AN INTRODUCTION TO VLSI PHYSICAL DESIGN

Computer Engineering

Senior Consulting Editors

Stephen W. Director, Carnegie Mellon University
C. L. Liu, University of Illinois Urbana-Champaign

McGraw-Hill Series in Computer Science

Senior Consulting Editor

C. L. Liu, University of Illinois at Urbana-Champaign

Consulting Editor

Allen B. Tucker, Bowdoin College

Fundamentals of Computing and Programming
Computer Organization and Architecture
Computers in Society/Ethics
Systems and Languages
Theoretical Foundations
Software Engineering and Database
Artificial Intelligence
Networks, Parallel and Distributed Computing
Graphics and Visualization
The MIT Electrical Engineering and Computer Science Series

AN INTRODUCTION TO VLSI PHYSICAL DESIGN

M. Sarrafzadeh

Northwestern University

C. K. Wong

IBM Thomas J. Watson Research Center and Chinese University of Hong Kong

The McGraw-Hill Companies, Inc.

New York   St. Louis   San Francisco   Auckland   Bogotá   Caracas
Lisbon   London   Madrid   Mexico City   Milan   Montreal   New Delhi
San Juan   Singapore   Sydney   Tokyo   Toronto

McGraw-Hill

A Division of The **McGraw·Hill** Companies

AN INTRODUCTION TO VLSI PHYSICAL DESIGN

Copyright ©1996 by The McGraw-Hill Companies, Inc. All rights reserved. Printed in the United States of America. Except as permitted under the United States Copyright Act of 1976, no part of this publication may be reproduced or distributed in any form or by any means, or stored in a data base or retrieval system, without the prior written permission of the publisher.

This book is printed on acid-free paper.

1 2 3 4 5 6 7 8 9 DOC DOC 9 0 9 8 7 6

ISBN 0-07-057194-5

This book was set in Times Roman by ETP/Harrison.
The editor was Eric M. Munson;
the production supervisor was Denise L. Puryear.
The design manager was Joseph A. Piliero.
Project supervision was done by ETP/Harrison (Portland, Oregon).
R.R. Donnelley & Sons Company was printer and binder.

Library of Congress Cataloging-in-Publication Data

Sarrafzadeh, Majid.
    An introduction to VLSI physical design / M. Sarrafzadeh and C. K. Wong.
        p.   cm.
    Includes index.
    ISBN 0-07-057194-5
    1. Integrated circuits—Very large scale integration—Design and construction—Data processing.   2. Computer-aided design.   I. Wong, C. K.   II. Title.
TK7874.75.S27   1996
621.39'5--dc20   95-47560

ABOUT THE AUTHORS

Majid Sarrafzadeh received his B.S., M.S., and Ph.D. in 1982, 1984, and 1987, respectively, all from the University of Illinois at Urbana-Champaign in the Electrical and Computer Engineering Department.

He joined Northwestern University as an Assistant Professor in 1987. Since 1991 he has been Associate Professor of Electrical Engineering and Computer Science at Northwestern University. His research interests lie in the area of design and analysis of algorithms and computational complexity, with emphasis in VLSI.

Dr. Sarrafzadeh is a fellow of IEEE. He received an NSF Engineering Initiation award in 1987, two distinguished paper awards in ICCAD-91, and the best paper award for physical design in DAC-93. He has served on the technical program committee of various conferences, for example, ICCAD, EDAC and ISCAS. He has published over 150 papers in the area of design and analysis of algorithms, is a co-editor of the book *Algorithmic Aspects of VLSI Layout*, a co-editor-in-chief of *The International Journal of High Speed Electronics*, and an associate editor of *IEEE Transactions on Computer-Aided Design*, 1993–present.

C. K. Wong received the B.A. degree (First Class Honors) in mathematics from the University of Hong Kong in 1965, and the M.A. and Ph.D. degrees in mathematics from Columbia University in 1966 and 1970, respectively. He joined the IBM T. J. Watson Research Center in 1969 as a Research Staff Member and was manager of the VLSI Design Algorithms group from 1985 to 1995. He was Visiting Associate Professor of Computer Science at the University of Illinois, Urbana, in 1972–73 and Visiting Professor of Computer Science at Columbia University in 1978–79. Currently he is Chair Professor and Chairman of the Department of Computer Science and Engineering at the Chinese University of Hong Kong on leave from IBM. His research interests include combinatorial algorithms, such as sorting, searching, and graph algorithms; computational geometry; and algorithms arising directly from industrial applications.

He holds four U.S. patents and has published close to 200 papers. He is author of the book *Algorithmic Studies in Mass Storage Systems*, published by Computer Science Press, 1983. He received an Outstanding Invention Award (1971), an Outstanding Technical Achievement Award (1988), and four Invention Achievement Awards (1977, 1980, 1983, 1989) from IBM.

Dr. Wong is a Fellow of IEEE and a Fellow of ACM. He was Chair of the IEEE Computer Society Technical Committee on VLSI from 1990 to 1991 and editor of *IEEE Transactions on Computers* from 1982 to 1985. He is the founding editor-in-chief of the international journal *Algorithmica*, and a founding member of the editorial board of *IEEE Transactions on VLSI Systems*. He is also on the editorial boards of the international journals, *Networks, Fuzzy Sets and Systems*, and *The Journal of Fuzzy Mathematics*.

To my mother, to the memory of my father,
and to my wife Marjan
Majid Sarrafzadeh

To Catherine, Henry and Andrew
C. K. Wong

CONTENTS

Preface   xv

1 Introduction   1
  1.1   VLSI Technology   2
  1.2   Layout Rules and Circuit Abstraction   5
  1.3   Cell Generation   7
        1.3.1   Programmable Logic Arrays   8
        1.3.2   Transistor Chaining (CMOS Functional Arrays)   9
        1.3.3   Weinberger Arrays and Gate Matrices   9
  1.4   Layout Environments   11
        1.4.1   Layout of Standard Cells   12
        1.4.2   Gate Arrays and Sea-of-Gates   12
        1.4.3   Field-Programmable Gate Arrays (FPGAs)   14
  1.5   Layout Methodologies   14
  1.6   Packaging   18
  1.7   Computational Complexity   19
  1.8   Algorithmic Paradigms   21
  1.9   Overview of the Book   25
        Exercises   25
        Computer Exercises   26

2 The Top-Down Approach: Placement   31
  2.1   Partitioning   31
        2.1.1   Approximation of Hypergraphs with Graphs   33
        2.1.2   The Kernighan-Lin Heuristic   34
        2.1.3   The Fiduccia-Mattheyses Heuristic   37
        2.1.4   Ratio Cut   39
        2.1.5   Partitioning with Capacity and I/O Constraints   43
        2.1.6   Discussion   45
  2.2   Floorplanning   47
        2.2.1   Rectangular Dual Graph Approach to Floorplanning   50

		2.2.2	Hierarchical Approach	54
		2.2.3	Simulated Annealing	57
		2.2.4	Floorplan Sizing	60
		2.2.5	Discussion	67
	2.3	Placement		69
		2.3.1	Cost Function	70
		2.3.2	Force-directed Methods	71
		2.3.3	Placement by Simulated Annealing	73
		2.3.4	Partitioning Placement	74
		2.3.5	Module Placement Based on Resistive Network	75
		2.3.6	Regular Placement: Assignment Problem	80
		2.3.7	Linear Placement	82
		2.3.8	Discussion	83
		Exercises		84
		Computer Exercises		86

3 The Top-Down Approach: Routing — 91

	3.1	Fundamentals		91
		3.1.1	Maze Running	91
		3.1.2	Line Searching	94
		3.1.3	Steiner Trees	96
	3.2	Global Routing		107
		3.2.1	Sequential Approaches	110
		3.2.2	Hierarchical Approaches	112
		3.2.3	Multicommodity Flow-based Techniques	114
		3.2.4	Randomized Routing	117
		3.2.5	One-Step Approach	118
		3.2.6	Integer Linear Programming	119
		3.2.7	Discussion	120
	3.3	Detailed Routing		121
		3.3.1	Channel Routing	123
		3.3.2	Switchbox Routing	133
		3.3.3	Discussion	138
	3.4	Routing in Field-Programmable Gate Arrays		140
		3.4.1	Array-based FPGAs	141
		3.4.2	Row-based FPGAs	145
		3.4.3	Discussion	149
		Exercises		151
		Computer Exercises		152

4 Performance Issues in Circuit Layout — 155

	4.1	Delay Models		155
		4.1.1	Gate Delay Models	156
		4.1.2	Models for Interconnect Delay	159
		4.1.3	Delay in RC Trees	161
	4.2	Timing-Driven Placement		163
		4.2.1	Zero-Slack Algorithm	163
		4.2.2	Weight-based Placement	169
		4.2.3	Linear Programming Approach	172
		4.2.4	Discussion	174

4.3		Timing-Driven Routing	175
	4.3.1	Delay Minimization	175
	4.3.2	Clock Skew Problem	180
	4.3.3	Buffered Clock Trees	184
4.4		Via Minimization	187
	4.4.1	Constrained Via Minimization	187
	4.4.2	Unconstrained Via Minimization	193
	4.4.3	Other Issues in Via Minimization	195
4.5		Power Minimization	196
4.6		Discussion and Other Performance Issues	198
		Exercises	201
		Computer Exercises	203

5 Single-Layer Routing and Applications — 207

5.1		Planar Subset Problem (PSP)	207
	5.1.1	General Regions without Holes	208
	5.1.2	PSP with a Fixed Number of Modules	210
5.2		Single-Layer Global Routing	212
5.3		Single-Layer Detailed Routing	215
	5.3.1	Detailed Routing in Bounded Regions	215
	5.3.2	Detailed Routing in General Regions	219
5.4		Wire-Length and Bend Minimization Techniques	222
	5.4.1	Length Minimization	222
	5.4.2	Bend Minimization	227
5.5		Over-the-Cell (OTC) Routing	230
	5.5.1	Physical Model of OTC Routing	230
	5.5.2	Basic Steps in OTC Routing	231
5.6		Multichip Modules (MCMs)	233
	5.6.1	An Overview of MCM Technology	234
	5.6.2	Requirements on MCM Routers	235
	5.6.3	Routing Problem Formulation and Algorithms	237
5.7		Discussion	241
		Exercises	242
		Computer Exercises	244

6 Cell Generation and Programmable Structures — 247

6.1	Programmable Logic Arrays	248
6.2	Transistor Chaining	253
6.3	Weinberger Arrays and Gate Matrix Layout	256
6.4	Other CMOS Cell Layout Generation Techniques	263
6.5	CMOS Cell Layout Styles Considering Performance Issues	265
6.6	Discussion	269
	Exercises	269
	Computer Exercises	270

7 Compaction — 271

7.1		1D Compaction	272
	7.1.1	Compression-Ridge Techniques	273
	7.1.2	Graph-based Techniques	276
	7.1.3	Wire-Length Minimization	279

	7.1.4	Compaction with Automatic Jogs	286
	7.1.5	Grid Constraints	289
7.2	2D Compaction		293
7.3	Discussion		295
	Exercises		298
	Computer Exercises		299

Appendix   302

A.1	Using DISPLAY	303
A.2	Examples	305
A.3	Running display	306
A.4	Bugs	306

Bibliography   307

Index   325

PREFACE

This book focuses on the VLSI physical design process. A large number of problems studied here are of a fundamental nature and have applications in many other scientific and engineering fields. In particular, the basic paradigms studied here apply to a number of large-scale computationally intensive optimization problems.

This book has evolved from the lecture notes used since 1991 for teaching a VLSI physical design class at Northwestern University (EECS-C57: "Design Automation in VLSI"), taken by both advanced undergraduate students and graduate students. Students learn about various optimization problems related to physical design and study the corresponding solution techniques. They also learn the underlying technologies and get a chance to develop computer-aided design (CAD) tools.

This book assumes that the readers have a basic understanding of computer data representation and manipulation (i.e., data structures), computer algorithms, and computer organization. An introductory course in algorithm design and a first course in logic design can serve as student prerequisites for a course using this book.

At Northwestern University, we teach two classes related to physical design. The first one (EECS-C57) is taken by juniors, seniors, and graduate students. In that course, this book is followed closely. The second course (EECS-D59) is a more advanced course, taken by graduate students. (In EECS-D59, interaction of physical design with higher-level phases such as logic design and behavioral design is also studied.) Several readings are assigned for that course, typically, this book, Lengauer [213], and recent papers.

The approach we have taken in writing this book is as follows. We first introduce commonly used algorithmic paradigms. For each topic we will discuss several basic approaches and elaborate on the more fundamental ones. Other techniques will be summarized at the end of the corresponding section or chapter. Some of the algorithms described might be very simple; however, they are widely used in industry and form the basis for understanding more advanced techniques. The concepts introduced in earlier chapters and sections are presented in a less formal manner. Once the reader understands the basic concepts, more formal (theoretical) issues are introduced in later chapters and sections. This manner of presentation has worked well in EECS-C57 at Northwestern University.

There are two types of exercises at the end of each chapter. The first type of exercise involves the design and analysis of algorithms for physical design problems. The second are computer exercises that can be used in conjunction with the display environment presented in the Appendix. Such exercises involve the implementation of the algorithms presented in the text or the implementation of new algorithms to be designed by students. Some of the computer exercises are meant to provide a complete understanding of tools development. They start with a simple model in the earlier chapters and evolve into more sophisticated (and practical) tools in later chapters.

We are grateful to our colleagues Prof. Wayne Dai (UC–Santa Cruz), Dr. Fook-Luen Heng (IBM T. J. Watson), Prof. Yoji Kajitani (TIT–Japan), Prof. Steve Kang (UI–Urbana), Prof. Ernie Kuh (UC–Berkeley), Dr. Jin-Fuw Lee (IBM T. J. Watson), Prof. Dave Liu (UI–Urbana), Prof. Margaret Marek-Sadowska (UC–Santa Barbara), and Dr. Gary Yeap (Motorola) for their valuable suggestions on an early draft of the book. Dr. Jin-Fuw Lee's insightful comments on Chapter 6, Compaction, have significantly improved the presentation of that chapter.

Collaboration with our colleagues in the past has provided new insights into the VLSI physical design problem. Among them are Prof. D. T. Lee (Northwestern University), Prof. Thomas Lengauer (GMD, Germany), and Prof. Franco Preparata (Brown University). We are also grateful to former and current CAD students at Northwestern University, Dr. D. S. Chen, Dr. Charles Chiang, Professor J. D. Cho, Amir Farrahi, Dr. Y. M. Huang, David Knol, Dr. K. F. Liao, Dr. Weiliang Lin, Salil Raje, Dr. Richard Sun, and Gustavo Tellez, for their valuable comments and suggestions. David Knol has helped design the cover of the book.

Majid Sarrafzadeh
C. K. Wong

AN INTRODUCTION TO VLSI PHYSICAL DESIGN

CHAPTER

1

INTRODUCTION

The size of present-day computing systems demands the elimination of repetitive manual operations and computations in their design. This motivates the development of automatic design systems. To accomplish this task, a fundamental understanding of the design problem and full knowledge of the design process are essential. Only then could one hope to efficiently and automatically fill the gap between system specification and manufacturing. Automation of a given (design) process requires an algorithmic analysis of it. The availability of fast and easily implementable algorithms is essential to the discipline.

In order to take full advantage of the resources in the very-large-scale integration (VLSI) environment, new procedures must be developed. The efficiency of these techniques must be evaluated against the inherent limitations of VLSI. Previous contributions are a valuable starting point for future improvements in design performance and evaluation.

Physical design (or layout phase) is the process of determining the physical location of active devices and interconnecting them inside the boundary of a VLSI chip (i.e., an integrated circuit). This book focuses on the layout problem that plays an important role in the design process of current architectures. The measure of the quality of a given solution to the circuit layout problem is the efficiency with which the circuit (corresponding to a given problem) can be laid out according to the formal (design) rules dictated by the VLSI technology. Since the cost of fabricating a circuit is a function of the circuit area, circuit layout techniques aim to produce layouts with a small area. Also, a smaller area implies fewer defects, hence a higher yield. These layouts must have a special structure to guarantee their wirability (using a small number of planes in the third dimension). Other criteria of optimality, for example, wire length length minimization, delay minimization, power minimization, and via minimization also have to be taken into consideration. In present-day systems, delay minimization is becoming more crucial. The aim is to design circuits that are fast while having small area. Indeed, in "aggressive" designs, used for example in the medical electronics industry, speed and reliability are the main objectives.

In the past two decades, research has been directed toward automation of layout process. Many invaluable techniques have been proposed. The early layout techniques, designed for printed circuit boards containing a small number of components, assumed a fixed position for the components. The earliest work along this line, proposed in [200], is a wiring algorithm based on the propagation of "distance wave." Its simplicity and effectiveness have been the reasons for its success. As the number of components increased and with the advent of VLSI technology, efficient algorithms for placing the components and effective techniques for component (or cell) generation and wiring have been proposed.

Because of the inherent complexity of the layout problem, it is generally partitioned into simpler subproblems, with the analysis of each of these parts providing new insights into the original problem as a whole. In this framework, the objective is to view the layout problem as a collection of subproblems; each subproblem should be efficiently solved, and the solutions of the subproblems should be effectively combined. Indeed, a circuit is typically designed in a hierarchical fashion. At the highest level of hierarchy, a set of ICs are interconnected. Within each IC, a set of modules, for example, memory units, ALUs, input-output ports, and random logic, are arranged. Each module consists of a set of gates, where each gate is formed by interconnecting a collection of transistors. Transistors and their interconnections are defined by the corresponding masks.

In this chapter VLSI technology is first reviewed and a brief overview of the layout rules is given. In Section 1.3, cell generation techniques are reviewed, followed by layout environments in Section 1.4, layout methodologies in Section 1.5, and VLSI packaging issues in Section 1.6.

1.1 VLSI TECHNOLOGY

The most prevalent VSLI technology is metal-oxide-semiconductor (MOS) technology. The three possibilities of functional cells (or subcircuits) are p-channel MOS (PMOS), n-channel MOS (NMOS), and complementary MOS (CMOS) devices. PMOS and NMOS are not used anymore. CMOS offers very high regularity and often achieves much lower power dissipation than other MOS circuits.

Although this is an overview of MOS technology, most concepts developed in this book are to a large extent technology-independent. CMOS is likely to be current for some time, as it satisfies VLSI system requirements. For more details see [247, 256, 381].

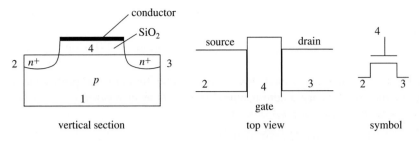

FIGURE 1.1
Geometry of an NMOS switch.

VLSI technology offers the user a new and more complex range of "off the shelf" circuits (i.e., predesigned circuits), but MOS VLSI design processes are such that system designers can readily design their own special circuits of considerable complexity. This provides a new degree of freedom for designers.

The geometry of an NMOS switch is shown in Figure 1.1. On a silicon substrate (1) of the p-type (i.e., doped with 3-valent atoms)—where positive carriers (holes) are available—two strips (2) and (3), separated by a narrow region (4), are heavily doped with 5-valent atoms. This modified material is called *diffusion*, with reference to the doping process. The two regions (2) and (3) are respectively called the *source* and *drain*, and region (4) is called the *channel*. Over the channel, a thin layer of silicon dioxide, SiO_2, is created, and a conductor plate is placed on top of it. The latter, called the *gate*, is typically realized in polysilicon.

Photolithography is used to pattern the layers of an integrated circuit. Photoresist (PR) is placed on the wafer surface and the wafer is spun at high speed to leave a very thin coating of PR. PR is a photosensitive chemical used with a mask to define areas of wafer surface by exposure to ultraviolet light. The mask consists of opaque and transparent materials patterned to define areas on the wafer surface. It is the pattern of each mask that an engineer designs.

MOS design is aimed at turning a specification into masks for processing silicon. Typical NMOS circuits are formed on three layers, diffusion, polysilicon and metal, that are isolated from one another by thick or thin silicon dioxide insulating layers. The thin oxide (thinox) region includes n-diffusion, p-diffusion, and transistor channels. Polysilicon and thinox regions interact so that a transistor is formed where they "cross" one another. Layers may be deliberately joined together where contacts, also called vias, are formed for electrical connection.

Typical processing steps are:

- Mask 1 defines the areas in which the deep p-well diffusions are to take place (similar to region 1 in Figure 1.1) on an n-type substrate.

- Mask 2 defines the thinox (or diffusion) regions, namely, those areas where the thick oxide is to be stripped and thin oxide grown to accommodate p- and n-transistors and wires (similar to regions 2–4 in Figure 1.1).

- Mask 3 is used to pattern the polysilicon layer that is deposited after the thin oxide.
- Mask 4 is a p-plus mask used to define all areas where p-diffusion is to take place.
- Mask 5 is usually performed using the negative form of the p-plus mask and defines those areas where n-type diffusion is to take place.
- Mask 6 defines contact cuts.
- Mask 7 defines the metal layer pattern.
- Mask 8 is an overall passivation layer that is required to define the openings for access to bonding pads.

CMOS gates are basic elements used in digital design. They act as switches. CMOS is discussed in more detail when we discuss cell generation (see Section 1.3). For now, it is sufficient to know about the layers (masks) involved in the process; for example, for the layout problem, the polysilicon layer and a collection of metal layers, typically between one and three in number.

Consider the circuit shown in Figure 1.2a. The power supply is typically at 5 volts (more recent technologies work with 3 volts—or even less). When the transistor (or switch) is on (i.e., logic value 1), the current flows from the power supply to the ground. A voltage drop of approximately 5 volts across the resistor brings the voltage of point f to 0. When the transistor is off, there is no current flowing. Thus, point f is at 5 volts, that is, logic value 1. The circuit works as an inverter. Similarly, the circuit shown in Figure 1.2b works as a NAND gate and the one in Figure 1.2c as a NOR gate. In NMOS circuits, a transistor replaces the resistor shown (as resistors are hard to realize in silicon). In CMOS circuits, the dual of the circuit "below" function f replaces the resistors. The lower circuit consists of NMOS transistors and the upper one consists of PMOS transistors.

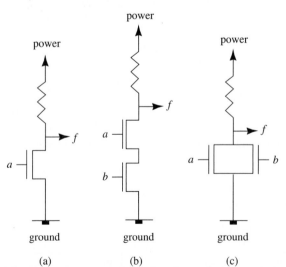

FIGURE 1.2
(a) Inverter, (b) NAND, and (c) NOR gates.

Depending on the input signals, either the upper or the lower circuit is on. The design of the circuit is similar in CMOS and NMOS—they are both based on classical switching theory. See a book on switching theory or logic design for more information on the topic.

1.2 LAYOUT RULES AND CIRCUIT ABSTRACTION

A circuit is laid out according to a set of *layout rules* (or *geometric design rules*). The layout rules, being in the form of minimum allowable values for certain widths, separations, and overlaps, reflect the constraints imposed by the current technology. These values are expressed as a function of a parameter λ that depends on the technology. The parameter λ is approximately the maximum amount of accidental displacement. (In the early 1980s, λ was about 3 microns; in the early 1990s, submicron fabrication became feasible.)

In realizing the interconnections, the following set of rules is adopted. Assume that layers $L_1, \ldots, L_v$ are available, ordered from 1 to v, so that L_i is below L_{i+1}. L_1 is typically polysilicon and $L_2, \ldots, L_v$ are metal. (There is a diffusion layer below L_1; however, it is not used for interconnection).

R1. *Wire width:* Each wire in layer L_i ($1 \le i \le v$) has a minimum width $w_i\lambda$ (see Figure 1.3a). Due to possible displacement λ for each edge of a wire in layer $L_1, w_1 > 2$. In this case, even if an edge of the wire displaces by λ, the width of the wire remains nonzero. Also since a wire in layer L_i runs over more wires than a wire in layer L_{i-1} (i.e., in upper layers the surface becomes less smooth), $w_i > w_{i-1}$.

R2. *Wire separation:* Two wires in layer L_i have a minimum separation of $s_i\lambda$ (see Figure 1.3b). Normally $s_1 = 3$ since there is a possible displacement of λ for each wire, and after possible displacement the two wires must be separated by λ units to avoid cross-talk. Also $s_i \ge s_{i-1}$.

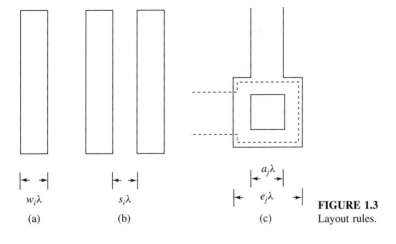

$w_i\lambda$ $s_i\lambda$ $a_j\lambda$ $e_j\lambda$

(a)   (b)   (c)

FIGURE 1.3
Layout rules.

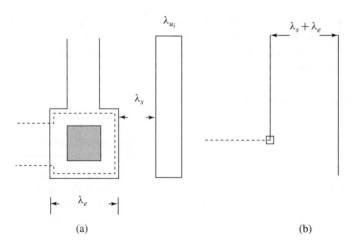

FIGURE 1.4
Abstract model. (a) Mask layout; (b) abstract layout.

R3. *Contact rule:* To connect two wires in layers L_i and L_j ($i < j$) a contact (via) must be established. The two wires must overlap for $e_j\lambda \times e_j\lambda$ units and the contact cut must be $a_j\lambda \times a_j\lambda$ units (see Figure 1.3c). Typically $e_i \geq e_{i-1}$ and $a_i \geq a_{i-1}$.

To facilitate the analysis of the layout problem, assume for every pair of layers: (A1) $w_i\lambda = w_v\lambda = \lambda_w$, (A2) $s_i\lambda = s_v\lambda = \lambda_s$ for $1 \leq i \leq v$, (A3) $e_j\lambda = e_k\lambda = \lambda_e$ and $a_j\lambda = a_k\lambda = \lambda_a$. Thus, in an abstract model, the wires are viewed as segments (i.e., they have zero width) on the plane with $\lambda_s + \lambda_e$ separation between two wires, as shown in Figure 1.4. Actually, in the situation shown in the figure, the separation is $\lambda_s + (\lambda_e + \lambda_w)/2$ between two wires. However, to make all separations uniform (for simplicity), it is assumed that all separations are caused by vias, and thus $\lambda_s + \lambda_e$ is used as the corresponding separation. When the number of layers is small (e.g., $v = 3$) the wasted area, due to assumptions A1–A3, is negligible. A layout conforming with the given set of design rules is called a *legal layout*. How to transform an abstract layout into a mask layout and how to effectively "compact" the mask layout is discussed in Chapter 7. Some layout systems deal directly with geometry of wires (i.e., do not go through the abstraction steps). Such systems/algorithms are called *gridless*.

The chip area, which must be minimized, is the smallest rectangle (IC packages are rectangular in shape) enclosing a legal layout of the circuit. In order to simplify design rule checking, consider a grid environment. A circuit, represented by a circuit graph (to be defined), will be mapped into or placed in a grid.

A formal definition of a grid follows. A plane figure is called a *tile* if the plane can be covered by copies of the figure without gaps and overlaps (the covering is then called a *tessellation*). A square tessellation is one whose tiles are squares. The dual of the tessellation is called a *(square) grid-graph*. The vertices (grid points) of the grid-graph are the centers of the tiles, and edges join grid points

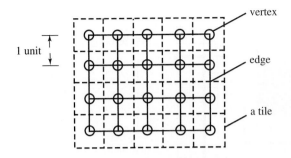

vertex

1 unit

edge

a tile

FIGURE 1.5
A square grid-graph.

belonging to neighboring tiles. The separation between two adjacent columns or two adjacent rows is 1 unit, that is, $\lambda_s + \lambda_e$ (see Figure 1.5). When a grid-graph is placed on the plane (a graph is a topology), we call it a *grid*.

A circuit $\mathcal{C} = \{\mathcal{M}, \mathcal{N}\}$ consists of a collection $\mathcal{M} = \{M_1, \ldots, M_m\}$ of modules—each module being a collection of active devices—and a set $\mathcal{N} = \{N_1, \ldots, N_n\}$ of nets. Each net specifies a subset of points on the boundary of the modules to be interconnected. A *circuit graph* G_C is a hypergraph associated with $\mathcal{C}$, where vertices correspond to the modules and hyperedges correspond to the nets. In certain problems, it is more convenient to deal with a circuit graph than with the circuit. A solution to the grid layout problem consists of embedding each module M_i ($1 \leq i \leq m$) of the circuit on the grid using a finite collection T_i of tiles and interconnecting the terminals of each net by means of wires in the region outside the modules. An example is shown in Fig. 1.6.

A *conducting layer* is a graph isomorphic to the layout grid. Assume that layers $L_1, \ldots, L_v$ are available, ordered from 1 to v, so that L_{i-1} is below L_i. Contacts (vias) between two distinct layers can be established only at grid points. There are various layout models (e.g., planar, Manhattan, knock-knee, overlap). See Chapter 3.

A *wiring* of a given two-dimensional layout (e.g., the 2D layout shown in Figure 1.6) is a mapping of each edge of wires to a conducting layer, where a wire is a tree interconnecting terminals of a net. A via is established between layers L_h and L_k ($h < k$) at a grid point, then layers $L_j, h < j < k$, cannot be used at that grid point. A layout W is v-*layer wirable* if there exists a v-layer wiring of W.

In this book, the terms routing, interconnection, and layout refer to a two-dimensional problem, and the terms wiring and layer-assignment refer to the mapping of the two-dimensional entities to the third dimension.

1.3 CELL GENERATION

In VLSI design, a logic function is implemented by means of a circuit consisting of one or more basic cells, such as NAND or NOR gates. The set of cells form a library that can be used in the design phase. Basic cells have a smaller size and better performance, for they have been optimized through experience. Thus, employing predesigned basic cells decreases design time and produces structured

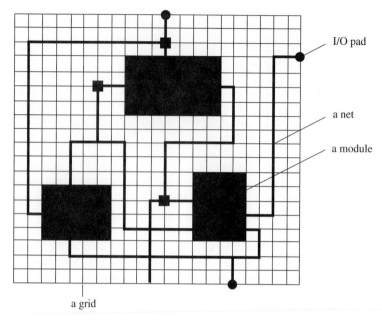

I/O pad

a net

a module

a grid

FIGURE 1.6
An example of the grid layout problem.

designs. In CMOS circuits, it is possible to implement complex Boolean functions by interconnecting NMOS and PMOS transistors.

Cell generation techniques are classified as random generation or regular style. A random generation technique is obtained by placing the basic components and interconnecting them. That is, there is no regular connection pattern. It is difficult to create a library of such cells because of their complexity. Thus, they must be designed from scratch. In contrast, the interconnection in a regular style technique admits a pattern. Compared to the regular cells (e.g., PLAs, ROMs, and RAMs), random logic cells occupy less silicon area, but take longer design time. Regular cells can be used to easily implement a set of Boolean expressions. The disadvantage of a regular cell, for example, a ROM-based cell, is that it takes a lot of area, for it uses many redundant transistors. Clearly, reducing the space required is important in designing functional cells. Several systematic layout methods to minimize the total area, for example, gate matrices and transistor chaining techniques, have been introduced [365, 214, 384, 261].

Some of the most commonly used cell structures are introduced below. They are elaborated on along with techniques for optimizing them in Chapter 6.

1.3.1 Programmable Logic Arrays

A programmable logic array (PLA) provides a regular structure for implementing combinational and sequential logic functions. A PLA may be used to take inputs and compute some combinational function of these inputs to yield outputs. Additionally some of the outputs may be fed back to the inputs through some flip flops, thus forming a finite-state machine.

Boolean functions can be converted into a two-level sum-of-product form and then be implemented by a PLA. A PLA consists of an AND-plane and an OR-plane. For every input variable in the Boolean equations, there is an input signal to the AND-plane. The AND-plane produces a set of product terms by performing AND operations. The OR-plane generates output signals by performing an OR operation on the product terms fed by the AND-plane. Reducing either the number of rows or the number of columns results in a more compact PLA. Two techniques have been developed, logic minimization for reducing the number of rows and PLA folding for reducing the number of columns. Using the technique, the number of product terms can be reduced while still realizing the same set of Boolean functions. Folding greatly reduces the area and is performed as a post-processing step.

1.3.2 Transistor Chaining (CMOS Functional Arrays)

Traversing a set of transistors in a cell dictates a linear ordering. Transistor chaining is the problem of traversing the transistors in an optimal manner. One can obtain a series-parallel implementation in CMOS technology, in which the PMOS and NMOS sides are dual of each other. If two transistors are placed side by side and the source or drain of one is to be connected to the source or drain of the other, then no separation (or space) is needed between the two transistors. Otherwise, the two transistors need to be separated. Thus an optimal traversal corresponds to a minimum separation layout. Since both the cell height and the basic grid size are a function of the technology employed, an optimal layout is obtained by minimizing the number of separations. Figure 1.7a shows a circuit and Figure 1.7b shows a corresponding chaining of the transistors. If the lower transistor is turned (or flipped) (Figure 1.7c), the first two columns can be merged, as shown in Figure 1.7d; thus a smaller width layout is obtained. Note that the resulting saving in width is not only due to no longer having a separation between columns but it is also due to the fact that transistor 2's drain and transistor 1's source (both in the n- and p-part) can use a common area.

1.3.3 Weinberger Arrays and Gate Matrices

Whereas the previous two layout styles (PLAs and transistor chaining) have been mainly used for AND-OR functions, the following two styles can be used to implement more general functions.

The first of these was introduced by Weinberger [379] and was one of the first methods for regular-style layout. The idea is to have a set of columns and rows, each row corresponding to a gate signal and the columns responsible for realization of internal signals (e.g., transistor to V_{dd} connection). The Weinberger

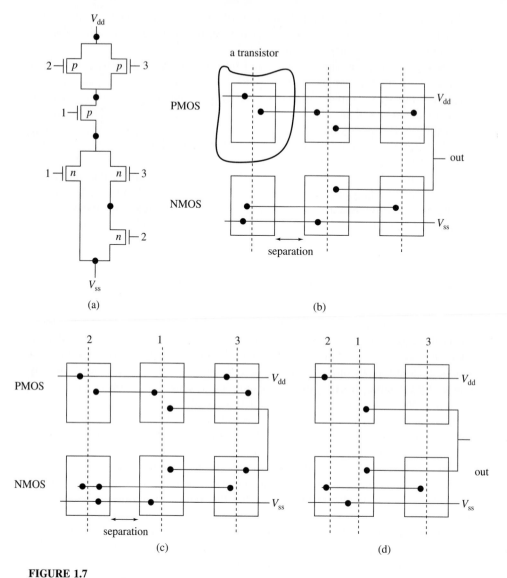

FIGURE 1.7
An example of transistor chaining. (a) A logic diagram; (b) a transistor chaining; (c) turning the lower left transistor; (d) a smaller width layout.

array style has led to other regular-style layouts, one of the most common ones being gate matrices.

Gate matrix layout was first introduced by Lopez and Law [228] as a systematic approach to VLSI layout. One of the objectives in the gate matrix layout problem is to find a layout that uses the minimum number of rows by permuting columns. The structure of the gate matrix layout is as follows. In a gate matrix

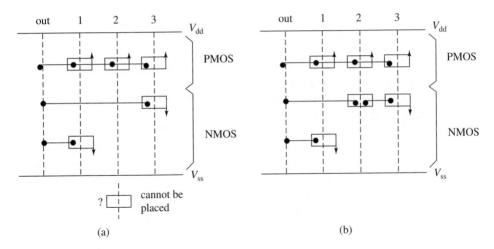

FIGURE 1.8
An example of gate matrix layout.

layout, a vertical polysilicon wire corresponding to a net is placed in every column. As polysilicon is of constant width and pitch, the number of columns is fixed. All transistors using the same gate signal are constructed along the same column; a transistor is formed at the intersection of a polysilicon column and a diffusion row. Transistors placed in adjacent columns of the same row are connected using shared diffusion. Transistors separated by one or more polysilicon columns are connected by metal lines (also called runners). Connections to power and ground are in the second metal layer.

The gate matrix structure allows a simple symbolic description for the layout. The size of a gate matrix layout is proportional to the product of the number of columns and rows. To minimize the area of a gate matrix layout, the number of rows must be reduced, since the number of columns is fixed to the number of nets in the circuit schematic. Because a row can be shared by more than one net, the number of rows depends heavily on both the column ordering and the net assignment to rows. A gate matrix layout of the circuit shown in Figure 1.7a is shown in Figure 1.8a. Note that it is not possible to place the last transistor because the two nets to transistor 3 and to signal out will overlap. However, if the columns are permuted, as shown in Figure 1.8b, a realizable layout is obtained.

1.4 LAYOUT ENVIRONMENTS

A circuit layout problem, as discussed before, involves a collection of cells (or modules). These modules could be very simple elements (e.g., a transistor or a gate) or may contain more complicated structures (e.g., a multiplier).

Layout architecture refers to the way devices are organized in the chip area. Different layout architectures achieve different trade-offs among speed, packaging density, fabrication time, cost, and degree of automation. The fabrication technology for these layout architectures are generally identical. The design rules are also independent of the layout architectures. The main difference lies in design production.

There are three styles of design production: full custom, semicustom, and universal. In fullcustom,, a designer designs all circuitry and all interconnection paths, whereas in semicustom, a library of predesigned cells is available. In universal circuitry, the design is more or less fixed and the designer programs the interconnections. Examples of universal circuitry are PLAs and FPGAs (to be described next). The designer chooses the appropriate ones and places them in the chip. In full custom designs there are no restrictions imposed on the organization of the cells. Thus it is time-consuming to design them, and it is difficult to automate them. However, area utilization is very good. In semicustom design, there are restrictions imposed on the organization of the cells (e.g., row-wise or gridwise arrangements). These circuits can be designed faster and are easier to automate, but area efficiency is sacrificed. Universal circuitries rely on programmable memory devices for cell functions and interconnections.

Full custom layout will be discussed in the next chapter, where general paradigms will be developed. Most of these paradigms can also be used to effectively solve semicustom layout problems. In what follows, we shall discuss standard cell layouts, gate arrays, sea-of-gates, and field-programmable gate arrays. After that, we will summarize the advantages and disadvantages of each.

1.4.1 Layout of Standard Cells

The layout of standard cells consists of rows of cells (see Figure 1.9). Here each cell is a simple circuit element such as a flip-flop or a logic gate. The cells are stored in a cell library and usually need not be designed. For each function, typically, there are several cells available with different area, speed, and power characteristics. The layout of standard cells is highly automated.

Manual layout of standard cells is perhaps the most tedious semicustom design style. All masks need to be fabricated for a standard cell design. Since each standard cell chip involves a new set of masks, the probability of failure is increased as compared to other design styles, and designers need to take care of nonfunctional aspects of chip design such as design rule check, latchup, power distribution, and heat distribution.

1.4.2 Gate Arrays and Sea-of-Gates

In a gate array, each cell is an array of transistors, as demonstrated in Figure 1.10. Each cell is capable of implementing a simple gate or a latch by interconnecting a set of transistors. There are also cell libraries which contain patterns for larger cells such as flip-flops, multiplexors or adders. The cells are arranged in rows to allow spaces for routing channels. Recently, gate arrays where the whole substrate is filled with transistors have been proposed. Routing is achieved using metals 1

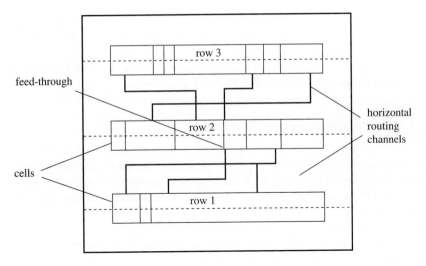

FIGURE 1.9
Organization of a standard cell.

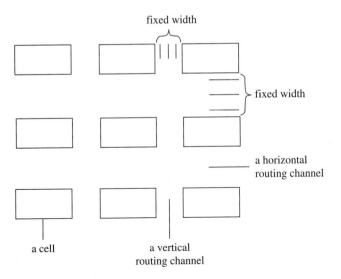

FIGURE 1.10
An example of a gate array.

and 2 (metal 1 is layer L_2, defined earlier, and metal 2 is layer L_3). This is called the *sea-of-gate* layout architecture or second generation gate arrays.

All transistors in gate arrays are prefabricated. The entire gate array substrate is prefabricated up to contact layer. The patterns of metals 1 and 2 define the cell functions and routing. Thus only a fraction of masks needs to be designed (typically three out of ten). This reduces the probability of failure. Also, since the early processing steps are prefabricated, processing time is shortened. Typically, it takes only one to two weeks to fabricate the chip when the masks are ready.

The prefabricated gate array substrates are called *masters*. Sometimes, regular structures, such as random access memory (RAM), programmable logic array (PLA), adders, and multipliers are also included in the chip.

1.4.3   Field-Programmable Gate Arrays (FPGAs)

All the circuit elements of a FPGA are prefabricated. The chip is already packaged and tested, like a PLA. In a PLA, only the connections are programmable. However in a FPGA, not only are the connections programmable, the cells are also programmable to achieve different functions.

Each cell of a FPGA typically contains flip-flops, multiplexors, and programmable functional gates. The gates can usually realize any function of a small set of inputs (say four). It may also contain testing circuitry for fabrication (not accessible to users). This is an advantage over the standard cells and gate arrays, where testing circuitry must be incorporated into the functional circuitry and the designer has to take into consideration the extra testing circuitry. The cells may be organized in one-dimensional (row-wise) or two-dimensional (gridwise) manner.

Using FPGAs involves programming the cells and interconnections to realize the circuits. Several types of routing resources are available. There are global wires that connect to every cell to provide global communication. Shorter wire segments are used for local signal communication. The programmable elements of FPGAs may be special devices that require extra processes. They may be permanently programmed or reprogrammed.

A typical programmable element (used by a number of leading manufacturers) is shown in Figure 1.11. It consists of logic elements, programmable interconnect points, and (programmable) switches, where each switch can realize various connections among the signals entering it.

A comparison of layout styles and architectures is given in Table 1.1. For example, it takes longer to fabricate a standard cell; however, the resulting chip operates at a higher speed. Thus, it makes sense to use standard cells when they are produced in large quantities.

1.5   LAYOUT METHODOLOGIES

The layout problem is typically solved in a hierarchical framework. Each stage should be optimized, while making the problem manageable for subsequent stages. Typically, the following subproblems are considered (Figures 1.12 shows each step):

- *Partitioning* is the task of dividing a circuit into smaller parts. The objective is to partition the circuit into parts, so that the size of each component is

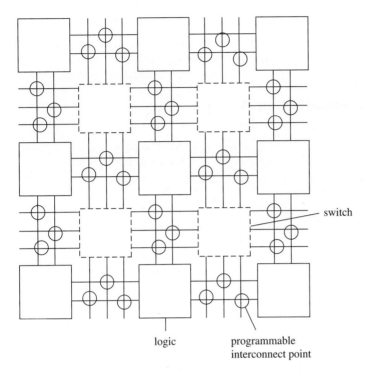

label: switch

label: logic

label: programmable
interconnect point

FIGURE 1.11
Architecture of an array-style FPGA.

TABLE 1.1
Comparison among various layout styles and architectures

	Full custom	Standard cell	Gate array	PLA	FPGA
Fabrication time	— —	— —	+	+	++
Packing density	++	+	−	— —	— —
Unit cost in large quantity	++	++	+	−	— —
Unit cost in small quantity	— —	— —	+	+	++
Easy design and simulation	— —	−	−	+	++
Remedy for erroneous design	— —	— —	−	++	++
Accuracy of timing simulation	−	−	−	+	+
Chip speed	++	++	+	−	−

\+ desirable
− not desirable

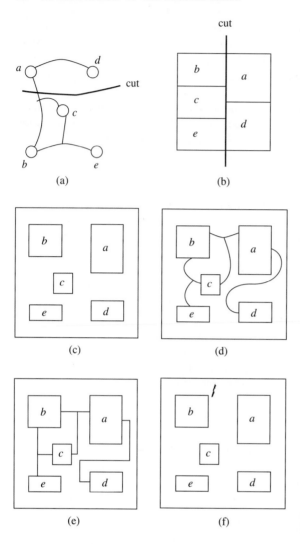

FIGURE 1.12
An example demonstrating hierarchical steps in the layout process.

within prescribed ranges and the number of connections between the components is minimized. Different ways to partition correspond to different circuit implementations. Therefore, a good partitioning can significantly improve circuit performance and reduce layout costs. A hypergraph and a partition of it is shown in Figure 1.12a. The cut (or general cuts) defines the partition.

- *Floorplanning* is the determination of the approximate location of each module in a rectangular chip area, given a circuit represented by a hypergraph—the shape of each module and the location of the pins on the boundary of each module may also be determined in this phase. The floorplanning problem in chip layout is analogous to floorplanning in building design, where we have a set of rooms (modules) and wish to decide the approximate location of each room based on some proximity criteria. An important step in floorplanning is to decide the relative location of each module. A good floorplanning algorithm should achieve many goals, such as making the subsequent routing phase easy, minimizing the total chip area, and reducing signal delays. The floorplan corresponding to the circuit shown in Figure 1.12a is shown in Figure 1.12b. Typically, each module has a set of implementations, each of which has a different area, aspect ratio, delay, and power consumption, and the best implementation for each module should be obtained.

- *Placement,* when each module is fixed, that is, has fixed shape and fixed terminals, is the determination of the best position for each module. Usually, some modules have fixed positions (e.g., I/O pads). Although area is the major concern, it is hard to control it. Thus, alternative cost functions are employed. There are two prevalent cost functions: wire-length-based and cut-based. The placement corresponding to the circuit shown in Figure 1.12a is shown in Figure 1.12c, where each module has a fixed shape and area.

- *Global routing* decomposes a large routing problem into small, manageable problems for detailed routing. The method first partitions the routing region into a collection of disjoint rectilinear subregions. This decomposition is carried out by finding a "rough" path (i.e., sequence of "subregions" it passes) for each net in order to reduce the chip size, shorten the wire length, and evenly distribute the congestion over the routing area. A global routing based on the placement shown in Figure 1.12c is shown in Figure 1.12d.

- *Detailed routing* follows the global routing to effectively realize interconnections in VLSI circuits. The traditional model of detailed routing is the two-layer Manhattan model with reserved layers, where horizontal wires are routed on one layer and vertical wires are routed in the other layer. For integrated circuits, the horizontal segments are typically realized in metal while the vertical segments are realized in polysilicon. In order to interconnect a horizontal and vertical segment, a contact (via) must be placed at the intersection points. More recently, the unpreserved layer model has also been discussed, where vertical and horizontal wires can run in both layers. A detailed routing corresponding to the global routing shown in Figure 1.12d is shown in Figure 1.12e; the Manhattan model is used. Recent designs perform multilayer detailed routing and over-the-cell (OTC) routing, as will be discussed in Chapter 5.

- *Layout optimization* is a post-processing step. In this stage the layout is optimized, for example, by compacting the area. A compacted version of the layout shown in Figure 1.12e is shown in Figure 1.12f.

- *Layout verification* is the testing of a layout to determine if it satisfies design and layout rules. Examples of design rules are timing rules. Layout rules were discussed in Section 1.2. In more recent CAD packages, the layout is verified in terms of timing and delay.

The above steps are followed in their entirety in full custom designs. In other layouts, such as standard cells and PLAs, some of the steps are not taken: this is due to standardization or prefabrication. Details of the steps taken in full custom designs and in the special cases will be discussed in subsequent chapters. We shall refer to the problems related to location of modules (i.e., partitioning, floor-planning, and placement) as the **placement** problem, and the problems related to interconnection of terminals (i.e., global and detailed routing) as the **routing** problem. In addition, there are other post-processing problems, such as via and bend minimization, that will be discussed.

1.6 PACKAGING

The previous sections have emphasized designing fast and reliable VLSI chips. These chips must be supported by an equally fast and reliable packaging technology. Packaging supplies chips with signals and powers, and removes the heat generated by circuitry. Packaging has always played an important role in determining the overall speed, cost, and reliability of high-speed systems such as supercomputers. In such high-end systems, 50% of the total system delay is usually due to packaging, and by the year 2000 the share of packaging delay may rise to 80% [13]. Moreover, increasing circuit count and density in circuits place further demands on packaging. A package is essentially a mechanical support for the chip and facilitates connection to the rest of the system. One of the earliest packaging techniques was dual-in-line packaging (DIP). An example is shown in Figure 1.13. A DIP has a small number of pins. Pin grid arrays (PGA) have more pins that are distributed around the packages (see Figure 1.13).

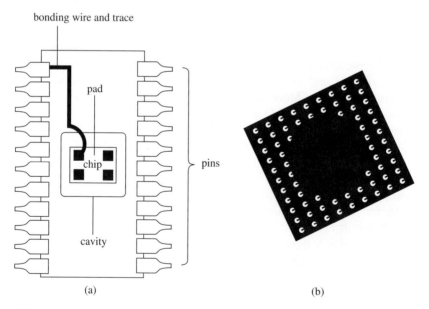

(a)   (b)

FIGURE 1.13
Typical DIP and PGA packages. (a) DIP; (b) PGA.

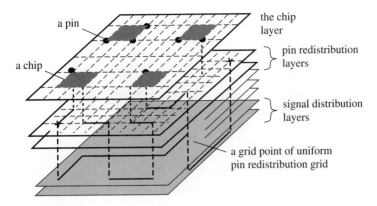

a pin

the chip
layer

a chip

pin redistribution
layers

signal distribution
layers

a grid point of uniform
pin redistribution grid

FIGURE 1.14
A typical MCM.

In order to minimize the delay, chips must be placed close together. Multichip module (MCM) technology has been introduced to significantly improve performance by eliminating packaging. An MCM is a packaging technique that places several semiconductor chips, interconnected in a high density substrate, into a single package. This innovation led to major advances in interconnection density at the chip level of packaging. Compared with single chip packages or surface mount packages, MCMs can reduce circuit board area by five to ten times and improve system performance by 20% or more. Therefore, MCM has been used in high performance systems as a replacement for the individual packages. An instance of a typical MCM is shown in Figure 1.14, where chips are placed and bonded on a surface at the top layer (called the *chip layer*). Below the chip layer, a set of *pin redistribution layers* is provided for distributing chip I/O pins for *signal distribution layers*. The primary goal of MCM routing is to meet high performance requirements, rather than overly minimizing the layout area. Techniques to be described in Chapters 2 and 3 can be employed, directly or with some modification, for placement and routing in MCMs.

1.7   COMPUTATIONAL COMPLEXITY

This book discusses the development of CAD tools for various layout problems. The algorithms used in such tools should be of *high quality* and *efficiency*, the two fundamental measures in CAD tool development.

Consider a problem of size n, for example sorting n numbers or deciding whether n small rectangles fit in one large rectangle. Let A be an algorithm solving the problem in time $T(n)$. Various values of n and $T(n)$ are shown in Table 1.2. Most problems in this text can be solved in (e.g., in $T(n) = 2^n$). However, the primary interest here is in solutions, that is, an algorithm with a running time expressed as a polynomial in n (e.g., in $T(n) = n^2$).

To simplify the analysis of algorithms, (big-O notation) is used here. This notation indicates the growth of a function; thus multiplicative and additive con-

TABLE 1.2
Comparison among various running times

	$n = 10$	$n = 20$	$n = 10^2$	$n = 10^3$	$n = 10^6$
$T(n) = n$	10	20	10^2	10^3	10^6
$T(n) = 3n$	30	60	3×10^2	3×10^3	3×10^6
$T(n) = n \log_{10} n$	10	26	2×10^2	3×10^3	6×10^6
$T(n) = n^2$	10^2	4×10^2	10^4	10^6	10^{12}
$T(n) = n^3$	10^3	8×10^3	10^6	10^9	10^{18}
$T(n) = 2^n$	10^3	10^6	10^30	10^{301}	∞
$T(n) = n!$	3×10^6	2×10^{18}	9×10^{157}	∞	∞

The numbers are approximate.
∞ means a number larger than 10^{500}.

stants and lower-order terms are ignored. For example, functions (n), $(3n + 2)$, $(.5n + \log n)$ are all of the same order. Formally, if a function $f(n)$ is of the same order as another function $g(n)$, we write $f(n) = O(g(n))$, if there exist constants c and n_0 such that $f(n) \leq cg(n)$ for all $n \geq n_0$.

Unfortunately, most problems encountered in VLSI layout are NP-complete or NP-hard. That is, (most probably) they require exponential time to be solved. For such problems, there are several strategies:

- Spend exponential time to solve the problem. This is, in general, an unfeasible alternative, for n is typically large in these problems.

- Instead of solving the problem optimally, solve the problem with high quality. An algorithm that does not guarantee an optimal solution is called a heuristic. This is often a good alternative. There is, obviously, a trade-off between the quality of the solution and the running time of the algorithm. Depending on the application, either the quality or the running time is favored.

- Solve a simpler (or restricted) version of the problem. Such an approach has two advantages. First, the simpler problem can reveal new insights into the complexity of the general problem. Second, the solution to the simpler problem can be used as a heuristic for solving the original problem.

There are two reasons for solving a problem. The first is to find a solution that will be used in designing a chip. In this application, quality is of crucial importance and running time, within tolerable limits, is of secondary importance. For example, once a placement has been obtained, a high-quality routing should be found to fabricate the chip. The second reason for solving a problem is to estimate the complexity of a problem. In this case, a reasonable solution is typically accepted, and such a solution should be obtained as fast as possible. For example, the solution to choosing the best of several possible placements might be to find a quick routing of each to decide which is easier to route. Once that is decided, a high-quality routing algorithm will be used to route the chip for fabrication.

Throughout this book, the quality and time complexity of various algorithms will be analyzed for a number of problems in VLSI layout. This will further illustrate the natural trade-off between quality and time complexity of algorithms.

1.8 ALGORITHMIC PARADIGMS

As discussed, most layout problems are NP-hard, that is, they (most probably) require exponential time to be solved exactly. Because of the size of the problems involved, exponential time is not affordable. Alternatively, suboptimal algorithms, those that are fast and produce good quality solutions, have been designed. Such algorithms are of *practical* importance.

- **Exhaustive search.** The most naive paradigm is exhaustive search. The idea is to search the entire solution space by considering every possible solution and evaluating the cost of each solution. Based upon this evaluation, one of the best solutions is selected. Certainly, exhaustive search produces optimal solutions. The main problem is that such algorithms are very slow—"very slow" does not mean that they take a few hours or a few days to run; it means they take a lifetime to run.

Alternatively, a number of efficient algorithmic techniques have been proposed. The following algorithmic paradigms have been employed in most subproblems arising in VLSI layout. These paradigms will be used throughout this book (see an introductory text in design and analysis of algorithms, e.g., [4, 75], for more details):

- **Greedy approach.** Algorithms for optimization problems go through a number of steps. At each step, a choice, or a set of choices, is made. In greedy algorithms the choice that results in a locally optimal solution is made. Typically, greedy algorithms are simpler than other classes of algorithms. However, they do not always produce globally optimal solutions. Even if they do, it is not always an easy task to prove them.

- **Dynamic programming.** A problem is partitioned into a collection of subproblems, the subproblems are solved, and then the original problem is solved by combining the solutions. Dynamic programming is applied when the subproblems are not independent. Each subproblem is solved once, and the solution is saved for other subproblems. In dynamic programming, first the structure of an optimal solution is characterized and the value of an optimal solution is recursively defined. Finally, the value of an optimal solution is computed in a bottom-up fashion.

- **Hierarchical approach.** In this approach, a problem is also partitioned into a set of subproblems. The subproblems are independent and the partitions are recursive. The sizes of the subproblems are typically balanced. The partition is usually done in a top-down fashion and the solution is constructed in a bottom-up fashion.

- **Mathematical programming.** In the mathematical programming approach, there is a set of constraints expressed as a collection of inequalities. The objective function is a minimization (or maximization) problem subject to a set of constraints. When the objective function and all inequalities are expressed as a linear combination of the involved variables then the system is called a .

- **Simulated annealing.** Simulated annealing is a technique used to solve general optimization problems. This technique used is especially useful when the solution space of the problem is not well understood. The idea originated from observing crystal formation of physical materials. A simulated annealing algorithm examines the configurations (i.e., the set of feasible solutions) of the problem in sequence. The algorithm evaluates the feasible solutions it encounters and moves from one solution to another.

- **Branch and bound.** This is a general and usually inefficient method for solving optimization problems. There is a tree-structured configuration space. The aim is to avoid searching the entire space by stopping at nodes when an optimal solution cannot be obtained. It is generally hard to claim anything about the running time of such algorithms.

- **Genetic algorithms.** At each stage of a genetic algorithm, a *population* of solutions (i.e., a solution subspace) to the problem is stored and allowed to *evolve* (i.e., be modified) through successive generations. To create a new generation, new solutions are formed by merging previous solutions (called *crossover*) or by modifying previous solutions (called *mutation*). The solutions to be selected in the next generation are probabilistically selected based on a *fitness* value. The fitness value is a measure of the competence (i.e., quality) of the solution.

There are other paradigms that will be introduced in later chapters, for example, neural-network algorithms. There are also paradigms that are not discussed in this text, for they have not been proven effective in VLSI layout problems, for example, artificial intelligence approaches.

The described paradigms can be demonstrated using the classic bin packing problem. As will be shown in the next chapter, bin packing is a subproblem and an easier version of the placement problem.

Example 1.1. In an instance of the bin packing problem, there is a collection of items $T = \{t_1, \ldots, t_n\}$, where t_i has (an integer) size s_i. There is a set B of bins each with a fixed (integer) size b. A subset of the items can be packed in one bin as long as the sum of the sizes of the items in the subset is less than or equal to b. The goal is to pack all the items, minimizing the number of bins used.

For example, assume there are seven items with sizes $1, 4, 2, 1, 2, 3, 5$ and $b = 6$. A solution in this instance is to place the items with sizes 1 and 5 in one bin, those with sizes 2 and 4 in one bin, and the rest (with sizes 1, 2 and 3) in the third bin. Three bins are used; this is optimal. Next the algorithmic paradigms are applied to this problem. Note that the solution presented within each paradigm is not unique. Also not every paradigm is effective for this problem: for any given problem the paradigms that best fit the problem must be chosen.

Solutions

- **Exhaustive search.** There are many ways to do exhaustive search. One approach is to find all possible partitions of the items (how many partitions are there for seven items?), then decide which partition is feasible, and among the feasible ones find one that uses the minimum number of bins.

- **Greedy approach.** First, sort the items based on their sizes to obtain the list $1, 1, 2, 2, 3, 4, 5$. Place the items one by one into the first bin until it is saturated. Next move to the next bin and repeat the procedure. Thus, items with sizes $1, 1, 2, 2$ will be placed in the first bin; the remaining three items each require a separate bin. Thus, a total of four bins is used.

- **Dynamic programming.** One strategy is to solve the problem for the first k items, then consider the $(k + 1)$st iteration, and determine the best way to place the $(k + 1)$st item in previous bins. In the first step, place the item with size 1 in the first bin. In the next step, considering the next item, note that there are two ways to place it—in the same bin as the item with size 1 or in a new bin. Certainly, the first choice is better. The next item has size 2 and requires a new bin. The next item has size 1 and can be placed in the first bin. Continue this procedure. Finally there are bins containing items $(1, 4, 1)$, $(2, 2)$, (3), and (5). A total of four bins is used.

- **Hierarchical approach.** Partition the problem into two subproblems, for example, $1, 4, 2, 1$ and $2, 3, 5$. Since neither of the subproblems can fit in one bin, partition them again to obtain $(1, 4)$, $(2, 1)$, and $(2, 3)$ (5). Now each subproblem can fit inside one bin; thus four bins are needed to fit them all. In this case the problem was divided into subproblems and the solutions to subproblems did not interact. In most hierarchical solutions, the subproblems do interact.

- **Mathematical programming.** First find a solution using $|B|$ bins (or decide there is no solution); then solve the problem for various values of $|B|$. Certainly, $|B| \leq n$. Formulate the decision problem as follows. Form a complete bipartite graph $G = (U, B, E)$, where U is the set of items, $U = \{u_1, u_2, \ldots, u_n\}$ and B is the set of bins, $B = \{b_1, b_2, \ldots, b_k\}$. The weight $w_{i,j}$ of an edge connecting one vertex u_i in U and one vertex b_j in B is set to $s(u_i)$, where $s(u_i)$ is the size of item u_i.

  The objective is to maximize $\sum_{(i,j) \in E} (w_{i,j} x_{i,j})$, subject to the following four constraints:

(1) [0,1] constraint: $x_{i,j} \in \{0, 1\}$,
(2) capacity constraint: $\sum(w_{i,j} x_{i,j} : \forall i \in U) \leq b_i, \forall j \in B$,
(3) assignment constraint: $\sum(x_{i,j} : \forall j \in B) = 1, \forall i \in U$.
(4) completeness constraint: the total number of $x_{i,j}$s whose value is 1 is equal to n.

  A possible solution is as follows.

Step 1. $|B|$ is set to the lower bound of the number of bins, that is, $\lceil \sum_{i=1}^{n} s(u_i) \rceil$ divided by the capacity of the bins.

Step 2. Formulate the above linear programming problem as to find a feasible solution satisfying the set of constraints.

Step 3. If the solution exists, the number of bins required is $|B|$. Then exit.

Step 4. Otherwise, we set $|B| = |B| + 1$ and repeat steps 1–4.

- **Simulated annealing.** Start with an initial feasible solution, for example, bins containing $(1, 4)$ $(2, 1)$ (2) (3) and (5), and then try to merge two of the bins, for example, $(1,4)$ $(2,1,3)$ (2) and (5). This improves the solution. At the beginning of this algorithm one may even accept (with certain probability) solutions that are worse than the original one. Thus, from the previous solution, the solution (1) (4) $(2, 1, 3)$ (2) and (5) can be obtained, where $(1, 4)$ has been split into (1) and (4). From this (1) $(4, 2)$ $(2, 1, 3)$ and (5), and, finally, $(1, 5)$ $(4, 2)$ and $(2, 1, 3)$ can be obtained. Moving from one solution to another is done in a probabilistic manner. The probability of moving from one solution to another depends on the difference between the cost of the previous solution and the new solution (i.e., the gain). It also depends on the stage of the algorithm—in early stages the probability of moving to higher-cost solutions is greater.

- **Branch and bound.** At the first level, consider all possible ways of partitioning the solution space. For example, consider all the possible ways of placing the item with size 5 in the solution: 5 is either in one bin by itself, in the same bin as an item with size 1, in the same bin as an item with size 2, or in the same bin as an item with size 3. Certainly, the last two possibilities may not happen; otherwise there will be a bin in which the total size of items is larger than 6. Thus, bound the solution and continue with the the the first two possibilities. This branching and bounding procedure continues until an optimal solution is obtained.

- **Genetic algorithms.** First obtain a few initial solutions: $S_1 = \{(1, 4), (2, 1), (2), (3), (5)\}$ and $S_2 = \{(1, 2), (4, 1), (2, 3), (5)\}$. As in simulated annealing, from S_1 other solutions are obtained by mutation; the same is true for S_2. For example from S_1 $S_3 = \{(1, 4)(2, 1, 2), (3), (5)\}$ is obtained, and from S_2 $S_4 = \{(1), (2), (4, 1), (2, 3), (5)\}$ is obtained. Then by crossover of S_1 and S_2 (e.g., taking some members of S_1 and some members of S_2 and merging them) $S_5 = \{(1, 2, 1), (4), (3, 2), (5)\}$ is obtained. Since it is not desirable to have a large population in one generation (because the space grows exponentially), from among $S_1, \ldots, S_5$ a subset is selected, for example, those that require the fewest bins. The task is repeated until either a good solution is obtained or time runs out.

As not every paradigm is suitable for any given problem, a paradigm should be selected based upon on quality and efficiency requirements. For example, (usually) if the quality of the solution is not extremely important but the running time of the algorithm is of crucial importance, select a greedy algorithm. It is also possible to combine more than one paradigm in a given algorithm. For example, one can use branch and bound to obtain a few partial solutions and then apply a greedy strategy to each of these partial solutions.

1.9 OVERVIEW OF THE BOOK

The remaining chapters of this book will focus on specific approaches to the physical design problem. Chapters 2 and 3 review the top-down approach to the circuit layout problem and elaborate on various subproblems (that is, partitioning, floorplanning/placement, global and detailed routing). Various layout environments such as standard cells and FPGAs will be discussed. An overview of other approaches to the layout problem is also given. Chapter 4, after introducing signal and gate delay models, elaborates on the timing-driven layout algorithms, and discusses other performance issues such as the via minimization problem and power minimization. Chapter 5 covers the single-layer layout problem and its application to over-the-cell routing and multichip modules. The problem of cell generation is discussed in Chapter 6. How to compact a given layout to obtain a layout with small area is shown in Chapter 7. In these chapters, a set of fundamental algorithms will be studied, with references to other techniques. Other books, monographs, and surveys that consider physical design problems are [213, 326, 153, 277, 264, 306].

EXERCISES

1.1. Design a circuit realizing the Boolean function $f = abc + bcd$ (see Figure 1.2 and use resistors as shown). How many transistors did you use? Design another circuit realizing the function $g = bc + e$. How many transistors did you use? Now, design one circuit realizing both functions f and g. How many transistors were used?

1.2. Some design rules are given as minimum requirements while others are expressed as maximum requirements. Explain the significance of the following parameters and decide whether minimum or maximum requirements are expected (see Figure 1.3): (a) w_i, (b) s_i, (c) a_j, (d) e_j.

1.3. Explain why e_j must be larger than a_j (see Figure 1.3c).

1.4. Find a minimum area layout of the circuit shown in Figure 1.6.

1.5. In certain layout styles, wires are allowed to be routed over the modules. Find a minimum area routing of the circuits shown in Figure 1.6 under this model. State the assumptions you made and justify their realistic nature.

1.6. Discuss the advantages and disadvantages of using a grid abstraction throughout the layout process.

1.7. Compare the top-down chip partitioning with bottom-up clustering.

1.8. Route the two-terminal subcircuits (also called a switchbox) shown in Figure E1.8 using edge-disjoint paths (each grid edge is to be used at most once). Find a layer assignment of your layout using the minimum number of layers.

1.9. Solve the sorting problem (sort a given list of integers) using each of the paradigms discussed in Section 1.8. Which paradigm is most effectively used?

1.10. What is the minimum number of bins needed in a one-dimensional bin packing problem (as a function of size of the items and size of the bins)?

1.11. Consider an instance of the one-dimensional bin packing problem. Assume the value B (i.e., the size of the bins) and the size of each item is less than or equal to a constant C. Design an optimal algorithm for solving this class of the one-dimensional bin packing problem. Analyze the time complexity of your algorithm.

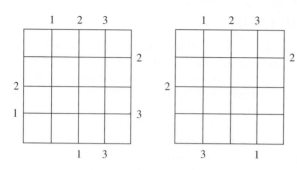

FIGURE E1.8
A switchbox.

1.12. What restricted (and nontrivial) classes of the one-dimensional bin packing problem can be optimally solved? Justify your answer.

1.13. Consider a two-dimensional version of the bin packing problem. Given a set of rectangular bins each with size $A \times B$, where A is the (horizontal) width and B is the (vertical) height of each bin, and a set of items $t_1, \ldots, t_n$, where item t_i has size $a_i \times b_i$; the problem is to pack the items in the bin trying to minimize the number of bins used. Neither the bins nor the items can be rotated. Design a greedy algorithm and a hierarchical algorithm for solving this problem. Analyze the time complexity of your algorithms.

1.14. Find an optimal solution to an instance of the two-dimensional bin packing problem (see the previous exercise) assuming:
$n = 10$, $A = 6$, $B = 4$,
$(a_1, b_1) = (2, 4)$, $(a_2, b_2) = (2, 3)$, $(a_3, b_3) = (1, 2)$, $(a_4, b_4) = (5, 2)$, $(a_5, b_5) = (4, 3)$, $(a_6, b_6) = (3, 3)$, $(a_7, b_7) = (1, 1)$, $(a_8, b_8) = (5, 3)$, $(a_9, b_9) = (2, 4)$, $(a_{10}, b_{10}) = (3, 5)$.
Solve the same problem with $A = 5$, $B = 4$.

1.15. Design a greedy algorithm for solving a restricted class of the two-dimensional bin packing problem where $A = B$ and $a_i = b_i$ for all i. Analyze the time complexity of your algorithm.

1.16. Design an optimal algorithm for solving a restricted class of the two-dimensional bin packing problem where $A = B$, and $a_i = b_i$ for all i, and either $a_i = a_j$ or $a_i = 2a_j$ or $a_j = 2a_i$.

1.17. Consider a set $\mathcal{I} = \{I_1, \ldots, I_n\}$ of intervals. Each interval $I_i = (l_i, r_i)$ is represented by its left point l_i and its right point r_i. Two intervals I_i and I_j are disjoint if $r_i < l_j$ or $r_j < l_i$. Design a greedy algorithm for finding the maximum number of pairwise disjoint intervals. Design a simulated annealing algorithm for the same problem. Analyze the time complexity of both algorithms.

1.18. Why is it necessary to redistribute the pins in MCM design?

COMPUTER EXERCISES

1.1. Given integers m and n and a two-terminal net N represented by terminals (a, b), (c, d) (coordinate $(1, 1)$ represents the northwest corner), and a set of obstacles at coordinates $(x_1, y_1), (x_2, y_2) \ldots (x_k, y_k)$; show an $m \times n$ grid graph and a routing interconnecting the terminals of N (or an appropriate "error" message). Minimize the total length. Report the resulting length. Describe your algorithm and discuss its performance.

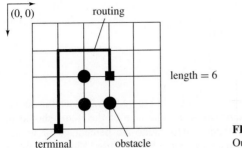

FIGURE CE1.1
Output format.

Input format. The first line contains the grid size (the first coordinate is the x-coordinate). The second line has coordinates of the source and the sink, and the rest of the lines contain coordinates of the obstacles. The northwest location is coordinate $(0, 0)$. The input format is:

> grid 5 4
> s-t (1,4) (3,2)
> obs (2,2) (2,3)
> (3,3)

Output format. The output format is shown in Figure CE1.1.

1.2. Consider a set of horizontal and vertical line segments in the plane. Design an algorithm for finding the maximum number of pairwise independent segments.

   Input format. Segments are given one by one and separated by a comma. x- and y-coordinate of the two endpoints of each segment is given. For each segment, the first two integers are x- and y-coordinates of one point of the segment and the following two integers are x- and y-coordinates of the other point of the segment (see Figure CE1.2a).

> 2 1 2 7, 1 2 9 2,
> 6 2 6 5, 8 2 8 5, 2 4 5 4

   Output format. The output format is shown in Figure CE1.2b.

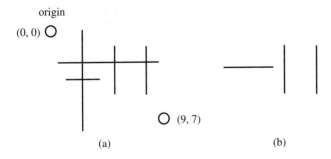

(a)   (b)

FIGURE CE1.2
(a) Input and (b) output formats.

1.3. Implement a greedy algorithm for the one-dimensional bin packing problem.

> **Input format.** The first line contains the size of the bins. The next lines contains the size of the items, separated by space.

> 5
> 1 4 3 2 5 2 1 2 3

> **Output format.** The output format is shown in Figure CE1.3. Each bin should be partitioned as dictated by the items inside it. The size of the items should be shown. The empty space should be shaded. Also write the percentage of wasted space: total wasted space divided by the sum of the sizes of used bins. Or send an error message indicating a solution does not exist.

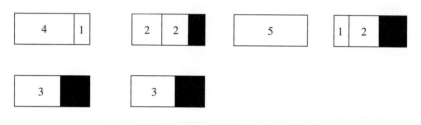

waste = 7/25 = 28%

FIGURE CE1.3
Output format, an instance of the bin packing problem.

1.4. Design and implement an improved greedy algorithm for the previous computer exercise. Introduce heuristics to improve the result. These heuristics can be of three types: preprocessing, post-processing, or heuristics during the algorithm.

1.5. Implement a hierarchical algorithm for the previous computer exercise.

1.6. Consider a set of vertical segments on the plane. Two segments are visible if there is an (imaginary) horizontal line segment intersecting those two segments and no other segments. These imaginary segments should not intersect the vertical segments at their endpoints. Given a set of vertical segments, design an algorithm for drawing (exactly) one horizontal segment for each pair of visible segments.

> **Input format.** Each segment is specified by the x- and y-coordinate of its upper point followed by its length. The specifications of two segments are separated by a comma.

> 3 1 6, 2 8 2, 6 2 3,
> 5 6 6, 7 7 3, 9 3 3

> **Output format.** The output format is shown in Figure CE1.6.

1.7. Consider a graph $G = (V, E)$. A coloring of G is a mapping of vertices into colors $(1, \ldots, r)$ such that two adjacent vertices (i.e., two vertices connected by an edge) are assigned different colors. The goal is to minimize the number of colors r. The minimum value r is called the *chromatic number* of the graph. Design an algorithm for obtaining a coloring of a given graph G.

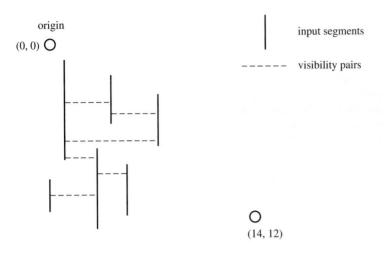

FIGURE CE1.6
Output format, visibility of vertical segments.

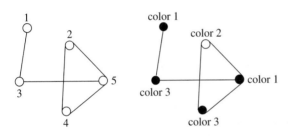

FIGURE CE1.7
Output format, coloring a graph.

Input format. The set of edges are given as input. An edge is specified by the two vertices it connects. Specifications of edges are separated by commas. An example of input follows.

3 1, 3 5, 2 4,
2 5, 4 5

Output format. The output format is shown in Figure CE1.7. You may either show color labels next to each vertex or actually color the vertices. (In the example, three colors are used and shown with different shades.)

1.8. Consider a graph $G = (V, E)$. An *independent set* $V_i \in V$ of G is a subset of vertices that are pairwise independent. A *maximum independent set* (MIS) is an independent set of maximum cardinality. That is, for every two vertices v_a and v_b in V_i there is no edge connecting v_a and v_b. Design and implement an algorithm that finds an MIS of a given graph G. Is your algorithm good? Verify your answer.

Input format. The input format is the same as the format given in Figure CE1.7.

Output format. The output should show the graph and highlight the selected vertices.

1.9. Consider a graph $G = (V, W)$. A *vertex cover* in G is a subset V_1 of vertices such that every edge of G has an end point in V_1. A *minimum vertex cover* (MVC) is a vertex cover of minimum cardinality.

Design a greedy algorithm for obtaining a MVC in a given graph G. Next, design a more sophisticated algorithm for the same problem. Compare the running time and quality of the two algorithms.

CHAPTER 2

THE TOP-DOWN APPROACH: PLACEMENT

In this chapter, we discuss the top-down approach to the VLSI layout problem. The circuit layout problem is partitioned into a collection of subproblems. Each subproblem should be solved efficiently to make subsequent steps easy. There are two major subproblems, placement and routing. This chapter focuses on the placement problem and the next chapter focuses on the routing problem.

There are three major problems related to the placement problem: partitioning, floorplanning, and placement. The focus is on minimizing the chip area and total wire length. In Chapter 4, these problems are dealt with considering other performance measures such as signal delays, number of vias, and power consumption.

Here the fundamental problems of circuit layout are discussed, and several techniques for each problem are described in order to familiarize the reader with more than one algorithm. At the end of each section, other known techniques, along with their advantages and disadvantages, are outlined.

2.1 PARTITIONING

Circuit partitioning is the task of dividing a circuit into smaller parts. The objective is to partition the circuit into parts such that the sizes of the components are within prescribed ranges and the number of connections between the components is minimized. Partitioning is a fundamental problem in many CAD problems. For example, in physical design, partitioning is a fundamental step in transforming a large problem into smaller subproblems of manageable size. Partitioning can be applied at many levels, such as the IC level (and the subIC level), the board level, and the system level.

Different partitionings result in different circuit implementations. Therefore, a good partitioning can significantly improve the circuit performance and reduce layout costs.

In order to explain the partitioning problem clearly, the following notations will be used. A graph $G = (V, E)$ consists of a set V of vertices, and a set E of edges. Each edge corresponds to a pair of distinct vertices (see Figure 2.1a). A hypergraph $H = (N, L)$ consists of a set N of vertices and a set L of hyperedges, where each hyperedge corresponds to a subset N_i of distinct vertices with $|N_i| \geq 2$ (see Figure 2.1b; e.g., the connection interconnecting vertices a, b, and c is a hyperedge). We also associate a vertex weight function $\omega : V \rightarrow N$ with every

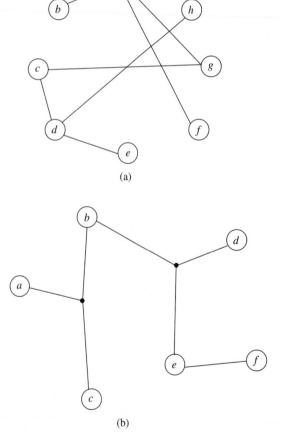

(a)

(b)

FIGURE 2.1
(a) Graph. (b) Hypergraph.

vertex, where N is the set of integers. Thus, a circuit can be represented by a graph or a hypergraph, where the vertices are circuit elements and the (hyper)edges are wires. The vertex weight may indicate the size of the corresponding circuit element.

A *multiway partition* of a (hyper)graph H is a set of nonempty, disjoint subsets of $N = \{N_1, \ldots, N_r\}$, such that $\cup_{i=1}^{r} N_i = N$ and $N_i \cap N_j = \emptyset$ for $i \neq j$. A partition is acceptable if $b(i) \leq \omega(N_i) \leq B(i)$, where $\omega(N_i)$ is the sum of the weight of vertices in N_i, $B(i)$ is the maximum size of part i and $b(i)$ is the minimum size of part i, for $i = 1, \ldots, r$; $B(i)$'s and $b(i)$'s are input parameters. A special case of multiway partitioning problem in which $r = 2$ is called the *bipartition* problem. In the bipartition problem, the bounds on the part size are usually chosen uniformly, $B(1) = B(2) = \lfloor \alpha \omega(N) \rfloor$, where $\omega(N)$ is the sum of the weight of the vertices, for some α, $\frac{1}{2} \leq \alpha < 1$, and $b(1) = b(2) = 1$. Typically, α is close to $\frac{1}{2}$. The number α is called the *balance factor*. The weight of either partition is no more than α times the total weight. The bipartition problem can also be used as a basis for heuristics in multiway partitioning.

In this section, several circuit partitioning techniques that are of fundamental importance in layout designs are discussed. Most techniques that have been proposed are variations of these. Since most algorithms work with graphs instead of hypergraphs, finding a suitable graph corresponding to a given hypergraph is discussed first.

2.1.1 Approximation of Hypergraphs with Graphs

The input circuit of a VLSI physical design problem is a hypergraph. Traditionally, it has been difficult to design efficient algorithms based on a hypergraph. Thus a circuit is usually transformed to a graph before subsequent algorithms are applied. The transformation process involves replacing a hyperedge with a set of edges such that the edge costs closely resemble the original hypergraph when processed in subsequent stages. In particular, it is best to have identical edge weight during partitioning.

Consider a hyperedge $e_a = (M_1, \ldots, M_n)$, $(n > 2)$ with n terminals. Let the weight of the hyperedge e_a be $w(e_a)$. One natural way to represent e_a is to put an edge between every pair of distinct modules (M_i, M_j) and assign weight $w(e_a)/n_0$ to each edge e_a in the graph, where n_0 is the number of added edges (i.e., $O(n^2)$). For $n = 3$, the cost of any bipartition of the modules M_i on the resulting graph is identical to that of the hypergraph (weights have to be set to 0.5) (see Chapter 2 Exercises). However, for $n \geq 4$ the cost of bipartition on the graph is an overestimate of the original cost in the hypergraph. In practice, a large number of nets have a small terminal count and the approximation works well. One possible solution is to put varying weight on the hyperedges. These weights should (probably) depend on the problem itself and the problem instance.

Recently, it was shown [162] that, in general, the weight of the edges cut by any bipartition in the graph is not equal to the weight of the hyperedges cut by the same bipartition in the hypergraph. However, the transformation can be done if the addition of dummy vertices and the use of positive and negative weights are allowed. For example, some CAD tools proceed as follows. In order to transform a hypergraph into a graph, each hyperedge is transformed into a complete graph. A minimum spanning tree of the complete graph is obtained. The result is a (in general, weighted) graph. Partitioning algorithms are applied to the resulting graph. An outline of the procedure follows.

> *Procedure:* Hypergraph-transform(H)
> > *begin-1*
> > > *for* each hyperedge $e_a = (M_1, \ldots, M_n)$ *do*
> > > > *begin-2*
> > > > > form a complete graph G_a with vertices $(M_1, \ldots, M_n)$;
> > > > > weight of an edge (M_i, M_j) is proportional
> > > > > > to the number of hyperedges of the circuit
> > > > > > between M_i and M_j;
> > > > > find a minimum-spanning tree T_a of G_a;
> > > > > replace e_a with the edges of T_a in the
> > > > > > original hypergraph;
> > > > *end-2*;
> > *end-1*.

An effective way of representing hyperedges with graph edges is still unknown. For the rest of this book, it is assumed that a graph representation of the given circuit is available. In some cases, a hypergraph is used directly.

2.1.2 The Kernighan-Lin Heuristic

To date, iterative improvement techniques that make local changes to an initial partition are still the most successful partitioning algorithms in practice. One such algorithm is an iterative bipartitioning algorithm proposed by Kernighan and Lin [181].

Given an unweighted graph G, this method starts with an arbitrary partition of G into two groups V_1 and V_2 such that $|V_1| = |V_2| \pm 1$, where $|V_1|$ and $|V_2|$ denote the number of vertices in subsets V_1 and V_2, respectively. After that, it determines the vertex pair (v_a, v_b), $v_a \in V_1$ and $v_b \in V_2$, whose exchange results in the largest decrease of the cut-cost or in the smallest increase if no decrease is possible. A cost increase is allowed now in the hope that there will be a cost decrease in subsequent steps. Then the vertices v_a and v_b are locked. This locking prohibits them from taking part in any further exchanges. This process continues, keeping a list of all tentatively exchanged pairs and the decreasing gain, until all the vertices are locked.

A value k is needed to maximize the partial sum $\sum_{i=1}^{k} g_i = \text{Gain}_k$, where g_i is the gain of ith exchanged pair. If $\text{Gain}_k > 0$, a reduction in cut-cost can be achieved by moving $\{v_{a_1}, \ldots, v_{a_k}\}$ to V_2 and $\{v_{b_1}, \ldots, v_{b_k}\}$ to V_1. After this is done, the resulting partition is treated as the initial partition, and the procedure is repeated for the next pass. If there is no k such that $Gain_k > 0$ the procedure halts. A formal description of the Kernighan-Lin algorithm is as follows.

> *Procedure:* Kernighan-Lin heuristic(G)
>   *begin-1*
>   bipartition G into two groups V_1 and V_2, with $|V_1| = |V_2| \pm 1$;
>   *repeat-2*
>   *for $i = 1$ to $n/2$ do*
>   *begin-3*
>   find a pair of unlocked vertices $v_{a_i} \in V_1$ and $v_{b_i} \in V_2$
>   whose exchange makes the largest decrease or
>   smallest increase in cut-cost;
>   mark v_{a_i} ,v_{b_i} as locked and store the gain g_i;
>   *end-3*
>   find k, such that $\sum_{i=1}^{k} g_i = Gain_k$ is maximized;
>   *if $Gain_k > 0$ then*
>   move $v_{a_1}, \ldots, v_{a_k}$ from V_1 to V_2 and $v_{b_1}, \ldots, v_{b_k}$
>   from V_2 to V_1;
>   *until-2 $Gain_k \leq 0$;*
>   *end-1.*

Figure 2.2a illustrates the algorithm. Assume an arbitrary partition of G into two sets $V_1 = \{a, b, c, d\}$ and $V_2 = \{e, f, g, h\}$ (see Figure 2.2a). Figure 2.2b shows a possible choice of vertex pairs step by step for the partition shown in Figure 2.2a. In this example, k is equal to 2. Exchanging the two vertex sets $\{c, d\}$ and $\{f, g\}$, yields a minimum cost solution. For example, in the figure, the final cost is equal to the original cost minus the gain of step 1 minus the gain of step 2, that is, $5 - 3 - 1 = 1$.

The for-loop is executed $O(n)$ times. The body of the loop requires $O(n^2)$ time. More precisely, step 1 takes $\frac{n}{2} \times \frac{n}{2}$ time and step i takes $(\frac{n}{2} - i + 1)^2$ time. Thus, the total running time of the algorithm is $O(n^3)$ for each pass of the repeat loop. The repeat loop usually terminates after several passes, independent of n. Thus, the total running time is $O(cn^3)$, where c is the number of times the repeat loop is executed.

In general, the bipartition obtained from the Kernighan-Lin algorithm is a local optimum rather than a global optimum. In their original paper, Kernighan and Lin gave further heuristic improvements of their algorithm (such as finding a good starting partition) that are aimed at finding better bipartitions. The extension of the Kernighan-Lin algorithm to multiway (i.e., r-way) partitioning can be easily done as follows. Start with an arbitrary partition of the graph into r equal-sized sets. Then, apply the algorithm to pairs of subsets so as to make the partition as close

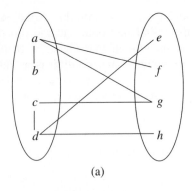

(a)

Step no.	Vertex pair	Gain	Cut-cost
0	–	0	5
1	{ d, g }	3	2
2	{ c, f }	1	1
3	{ b, h }	−2	3
4	{ a, e }	−2	5

(b)

FIGURE 2.2
An example demonstrating the Kernighan-Lin algorithm.

as possible to being pairwise optimal. There are $\binom{r}{2}$ pairs of subsets to consider, so the time complexity for one pass through all pairs is $\binom{r}{2}\left(\frac{n}{r}\right)^3$, that is, $O\left(\frac{n^3}{r}\right)$.

Consider the circuit shown in Figure 2.3a. Net 1 connects modules 1, 2, 3, 4; net 2 connects modules 1, 5; and net 3 connects modules 4, 5. The usual representation of the circuit as a graph is shown in Figure 2.3b (each hyperedge interconnecting a collection of terminals is replaced by a complete graph on the same set of terminals). However, the effect of this edge-cut model, which is used in the Kernighan-Lin algorithm, is to exaggerate the importance of any net with

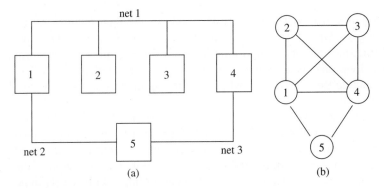

(a)   (b)

FIGURE 2.3
Example of a circuit and the corresponding graph.

more than two terminals. (Generally, the edge-cut model treats a k-terminal net as $(k-1) + (k-2) + \ldots + 1$ two-terminal nets.) A net-cut model was proposed by Schweikert and Kernighan [317] to handle multiterminal cases. The following statement is the essential part of this model: a net which is divided by the partition requires exactly one wire to connect its terminals from one set to the other. In the bipartitioning algorithm to be described next, the net-cut model will be applied to deal with hyperedges.

2.1.3 The Fiduccia-Mattheyses Heuristic

Fiduccia and Mattheyses [100] improved the Kernighan-Lin algorithm by reducing the time complexity per pass to $O(t)$, where t is the number of terminals in G. For example, in Figure 2.1b, t is equal to eight. They introduced the following new elements to the Kernighan-Lin algorithm:

1. Only a single vertex is moved across the cut in a single move.
2. Vertices are weighted.
3. A special data structure is used to select vertices to be moved across the cut to improve running time (the main feature of the algorithm).

Consider (3), the data structure used for choosing the (next) vertex to be moved. Let the two partitions be A and B. The data structure consists of two pointer arrays, list A and list B indexed by the set $[-d_{max} \cdot w_{max}, d_{max} \cdot w_{max}]$ (see Figure 2.4). Here d_{max} is the maximum vertex degree in the hypergraph, and w_{max} is the maximum cost of a hyperedge. Moving one vertex from one set to the other will change the cost by at most $d_{max} \cdot w_{max}$. Indices of the list correspond to possible (positive or negative) gains. All vertices resulting in gain g are stored in the entry g of the list. Each pointer in the array list A points to a linear list of unlocked vertices inside A with the corresponding gain. An analogous statement holds for list B.

Since each vertex is weighted, a maximum vertex weight W is defined such that the *balanced partition* is maintained during the process. W must satisfy $W \geq w(V)/2 + \max_{v \in V}\{w(v)\}$, where $w(v)$ is the weight of vertex v. An initial balanced partition is one with either side of the partition having a total vertex weight of at most W, that is, $w(A), w(B) \leq W$. An initial balanced partition can be obtained by sorting the vertex weights in decreasing order, and placing them in A and B alternately.

This algorithm starts with a balanced partition A, B of G. Note that a move of a vertex across the cut is allowable if such a move does not violate the balance condition. To choose the next vertex to be moved, consider the maximum gain vertex a_{max} in list A or the maximum gain vertex b_{max} in list B, and move both across the cut if the balance condition is not violated. As in the Kernighan-Lin algorithm, the moves are tentative and are followed by locking the moved vertex. A move may increase the cut-cost. When no moves are possible or if there are

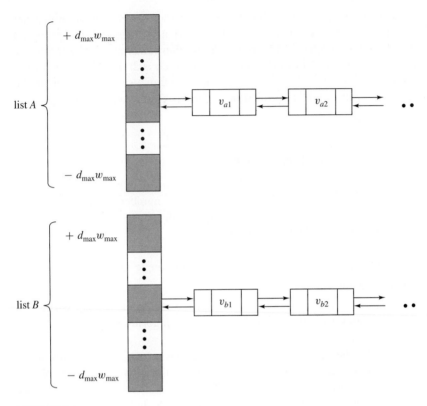

FIGURE 2.4
The data structure for choosing vertices in the Fiduccia-Mattheyses algorithm.

no more unlocked vertices, choose the sequence of moves such that the cut-cost is minimized. Otherwise the pass is ended.

Theorem 2.1. The running time of the Fiduccia-Mattheyses algorithm is $O(t)$, where t is the number of terminals.

Proof. Data structures list A and list B are easily initialized by traversing all hyperedges, one by one. Assume some of the hyperedges $e_1, \ldots, e_{a-1}$ have been visited and the lists have been initialized based on them. That is, the list is correctly initialized for the hypergraph induced by $e_1, \ldots, e_{a-1}$. Now consider $e_a = \{M_1, \ldots, M_d\}$. Let num$A$ be the number of modules connected by e_a that are in A. If num$A = 0$, the gain of each module $M_i \in \{M_1, \ldots, M_d\}$ (they are all in B) is decreased by $w(e_a)$. When M_i is moved to the other partition, the cut-cost is increased by $w(e_a)$. If num$A = 1$, the gain of the one module $M_i \in \{M_1, \ldots, M_d\}$ that is in A is increased by $w(e_a)$. If num$A > 1$, no actions are taken unless num$B = 0$ or num$B = 1$. A similar case analysis holds for num$B$, where num$B$ is the number of modules connected by e_a that are in B. Note that the cases for num$A$ and num$B$ are not necessarily disjoint. Both must be examined and then actions taken as dictated above. $O(d)$ time is spent processing e_a. Note that summation of d's over all hyperedges is simply the total number of terminals. Finding the modules with maximum gain in list A and list B takes a constant time, if this value is maintained and updated while processing e_a. Thus the time complexity of building data structures list A and list B is $O(t + n)$, where t is the number of terminals in G and n is the number of modules.

In one pass of the algorithm, find a module with maximum gain. Finding such a module is easy if doubly linked pointers are kept from the max gain in list A to the second gain (i.e., the second nonempty element in list A), from the second gain to the third, and so on (and similarly for list B).

Without loss of generality, assume this module is in A (it is being moved to B), and denote it by M_i. Repeat the following for each hyperedge that is connected to M_i. Let e_a be one of the hyperedges that is connected to M_i.

Case 1. If all modules of e_a are in A (that is, right before moving M_i) then the gain of all these modules will be increased by $w(e_a)$. This is the case because M_i will be locked in B and moving the other modules of e_a cannot change the gain (unless all of them are moved).

Case 2. If all modules of e_a are in B (that is, right after moving M_i) then the gain of each module of e_a will be decreased by $w(e_a)$.

For each hyperedge e_i it is necessary to modify the gains only in Case 1 and Case 2. It takes $O(d_a)$ time to update the gains in each case, where d_a is the number of modules connected to e_a, that is, the number of terminals of e_a. Since each hyperedge will be considered twice (once in Case 1 and once in Case 2), it takes time proportional to the number of terminals to update the lists. The total time complexity of the algorithm is $O(ct)$, where c is the number of passes and is usually a small constant (independent of t).

Note that if a move cannot be made, the algorithm terminates.

Note that the Fiduccia-Mattheyses algorithm chooses arbitrarily between vertices that have equal gain and equal weight. A further improvement was proposed by Krishnamurthy [190]. He introduced a look-ahead ability into the algorithm. Thus the best candidate among such vertices can be selected with respect to the gains they make possible in later moves. Subsequently, Sanchis [301] extended Krishnamurthy's algorithm to handle multiway network partitioning and showed that the technique was especially useful for partitioning a network into a large number of subsets.

2.1.4   Ratio Cut

Almost all circuit layout techniques have the tendency to put components of similar functionality together to form a strongly connected group. Predefining the partition size, as the Kernighan-Lin algorithm does, may not be well-suited for hierarchical designs because there is no way to know cluster size in circuits before partitioning. The notion of *ratio cut* was, presented to solve the partitioning problem more naturally. Earlier work on ratio cut, establishing the effectiveness of the technique, appeared in [286]. An approximation algorithm was proposed in [208]. The approach is based on a current flow formulation. The technique was

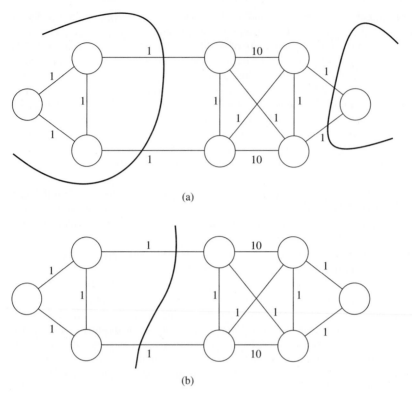

FIGURE 2.5
An eight node example: (a) the Kernighan-Lin approach; (b) the ratio cut approach.

extended to weighted nodes and hypergraphs in [236, 356]. Here the ratio cut algorithm proposed in [378] will be presented.

The ratio cut approach concept can be described as follows. Given a graph $G = (V, E)$, (V_1, V_2) denotes a cut that separates a set of nodes V_1 from its complement $V_2 = V - V_1$. Let c_{ij} be the cost of an edge connecting node i and node j. The *cut-cost* is equal to $C_{V_1 V_2} = \sum_{i \in V_1} \sum_{j \in V_2} c_{ij}$. The *cut-cost ratio* is defined as $R_{V_1 V_2} = C_{V_1 V_2}/(|V_1| * |V_2|)$, where $|V_1|$ and $|V_2|$ denote the size of subsets V_1 and V_2, respectively. The objective is to find a cut that generates the minimum ratio among all cuts in the graph. Figure 2.5 shows an eight node example, which will be partitioned as Figure 2.5a if the Kernighan-Lin algorithm is applied, with the subset size predefined as one-half of its total size; the cost is equal to four. A better partition, with cost equal to two, is obtained by applying the ratio cut algorithm as shown in Figure 2.5b.

Like many other partitioning problems, finding an optimal ratio cut in general graphs is NP-complete. Therefore, a good and fast heuristic algorithm is needed for dealing with complex VLSI circuits. The ratio cut algorithm consists of three major phases: initialization, iterative shifting, and group swapping. These phases are discussed in more detail below.

- **Initialization**
  1. Randomly choose a node s. Find the longest path starting from s (i.e., find a node farthest from s). The node at the end of one of the longest paths is called t. Let $X = \{s\}$, and $Y = V - \{s, t\}$.
  2. Choose a node i in Y whose movement to X will generate the best ratio among all the other competing nodes. Move node i from Y to X; update $X = X \cup \{i\}$, and $Y = Y - \{i\}$.
  3. Repeat step 2 until $Y = \phi$.
  4. The cut giving the minimum ratio found in the procedure forms the initial partitioning.

  Suppose seed s is fixed at the left end, and seed t at the right end. A *right-shifting operation* is defined as shifting the nodes from s toward t. The partition with the best resultant ratio is recorded. A left-shifting operation is similarly defined.

- **Iterative shifting.** Once an initial partitioning is made, repeat the shifting operation in the opposite direction (that is, start moving the nodes from X to Y) to achieve further improvement. Once $X = \{s\}$, switch direction again and move nodes from Y to X. The same procedure is applied again, starting with the best partition obtained in the previous application of iterative shifting.

  Figure 2.6 is a two-terminal net example to show the effect of iterative shifting operations after the initialization phase. Suppose that two seeds s and t have been chosen. From s to t, a cut with the ratio $\frac{1}{36}$ (that is $4/(12 \times 12)$) can be found in Figure 2.6a; from t to s, a cut with the ratio $\frac{1}{28}$ ($= 3/(10 \times 14)$) is shown in Figure 2.6b. Since the ratio of the first cut is lower than that of the second cut, the first cut is used as the initial partitioning. Then, applying a left-shifting operation on the cut in Figure 2.6a, find a new cut with better ratio $3/(8 \times 16)$, as illustrated in Figure 2.6c. That is followed by the right-shifting operations in the opposite direction as shown in Figure 2.6c. We finally reach a cut with the ratio $3/(12 \times 12) = \frac{1}{48}$, which is the optimal ratio cut in this example.

- **Group swapping.** After the iterative shifting is complete, a local minimum is reached because a single node movement from its current subset to the other cannot reduce the ratio value. A group swapping is utilized to make further improvement.

  The *ratio gain* $r(i)$ of a node i is defined as the decrease in ratio if node i were to move from its current subset to the other. The ratio gain is not defined for the two seeds s and t. The ratio gain could be a negative real number if the node movement increases the ratio value of the cut. The process is as follows.

  1. Calculate the ratio gain $r(i)$ for every node i, and set all nodes to be in the unlocked state.
  2. Select an unlocked node i with the largest ratio gain from two subsets.
  3. Move node i to the other side and lock it.
  4. Update the ratio gains for the remaining affected and unlocked nodes.

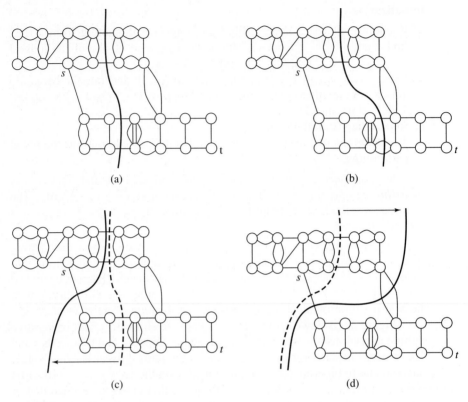

FIGURE 2.6
An example of the first two phases. (a) The initialization phase from s to t; (b) the initialization phase from t to s; (c) left-iterative shifting; (d) right-iterative shifting.

5. Repeat steps 2–4 until all nodes are locked.
6. If the largest accumulated ratio gain during this process is positive, swap the group of nodes corresponding to the largest gain, and go to step 1; otherwise, output the previous partition and stop.

Theorem 2.2. The running time of the ratio-cut algorithm is $O(|E|)$ time.

Proof. Examine the time complexity of each phase separately. In the initialization step, randomly find a node s. Node t (the node farthest from s) can be obtained by breadth-first search in $O(|E|)$ time. There are two lists, list A and list B, to store nodes in X and Y, respectively. As in the Fiduccia-Mattheyses algorithm, each list is indexed by the set $[-d_{max} \cdot w_{max}, d_{max} \cdot w_{max}]$. As before, d_{max} is the maximum vertex degree in the graph, and w_{max} is the maximum cost of an edge. Finding a node with maximum gain is readily done. There is one difference between this algorithm and that of Fiduccia-Mattheyses. In both cases, the gain is stored. However, in this algorithm, the ratio cost (i.e., cut-cost divided by the product of the sizes of the partitions) must be considered. The best ratio cost is obtained either by a node in list A with maximum gain or by a node in list B with maximum gain. Thus both nodes must be examined and the one that results in the best ratio cost should be moved. The initialization step takes $O(|E|)$ time as discussed in the proof of the Fiduccia-Mattheyses algorithm.

The next phase is iterative shifting. Since nodes were moved from B to A, now (in the first round) move the node with maximum gain in list A to list B. Repeat until $X = \{s\}$. Then the second round starts. In this round, move the node with maximum gain in list B to list A. These rounds are repeated for a constant number of steps or until no further gain is obtained. Each round takes $O(|E|)$ time.

The last phase is group swapping. First find a node in list A or a node in list B whose move to the opposite partition results in the best ratio cost. To accomplish this, examine only the node with maximum gain in list A and the node with maximum gain in list B. Calculation of gains (step 1) are done in list A and list B. Each round (steps 2–4) takes $O(|E|)$ time. In practice it is sufficient to repeat these steps a constant number of times, independent of the problem size.

Based on the proof of Theorem 2.1, the time complexity of the Fiduccia-Mattheyses algorithm, and the proof of Theorem 2.2, the following is established.

Theorem 2.3. The running time of the ratio-cut algorithm in hypergraphs is $O(t)$ time, where t is the number of terminals.

The concept of ratio cut has been applied to find a two-way stable partition [51], that is, a partition whose final result has little relationship to the choice of initial partition. First, using the ratio-cut algorithm, the graph is partitioned into a collection of smaller, and highly connected, subgraphs. Next, the subgraphs are rearranged into two subsets that match the specified size constraints. Finally, the Fiduccia-Mattheyses algorithm is employed to further improve the result. Good experimental results have been obtained for both cut-weight and CPU time with this approach.

Other approaches for finding a ratio cut have been proposed. One of the more interesting approaches is based on the *spectral* method, which use eigenvalues and eigenvectors of the matrices derived from the netlist. Early studies link the spectral properties and the partition properties of the graph. The algorithm proposed in [123] establishes a connection between graph spectra and ratio cuts and proposes an algorithm for solving the ratio-cut problem based on the underlying spectral properties.

2.1.5 Partitioning with Capacity and I/O Constraints

In traditional partitioning schemes, the main emphasis is on minimizing the number of edges between the set of partitions and balancing the size of the two partitions. In many applications, for example, in multichip modules (MCMs), a capacity constraint is imposed on each set; this is due to the area constraint of the

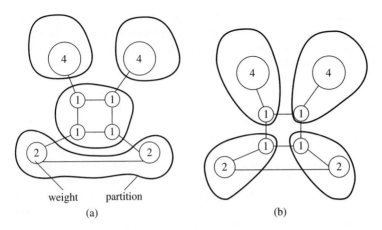

FIGURE 2.7
An example of partitioning with capacity and I/O constraints. (a) Capacity constraint = 4, I/O constraint = 4. (b) Capacity constraint = 5, I/O constraint = 3.

module (e.g., in MCMs, each chip defines a module). Furthermore, the number of edges connected to each partition should be constrained; this is due to the constraints on the number of I/O pads in each module.

A technique for partitioning with capacity and I/O constraints was proposed in [328]. Consider the two partitions of a given circuit shown in Figure 2.7 (the example is from [328]). The partition shown on the left has a capacity constraint of 4 and an I/O constraint of 4, whereas the partition shown on the right has a capacity constraint of 5 and an I/O constraint of 3. As the example demonstrates, in general there is a trade-off between capacity and I/O constraints.

Note that if the aim is to minimize the sum of the edges cut by the partition, the capacity constraints may not be satisfied. For example, if a partition is made as in Figure 2.7a, the result is a total of four edges cut. However, such a partition may not satisfy the given I/O constraints.

The algorithm proposed in [328] for solving the problem is based on mathematical programming. A 0/1 variable x_{ij} indicates if module M_i is to be assigned to partition P_j. Since each module can be assigned to only one partition,

$$\sum_j x_{ij} = 1 \quad \text{for all } i.$$

Since the capacity of partition P_j is bounded by a constant c_j,

$$\sum_i w_i x_{ij} \le c_j \quad \text{for all } j,$$

where w_i is the size of module M_i. Since the I/O of partition P_j is bounded by a constant p_j,

$$\sum_{i_1} \sum_{i_2} a_{i_1 i_2} [x_{i_1 j}(1 - x_{i_2 j}) + x_{i_2 j}(1 - x_{i_1 j})] \le p_j \quad \text{for all } j,$$

where $a_{i_1 i_2}$ is the number of connections from M_{i_1} to M_{i_2}.

A number of objective functions in the above formulation may be included. For example, the number of all nets cut by the partitions may be minimized. The above formulation is a special case of the *quadratic Boolean programming* problem. A heuristic for solving the general problem was proposed in [35]. Experimental results, reported in [328], are very promising.

2.1.6 Discussion

Many other approaches have been proposed for solving circuit partitioning problems. Perhaps the oldest techniques are the *constructive methods* that follow the greedy paradigm. The goal is to start placing the nodes in one of the two partitions. At each stage, the goal is to minimize the cost of the subproblem considered so far. This is also called the *aggregation algorithm*. For more details see [131, 195].

Clustering [43, 315] is an intuitive method for building up clusters based on interconnections among components. Usually, certain seeds are given in input specifications as starting points. Neighboring components with strong connections to each seed are merged with that seed. This approach is limited by the lack of a global view of the connectivity of the entire system. Therefore, clustering is commonly used to yield an initial partitioning.

The *maximum-flow minimum-cut algorithm* was proposed by Ford and Fulkerson [101]. They transformed the minimum cut problem into the maximum flow problem. In order to separate a pair of nodes into two subsets, the minimum number of crossing edges is equal to the maximum amount of flow from one node to the other. Although this algorithm can find the optimal solution between any pair of vertices in a network, there is no constraint on the sizes of resultant subsets. In practice, the result will be useless if two very unevenly sized subsets are generated. Other techniques related to this method can be found in [171, 207].

Eigenvector decomposition [24, 104] is a method for finding an allocation by some metric other than the graph structure, reflecting the connectivity of the graph. It requires the transformation of every multiterminal net into several two-terminal nets first. The connections are represented in a matrix, and the eigenvectors of the matrix define the locations of all components and thus derive partitioning results.

Simulated annealing [186, 246] is another successful example of the iterative improvement method. The objective function in simulated annealing is analogous to energy in a physical system, and the moves are analogous to changes in the energy of the system. Moves are selected randomly, and the probability that a move is accepted is proportional to the system's current temperature. Accepting moves that increase the solution's cost stochastically allows simulated annealing to explore the design space more thoroughly and extricate the solution from a local optimum. This technique usually produces good results at the expense of very long running time. The structure of the simulated annealing paradigm will be discussed in a later section.

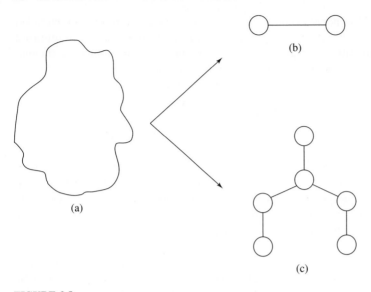

FIGURE 2.8
The MCTP model compared to standard min-cut bipartitioning model. (a) Hypergraph G. (b) Standard min-cut bipartitioning. (c) MCTP partitioning.

Recently, a generalization of the min-cut bipartitioning problem, in which a single edge e is extended to an arbitrary tree T (see Figure 2.8), called *min-cost tree partitioning* (MCTP) [374], was introduced. The MCTP problem is one of finding a feasible mapping of nodes of a hypergraph G onto the vertices of a tree such that when the hyperedges of G are globally routed on the edges of the tree, the sum of the "densities" or "flows" on the edges is minimized. As before, a mapping is said to be feasible if the sum of the sizes of the nodes of G mapped to a vertex (or partition set) of T does not exceed the capacity of the vertex. The standard min-cut bipartitioning problem is when the tree T is a single edge connecting two vertices. Note that each edge in the tree corresponds to a cut in hypergraph G.

MCTP can be used to model different kinds of partitioning problems in VLSI layout design, for example, simultaneous multiway partitioning, and partitioning within nonregular regions. The running time of this approach is $O(P \times D \times t^3)$, where P is the total number of terminals in the hypergraph G, D is the number of nodes in the hypergraph G, and t is the number of vertices in the tree T.

In summary, the Kernighan-Lin algorithm and its variants are simple and have faster running times compared with other methods however, they may fall into local minimum solutions, although experiments show that this does not happen often. Simulated annealing allows a designer to come arbitrarily close to the optimal solution statistically, but the user needs to trade off quality and computation time. Eigenvector decomposition is considerably more costly in computing time than other approaches, since it requires either the calculation of eigenvalues or the inversion of matrices. However, it often gives superior results. Very recently, some hybrid algorithms, such as finding the relationship between ratio cut and eigenvalue first and then solving the problem by eigenvector decomposition [68, 122], or incorporating cluster properties into simulated annealing [297], have been proposed to further improve the partitioning solutions.

The generalization of the bipartitioning, multiway partitioning, is also an important issue in VLSI design. However, it is known that the graph and network multiway partitioning problems are in general NP-hard [113]. Therefore, heuristics are needed to find good approximate solutions in polynominal time. Several approaches have been proposed [16, 181, 207]. Most of them adopt a bipartition algorithm iteratively to find a multiway partition. For example, for a four-way partition, we would use the bipartition algorithm to partition the nodes into two blocks, and then partition each of these blocks into two sub-blocks. As pointed out in [181], a bad result in the first partition may bias the second one (and the rest), if a bipartition algorithm is used hierarchically.

An alternative way to reduce the multiway partition problem into several bipartition problems is to attempt to improve the partition uniformly at each step. The author in [190] introduced the concept of level gains in computing gains in the cutset size. Sanchis [301] adapted the level gain concept to improve the partition uniformly with respect to all blocks as opposed to making repeated or hierarchical uses of bipartition algorithm. Experiments showed that the concept is especially useful for partitioning into a large number of subsets.

2.2 FLOORPLANNING

Given a circuit $C = (\mathcal{M}, \mathcal{N})$ represented by a hypergraph, the *floorplanning* problem is to determine the approximate location of each module in a rectangular chip area. The floorplanning problem in chip layout is analogous to floorplanning in building design where there is a set of rooms (modules) and the approximate location of each room must be determined based on some proximity criteria.

An important step in floorplanning is to decide the relative location of each module. A good floorplaning algorithm should:

- minimize the total chip area;
- make the subsequent routing phase easy; and
- improve performance, by, for example, reducing signal delays.

In general, it is difficult to achieve all these goals simultaneously since they mutually conflict. Furthermore, it is algorithmically difficult to consider all (or several) objectives simultaneously.

To facilitate subsequent routing phases, modules with relatively high connections should be placed close to one another. The set of nets $\mathcal{N}$ thus defines the closeness of the modules. Placing highly connected modules close to each other reduces routing space. Various floorplanning approaches have been proposed based upon the measure of closeness.

A floorplan is usually represented by a *rectangular dissection*. The border of a floorplan is usually a rectangle since this is the most convenient shape for chip

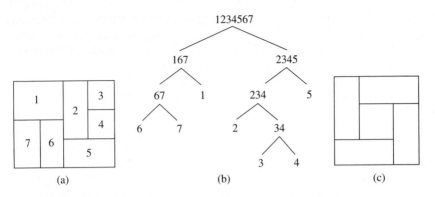

FIGURE 2.9
(a) A sliceable floorplan and (b) its corresponding tree. (c) A nonsliceable floorplan.

fabrication. The rectangle is dissected with several straight lines which mark the borders of the modules. The lines are usually restricted to horizontal and vertical lines only. Often, modules are restricted to rectangles to facilitate automation.

The restriction on the module organization of a floorplan is an important issue. Although in the most general case, the modules can have arbitrary organization, imposing some restrictions on the floorplan may have advantages. Several restricted floorplans and their characteristics will be discussed.

A *sliceable floorplan* is one of the simplest types of floorplans. Formally, a sliceable floorplan is recursively defined as follows.

1. A module, or

2. A floorplan that can be bipartitioned into two sliceable floorplans with a horizontal or vertical cutline.

Figure 2.9a shows a sliceable floorplan. By definition, a sliceable floorplan can be represented by a binary tree, as shown in Figure 2.9b. Each branch of the tree corresponds to a module and each internal node corresponds to a partial floorplan with the corresponding slice. To contrast, a smallest nonsliceable floorplan (i.e., one with the minimum number of modules) is shown in Figure 2.9c.

A sliceable floorplan is a special case of a *hierarchically defined floorplan*. Loosely speaking, a hierarchically defined floorplan is a floorplan that can be described by a floorplan tree. The leaves of the tree correspond to modules and an internal node defines how its child floorplans are combined to form a partial floorplan. Various restrictions can be imposed on the internal nodes, such as:

1. The number of child floorplans cannot exceed a certain limit.

2. The shape of the floorplan corresponding to the internal node is limited, for example, to rectangles or other simple shapes.

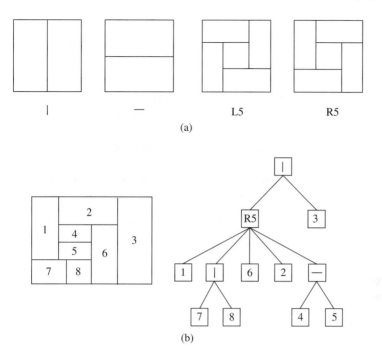

FIGURE 2.10
Hierarchical floorplan of order 5. (a) Templates. (b) An example, floorplan and tree.

A well-known example of a hierarchically defined floorplan is the hierarchical floorplan of order 5. There are four types of internal nodes, as shown in Figure 2.10a. An example is shown in Figure 2.10b. This class of floorplans offers a way to implement simple nonsliceable floorplan configurations in a hierarchical framework.

Rectangular floorplans are the most general class of floorplans considered here. There is no restriction with respect to the organization of the modules in this type of floorplan. The only restriction is that all modules and boundaries of the floorplans must be rectangular. In general, a module could be allowed to have a shape other than a rectangle; however, the usual practice is to restrict it to rectilinear shapes.

The complexity of a rectilinear shape can be measured by the number of concave corners. A rectangle has zero concave corners. An L-shaped module has one concave corner. There are four types of rectilinear shapes with two concave corners, if rotation symmetry is disregarded. The number of convex corners minus the number of concave corners is always four. The number of rectilinear shapes grows exponentially with the number of concave corners.

A sliceable floorplan is one of the simplest floorplan configurations. Because of its inherent binary tree structure, it facilitates efficient algorithm design. For example, a floorplan-sizing problem was shown to be NP-complete for general

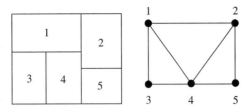

FIGURE 2.11
A rectangular floorplan and its dual graph.

(nonsliceable) floorplans but optimally solvable in polynomial time for sliceable floorplans [270, 344]. Also the routing channel ordering problem for sliceable floorplan is trivial. A hierarchically defined floorplan offers more flexibility in positioning the modules but increases the computational complexity. A hierarchical approach to floorplanning was reported in [235], and research on this class of floorplans provides some encouraging results [376, 391]; however, evidence suggests that the sizing optimization problem is intractable [377].

2.2.1   Rectangular Dual Graph Approach to Floorplanning

The rectangular dual graph approach to floorplanning is based on the proximity relation of a floorplan. A *rectangular dual graph* (*dual graph*, for short) of a rectangular floorplan is a plane graph $G = (V, E)$ where V is the set of modules and $(M_i, M_j) \in E$ if and only if modules M_i and M_j are adjacent in the floorplan. Figure 2.11, shows a rectangular floorplan and its corresponding rectangular dual graph.

Without loss of generality, it can be assumed that a rectangular floorplan contains no cross junctions. A cross junction, as shown in Figure 2.12, can be replaced by two T-junctions with a sufficiently short edge e. Under this assumption, the dual graph of a rectangular floorplan is a *planar triangulated graph* (PTG). Each T-junction of the floorplan corresponds to a triangulated face of the dual graph.

Every dual graph of a rectangular floorplan is a PTG. However, not every PTG corresponds to a rectangular floorplan. Consider the PTG in Figure 2.13a. By inspection, one can see that it is impossible to satisfy the adjacency requirements of edges (a, b), (b, c) and (c, a) simultaneously with three rectangular modules when the cycle (a, b, c) is not a face, as shown in Figure 2.13b. A cycle (a, b, c) of length 3 which is not a face in the dual graph is called a *complex triangle*.

If a PTG contains a complex triangle as a subgraph, a corresponding rectangular floorplan does not exist. It turns out that this is the only forbidden pattern for a rectangular floorplan; a rectangular floorplan exists if and only if

FIGURE 2.12
Replacing a cross junction with two T-junctions.

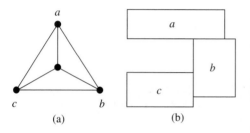

(a)

(b)

FIGURE 2.13
(a) A PTG containing a complex triangle.
(b) Forbidden pattern in rectangular dual graph.

the extended dual (defined later) of the floorplan contains no complex triangles [189]. The algorithm for constructing a rectangular floorplan, if one exists, ran in $O(n^2)$ time for a floorplan with n modules; this was later improved to linear time [19].

Assume that the floorplan F is enclosed by four infinite regions r, u, l, b, as shown in Figure 2.14b by the dual graph of the floorplan. Normally, this corresponds to a planar triangulated graph G. Appending the vertices r, u, l, b and some edges will produce a graph G_e, which has the general configuration shown in Figure 2.14a. Such a graph is called an *extended dual*.

Let n be the number of vertices of an arbitrary extended dual $G_e(n)$. By induction, it is hypothesized that it is possible to generate a floorplan for any dual graph $G_e(k)$ for $k < n$. If the external vertices r, u, l, b of graph G_e are examined, there are two possible cases: a) some of the vertices of r, u, l, b have degree 3, or b) none of the vertices has degree 3.

Consider the first case. Without loss of generality, it can be assumed that vertex r has degree 3. Since $(r, u), (r, b) \in G_e(n)$, there is exactly one edge (r, v) where $v \notin \{r, u, l, b\}$. The configuration of graph $G_e(n)$ is shown in Figure 2.15a. Remove vertex r from $G_e(n)$ and rename vertex v as r. The resulting graph $G_e(n-1)$ (Figure 2.15b) satisfies the hypothesis, and a floorplan $F(n-1)$ can be obtained (Figure 2.15c). The floorplan $F(n)$ of $G_e(n)$ is obtained by placing $F(n-1)$ on the left and module v on the right, as shown in Figure 2.15d.

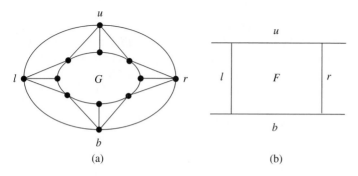

(a)

(b)

FIGURE 2.14
(a) Extended dual. (b) Floorplan and dual enclosed by four infinite regions.

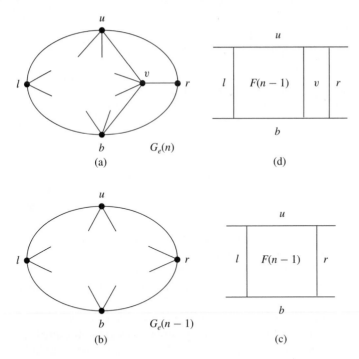

FIGURE 2.15
Dual decomposition and floorplan merging of rectangular graph.

Now consider the case where none of the vertices r, u, l, b has degree 3. In this case, find a path $P_V = \{u = p_1, p_2 \ldots, p_k = b\}$ in $G_e(n)$ from u to b with the following properties:

1. $p_2, \ldots, p_{k-1} \notin \{r, u, l, b\}$;
2. $(p_i, p_j) \notin G_e(n)$, for all distinct i, j; and
3. $(p_i, l) \notin G_e(n)$, for some i; and $(p_j, r) \notin G_e(n)$, for some j.

Such a path P_V is called a vertical *splitting path*. Similarly, a horizontal splitting path from l to r can be defined. It can be shown that at least one such path exists if $G_e(n)$ contains no complex triangles. $G_e(n)$ decomposes along P_V to obtain two subgraphs G_l and G_r. G_l consists of vertices to the left of (and including) P_V, and similarly G_r consists of vertices to the right of (including) P_V. Vertices of path P_V are duplicated so that they appear in both subgraphs. A vertex l' and edges (p_i, l'), $1 \le i \le k$, are added to obtain G_{er}, and similarly a vertex r' and edges (p_i, r'), $1 \le i \le k$, are added to obtain G_{el}. The decomposition is illustrated in Figure 2.16a. From the properties of P_V, it can be shown that graphs G_{el} and G_{er} both contain fewer than n vertices. Thus, the hypothesis applies and floorplans F_l and F_r are produced. The floorplans F_l and F_r are merged to obtain the floorplan of G_e as depicted in Figure 2.16b.

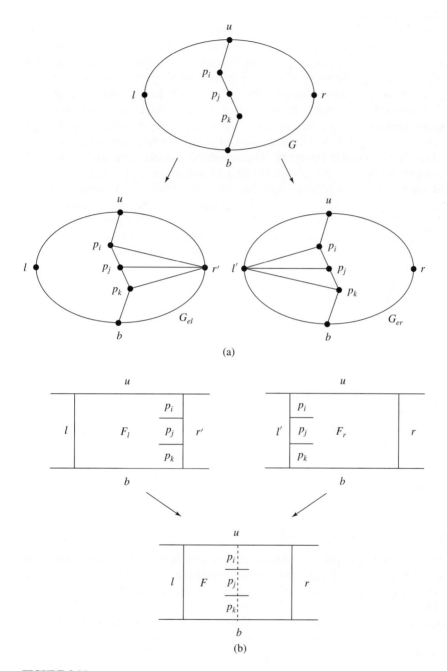

FIGURE 2.16
Dual decomposition and floorplan merging of rectangular graph. (a) Decomposition of G_e to G_{el} and G_{er}. (b) Merging floorplans of G_{el} and G_{er} to obtain the floorplan of G_e.

The rectangular dual graph approach is a new area in floorplanning that has not been widely accepted yet. Many problems in this approach are still unsolved, especially the quantitative aspect of the problems. Also, present dualization algorithms are too complicated to be implemented in commercial packages and research is currently underway to develop better and simpler algorithms.

One of the problems concerns the existence of the rectangular dual, that is, the elimination of complex triangles. Select a minimum set E of edges such that each complex triangle has at least one edge in E. A vertex can then be added to each edge of E to eliminate all complex triangles, resulting in a graph that admits a rectangular floorplan. The weighted complex triangle elimination problem was shown to be NP-complete [400]. However, an approximation solution is available [362], and an efficient solution exists for some special classes of planar triangulated graphs [400]. The unweighted complex triangle elimination problem is still open. The techniques for generating special classes of floorplans using the rectangular dual approach are not well studied either. The characterization of the dual graph of a sliceable floorplan is an open problem. Most of the problems related to the rectangular duals of sliceable floorplans eventually reduce to this important question: Given a planar triangulated graph, does a sliceable floorplan exist? All rectangular dual approaches to floorplanning require a PTG as input, but a circuit is basically a hypergraph. It is thus necessary to convert the hypergraph to a weighted PTG that can capture the information of the input circuit. This process is not well defined and needs more investigation.

The rectangular dual graph approach to floorplanning was first studied by Kozminski and Kinnen [189]. They also showed methods to transform one floorplan to another, where the rectangular duals are identical. Lai and Leinwand [196] later introduced another approach using graph matching where the horizontal and vertical orientation of each edge can be assigned in a consistent manner to obtain the floorplan. Bhasker and Sahni [19] reported a linear time algorithm to find a rectangular dual of a planar triangulated graph. Methods to enumerate all the floorplans of a planar triangulated graph were also reported in [352, 364]. An approximation algorithm for complex triangle elimination was reported in [362].

2.2.2 Hierarchical Approach

The hierarchical approach to floorplanning is a widely-used methodology. The approach is based on a divide and conquer paradigm, where at each level of the hierarchy, only a small number of rectangles are considered [79].

Consider a circuit C as shown in Figure 2.17. Since there are only three modules a, b, and c (and three nets ab, ac, and bc), it is possible to enumerate all floorplans for the given connection requirements. After an optimal configuration for the three modules has been determined, they are merged into a larger module. The vertices a, b, c are merged into a super vertex at the next level and will be treated as a single module. The number of possible floorplans increases exponentially with the number of modules d considered at each level. Thus, d is limited to a small number (typically $d \leq 5$). Figure 2.18 shows all possible floorplans for $d = 2$ and $d = 3$.

FIGURE 2.17
All possible floorplans of a small circuit, C.

BOTTOM-UP APPROACH. The hierarchical approach works best in bottom-up fashion using the clustering technique. The modules are represented as a graph where the edges represent the connectivity of the modules. Modules with high connectivity are clustered together while limiting the number in each cluster to d or less. An optimal floorplan for each cluster is determined by exhaustive enumeration and the cluster is merged into a larger module for higher-level processing.

A greedy procedure to cluster the modules is to sort the edges by decreasing weights. The heaviest edge is chosen, and the two modules of the edge are clustered in a greedy fashion, while restricting the number in each cluster to d or less. In the next higher level, vertices in a cluster are merged, and edge weights are summed up accordingly. One of the problems with this simple approach is that some lightweight edges are chosen at higher levels in the hierarchy, resulting in adjacency of two clusters of highly incompatible areas [79]. An example is illustrated in Figure 2.19. If the greedy clustering strategy is applied to the graph in Figure 2.19a for $d = 2$, the floorplan in Figure 2.19b is generated. Since module e has lightweight connections to other modules, its edges will be selected at a higher (later) level of the hierarchy. However, at higher levels, the size differences of the partial floorplans are much larger. When module e is chosen, it is paired up with a sibling of much larger size. The incompatibility of the areas results in a large wasted area in the final floorplan. The problem can be solved by arbitrarily assigning a small cluster to a neighboring cluster when their size will be too small for processing at higher a level of the hierarchy, as shown in Figure 2.19c.

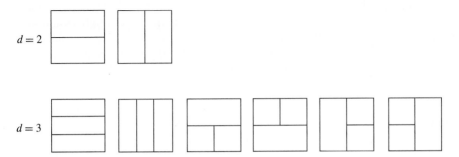

FIGURE 2.18
Floorplan enumeration.

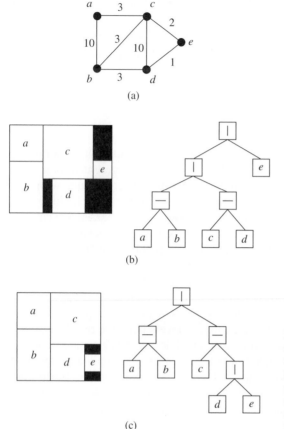

FIGURE 2.19
Size incompatibility problem in bottom-up greedy clustering of hierarchical floorplans. (a) Circuit connectivity graph. (b) Floorplan obtained by greedy clustering. (c) Limiting the size of a cluster at high level.

To evaluate the best floorplan at each level, some estimation techniques have to be introduced. The cost of a floorplan is usually estimated from the connections and the area of the floorplan. The area can easily be estimated because the dimensions of each cluster can be passed up from the bottom-up clustering. The area of a particular choice can thus be computed for each candidate floorplan. The routing cost can also be estimated by summing up the edge weights multiplied by the distance between the centers of the clusters. For example, consider two candidate floorplans in Figure 2.20 with the weights given. The routing costs for the corresponding floorplans are as given in the figure.

TOP-DOWN APPROACH. A hierarchical floorplan can also be constructed in a top-down manner. The fundamental step in the top-down approach is the partitioning of modules. Each partition is assigned to a child floorplan, and the partitioning is recursively applied to the child floorplans. One simple partitioning method is to apply the min-cut max-flow algorithm that can be computed in polynomial time. However, this method may result in two partitions of incompatible size. Many

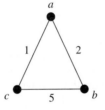

distance(a,b) = distance(a,c) = 2.24
distance(b,c) = 2.0
cost = (2.24 × 2) + (2.24 × 1) + (2.0 × 5) = 16.72

distance(a,b) = distance(b,c) = 2.24
distance(a,c) = 2.0
cost = (2.24 × 5) + (2.24 × 2) + (2.0 × 1) = 17.68

FIGURE 2.20
Estimating routing costs of two candidate floorplans.

heuristics have been proposed for balanced graph bipartitioning, as discussed in earlier sections. A generalization of bipartitioning is k-way partitioning ($k \geq 2$) which has also been widely studied [301].

It should be noted that one can combine the top-down and the bottom-up approaches. That is, for several steps apply the bottom-up technique to obtain a set of clusters. Then, apply the top-down approach to these clusters.

2.2.3 Simulated Annealing

Simulated annealing is a technique used to solve general optimization problems, floorplanning problems being among them. This technique is especially useful when the solution space of the problem is not well understood. The idea originated from observations of crystal formation. As a material is heated, the molecules move around in a random motion. When the temperature slowly decreases, the molecules move less and eventually form a crystal structure. When cooling is done more slowly, more of the crystal is at a minimum energy state and the material achieves a stronger crystal lattice. If the crystal structure obtained is not acceptable, it may be necessary to reheat the material and cool it at a slower rate.

Simulated annealing examines the *configurations* of the problem in sequence. Each configuration is actually a feasible solution of the optimization problem. The algorithm moves from one solution to another, and a global cost function is used to evaluate the desirability of a solution. Conceptually, it is possible to define a *configuration graph* where each vertex corresponds to a feasible solution and a directed edge (v_i, v_j) represents a possible movement from solution v_i to v_j.

The annealing process moves from one vertex (feasible solution) to another vertex following the directed edges of the configuration graph. The random motion of the molecules at high temperature is simulated by randomly accepting moves during the initial phases of the algorithm. As the algorithm proceeds, the temperature decreases and it accepts less random movements. Regardless of the temperature, the algorithm will accept a move (v_i, v_j) if $cost(v_j) \leq cost(v_i)$. When a local minimum is reached, all small moves lead to a higher cost solution. To avoid being trapped in a local minimum, simulated annealing accepts a movement to a higher cost when the temperature is high. As the algorithm cools down, such movement is less likely to be accepted. The best cost among all the solutions found by the process is recorded. When the algorithm terminates, hopefully it has examined enough solutions to achieve a low cost solution. Typically, the number of feasible solutions is an exponential function of the problem size. Thus, the movement from one solution to another is restricted to a very small fraction of the total configurations.

A pseudocode for simulated annealing is as follows:

Algorithm Simulated Annealing
Input: An optimization problem.
Output: A solution s with low cost.
begin-1
    s := random initialization;
    T := T_0; (* initial temperature *);
    while not frozen(T) do
        begin-2
        count := 0;
        while not equilibrium(count, s, T) do
            begin-3
            count := *count* +1;
            nexts := *generate(s)*;
            if (cost(next(s)) < cost(s)) or
                (f (cost(s),cost(next(s)),T) > random(0, 1)) then
                    s := *next(s)*;
        end-3;
        update(T);
    end-2;
end-1.

In the above, *generate()* is a function that selects the next solution from the current solution s following an edge of the configuration graph. *cost()* is a function that evaluates the global cost of a solution. $f()$ is a function that returns a value between 0 and 1 to indicate the desirability of accepting the next solution, and *random()* returns a random number between 0 and 1. A possible candidate function f is the well-known Boltzmann probability function $e^{\Delta C / k_B T}$, where ΔC is the cost change and k_B is the Boltzmann constant. The combined effect of $f()$ and *random()* is to have a high probability of accepting a high-cost movement at high temperature. *equilibrium*() is used to decide the termination condition of the random movement, and *update*() reduces the temperature to cool down the algorithm. *frozen*() determines the termination condition of the algorithm. The algorithm is usually frozen after an allotted amount of computation time has been consumed, a sufficiently good solution has been reached, or the solutions show no improvement over many iterations.

FLOORPLANNING BASED ON SIMULATED ANNEALING. This section describes a simulated annealing floorplanning algorithm proposed in [390]. Assume that a set of modules are given and each module can be implemented in a finite number of ways, characterized by its width and height. Some of the important issues in the design of a simulated annealing optimization problem are as follows:

1. The solution space.
2. The movement from one solution to another.
3. The cost evaluation function.

The solution is restricted to sliceable floorplans only. Recall that a sliceable floorplan can be represented by a tree. To allow easy representation and manipulation of floorplans, Polish expression notation is used. A Polish expression is a string of symbols that is obtained by traversing a binary tree in postorder. The branch cells correspond to the operands and the internal nodes correspond to the operators of the Polish expression. A binary tree can also be constructed from a Polish expression by using a stack. Examples are shown in Figure 2.21. The simulated annealing algorithm moves from one Polish expression to another.

A floorplan may have different slicing tree representations. For example, both trees in Figure 2.21 represent the given floorplan. The problem arises when there are more than one slice that can cut the floorplan. If the problem is ignored, both tree representations in our solution space could be accepted. But this leads to a larger solution space and some bias towards floorplans with multitree representations, since they have more chances to be visited in the annealing process.

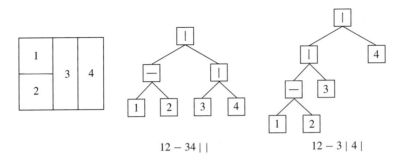

FIGURE 2.21
Two tree representations of a floorplan, with the corresponding Polish expressions.

The problem can be easily solved by giving priority to the slices, for example, from right to left and bottom to top. When there is more than one slice, the right-(bottom-) most slice is chosen. This property, when translated to the slicing tree, means that the orientation of a parent node must be different from that of its right child (if the child node is an internal node). In the corresponding Polish expression of this slicing tree, no two consecutive operators are identical. This is called a *normalized* Polish expression. There is a one-to-one correspondence between a floorplan and its normalized Polish expression.

Three types of movement are defined to move from one floorplan to another. They operate on the Polish expression representation of the floorplan.

OP1. Exchange two operands when there are no other operands in between.

OP2. Complement a series of operators between two operands.

OP3. Exchange adjacent operand and operator if the resulting expression is a normalized Polish expression.

It is obvious that the movements will generate only normalized Polish expressions. Thus in effect, the algorithm moves from one floorplan to another. One important question is the reachability defined by the movements: starting from an arbitrary floorplan, is it possible to visit all the floorplans using the movements defined above? If some floorplans cannot be reached, there is a danger of losing some valid floorplans in the solution. Fortunately the answer is yes. It can be shown that starting from any floorplan, we can move to the floorplan represented by $12|3|\dots|n|$ and vice versa by applying the above movements.

The cost function is a function of the floorplan or, equivalently, the Polish expression. There are two components for the cost function, area and wire length. The area of a sliceable floorplan can be computed easily using the floorplan sizing algorithm discussed in Section 2.2.4. The wire-length cost can be estimated from the perimeter of the bounding box of each net, assuming that all terminals are located on the center of their module. In general, there is no universally agreed-upon method of cost estimation. For simulated annealing, the cost function is best evaluated easily because thousands of solutions need to be examined.

Figure 2.22 shows a series of movements which lead to a solution. Simulated annealing has also been widely studied for various other applications, see for example [186, 373, 389].

2.2.4 Floorplan Sizing

In VLSI design, the circuit modules can usually be implemented in different sizes. Furthermore, each module usually has many implementations with a different size, speed, and power consumption trade-off. A poor choice of module implementation may lead to a large amount of wasted space. The floorplan-sizing problem is to determine the appropriate module implementation. If a cell has only one implementation, it is called a *fixed cell*; otherwise, it is a *variable cell*. A rotated cell is considered different from the original cell. Thus, if a cell of 2×4 is allowed to rotate, the rotated implementation is 4×2.

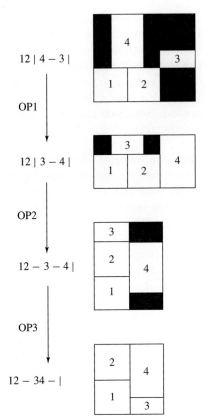

$12 \mid 4 - 3 \mid$

OP1

$12 \mid 3 - 4 \mid$

OP2

$12 - 3 - 4 \mid$

OP3

$12 - 34 - \mid$

FIGURE 2.22
A series of movements in a simulated annealing floorplanning.

Assume a fixed cell environment. Consider a floorplan of Figure 2.23a. Define a *horizontal dependency graph* $A_h(V, E)$, where V is the set of maximal vertical line segments in the floorplan. Each module of the floorplan is associated with a directed edge (v_i, v_j), where v_i (v_j) is the vertical segment to the left (or right) of the module. The graph A_h is a directed acyclic graph (with no directed

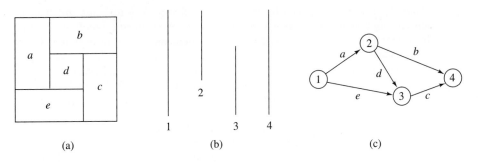

(a)   (b)   (c)

FIGURE 2.23
(a) Floorplan. (b) Maximal vertical segments. (c) Horizontal dependency graph A_h.

cycles); see Figure 2.23c. The length of each edge is defined as the width of the corresponding module in the floorplan. Using topological ordering on A_h, the distance of each line segment (vertex) from the leftmost segment can be computed. The longest path from the leftmost segment to the rightmost segment is the minimum width of the chip. Similarly, define a *vertical dependency* to find the height of the chip. Thus, the width and height of a fixed cell floorplan can be easily determined, and the area can be computed. Next, two classes of sizing techniques are described, hierarchical and nonhierarchical.

HIERARCHICAL FLOORPLAN SIZING. In a floorplan sizing problem developed by Otten and Stockmeyer [270, 344], the algorithm finds an area-optimal implementation of variable cells on a sliceable floorplan. Note that the horizontal and vertical dependency graphs of a sliceable floorplan are *series-parallel graphs*.

Given a sliceable floorplan F and, for each module M_i, a set of possible implementations of the module, where each implementation is described by a pair of real numbers (w_k, h_k) representing the width and height of a module for the corresponding implementation. It can be assumed that the w_i values of a given module are distinct; otherwise, for all the implementations with identical width, only the one with least height is selected. Without loss of generality, it can also be assumed that the set is sorted such that $w_p < w_q$ for all $p < q$. Since the goal is to find the most area-efficient implementation, claim that $h_p > h_q$ for $p < q$. If it is not, exclude (w_q, h_q) from the choices because it is inferior to (w_p, h_p) in width as well as height. For example, if there is a choice between $(3, 4)$ and $(4, 5)$, $(4, 5)$ can be immediately excluded from the solution. Summarizing the above observations gives

$$w_1 < w_2 < \ldots < w_{s_i}$$
$$h_1 > h_2 > \ldots > h_{s_i} \tag{2.1}$$

for each module M_i.

Lemma 2.1. Given two subfloorplans corresponding to two subtrees of a node v, one with t and the other with s nonredundant implementations, then v has at most $s + t - 1$ nonredundant implementations.

Proof. Consider a horizontal bipartition of the floorplan in a slicing tree with its left and right subtrees, as shown in Figure 2.24. (A symmetric proof can be used for a vertical bipartition.) Suppose the possible implementations of the left and right subtrees are $\{ (a_1, b_1), \ldots, (a_s, b_s) \}$ and $\{ (x_1, y_1), \ldots, (x_t, y_t) \}$, respectively. It can be assumed that the sets are sorted and exhibit the properties described in Equation 2.1. There are $s \times t$ possible implementations for node V in Figure 2.24, that is, by choosing one possible implementation from each list. Such an arbitrary choice of implementation results in exponential growth in the number of possible implementations at the root node. For example, if each module has two implementations and there are n modules, there will be 2^n possible implementations at the root node. Thus, exhaustive enumeration of the implementations is not a practical solution.

Next, examine the implementations at node V. Let (a_i, b_i) and (x_j, y_j) be the pairs selected to implement the left and right subtrees, respectively. The dimensions

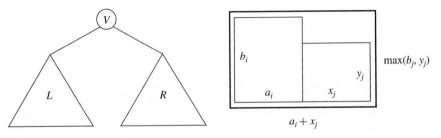

FIGURE 2.24
A horizontal bipartition of a floorplan.

of node V are given by $(a_i + x_j, \max(b_i, y_j))$. Consider the case where b_i is the dominant height, that is, $\max(b_i, y_j) = b_i$. Consider all pairs (x_k, y_k) with $k > j$. It is clear that all choices of (a_i, b_i) and (x_k, y_k) are inferior because (a_i, b_i) and (x_k, y_k) will not decrease the height but will increase the width (since $b_i \geq y_j > y_k$ and $x_k > x_j$). Thus, in the case when b_i is the dominant height of (a_i, b_i) and (x_j, y_j), all the implementations of (a_i, b_i) and (x_k, y_k) can be ignored for $k > j$. On the other hand, if y_j is the dominant height, skip the implementations of (a_l, b_l) and (x_j, y_j) for $l > i$. Thus not all the possible pairings need to be considered at node V.

Based on the previous lemma, an efficient algorithm can easily be designed for the generation of all the nonredundant implementations of a node. The procedure for generating the pairing of a node V is as follows:

Procedure Vertical Node Sizing
Input: Two sorted lists $L = \{(a_1, b_1), \dots, (a_s, b_s)\}$, $R = \{(x_1, y_1), \dots, (x_t, y_t)\}$
    where $a_i < a_j, b_i > b_j, x_i < x_j, y_i > y_j$ for all $i < j$.
Output: A sorted list $H = \{(c_1, d_1), \dots, (c_u, d_u)\}$ where $u \leq s + t - 1$,
    $c_i < c_j, d_i > d_j$ for all $i < j$.
begin-1
    $H := \emptyset$;
    $i := 1, j := 1, k = 1$;
    while $(i \leq s)$ *and* $(j \leq t)$ *do*
    begin-2
        $(c_k, d_k) := (a_i + x_j, \max(b_i, y_j))$;
        $H := H \cup \{(c_k, d_k)\}$;
        $k := k + 1$;
        if $\max(b_i, y_j) = b_i$ *then*
            $i := i + 1$;
        if $\max(b_i, y_j) = y_j$ *then*
            $j := j + 1$;
    end-2;
end-1.

Theorem 2.4. The sizing algorithm runs in $O(dn)$ time, where n is the number of modules and d is the height (or depth) of the slicing tree.

Proof. First, notice that at a node V, the number of implementations (i.e., possibilities) increases as the sum of s and t. At each iteration of the while loop, either i or j (or both) increases by one. The worst case scenario is when i and j increase alternately. Thus the loop is executed at most $s + t - 1$ times which is also the upper bound on the length of the output list. The algorithm works in bottom-up fashion until it reaches the root node. At the root node, the length of the list is the sum of the list lengths of all the modules. All pairs in the list of the root node are examined and one that achieves minimum area is selected. This is an optimal area implementation of the given sliceable floorplan.

Assume that there are n modules where each module has a constant number of implementations independent of n. At each node N of the slicing tree, the size of the list is proportional to the number of leaves of the subtree rooted at N. Let the node label denote the length of the implementation list of the partial floorplan represented by that node. A leaf node is labeled by the length of its implementation list. The label of an internal node is the sum of the two labels of its child nodes. Since the list of implementations has to be constructed at each node, the time complexity of the algorithm is the sum of all the node labels in the tree. Let d be the depth of the slicing tree. Consider all the nodes at an identical level of the tree. The sum of all the node labels at the level is bounded by the sum of all the leaf node labels, which is $O(n)$. Thus, at each level of the tree, a total of $O(n)$ time is spent processing the lists, and the total time complexity of the algorithm is $O(dn)$. For a balanced binary tree, $d = O(\log n)$ and the time complexity is $O(n \log n)$. However, for an unbalanced tree, the time complexity may become $O(n^2)$, as observed from summing up the labels of all the nodes in the tree shown in Figure 2.25.

The above algorithm operates within the framework of a binary hierarchical floorplan. Recently, the technique has been generalized to solve floorplan-sizing problems for hierarchical floorplans of order 5 [376, 391]. The idea is also based on enumerating all nonredundant floorplan implementations using the bottom-up technique. Experimentally, the technique has been shown to be effective for floorplans with a small number of modules. However the computational complexity of the algorithm is not known to be polynomial. In fact, the complexity of floorplans more general than sliceable floorplan has not been widely reported. For general nonsliceable floorplans, the sizing problem was shown to be NP-complete [344].

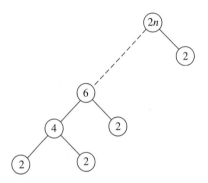

FIGURE 2.25
Analyzing the time complexity of unbalanced tree in floorplan sizing.

NONHIERARCHICAL FLOORPLAN SIZING. The efficient floorplan-sizing algorithm described in the last section relies on the hierarchical nature of the floorplan. This section describes a general floorplan-sizing algorithm which places no restriction on the organization of the modules. The approach is based on mixed integer linear programming. Linear programming (LP) has been used to solve a wide range of problems. The focus of most LP-based approaches is the formulation of LP equations, as in an LP floorplanning in [347] that uses the following notation:

w_i, h_i : width and height of module M_i.

(x_i, y_i) : coordinates of the lower left corner of module M_i.

x, y : the width and height of the final floorplan.

a_i, b_i : min and max values of the aspect ratio w_i/h_i of module M_i.

Given a set of modules $\mathcal{M}$, each module M_i has an unknown width w_i, height h_i and a prescribed *aspect ratio* $a_i \leq w_i/h_i \leq b_i$. Usually the constraint is symmetric, that is, $a_i = 1/b_i$. The solution is to find x_i, y_i, w_i, and h_i for each module such that all constraints are satisfied and xy is minimized. The constraints for a feasible LP solution follow.

Nonoverlap constraint. For simplicity, assume w_i and h_i are known constants. This restriction will be relaxed later. In a legal floorplan no two modules M_i, M_j can overlap. If M_i is to the left of M_j, the nonoverlap constraint requires that $x_i + w_i \leq x_j$. Including this with the other three cases (right, top, and bottom) gives:

$$\begin{aligned} x_i + w_i &\leq x_j \\ x_j + w_j &\leq x_i \\ y_i + h_i &\leq y_j \\ y_j + h_j &\leq y_i \end{aligned} \qquad (2.2)$$

At least one of the constraints in Equation 2.2 must be satisfied to obtain a legal solution, but not all of them need to be satisfied. The selection can be solved by introducing binary selection variables. For each pair of modules M_i and M_j, two variables p_{ij} and q_{ij} are introduced that assume only value 0 or 1. We also introduce two large numbers W and H, which are the upper bounds of the width and height of the feasible solution region. The following system of inequalities solves the nonoverlap constraint:

$$\begin{aligned} x_i + w_i &\leq x_j + W(p_{ij} + q_{ij}) \\ x_j + w_j &\leq x_i + W(1 - p_{ij} + q_{ij}) \\ y_i + h_i &\leq y_j + H(1 + p_{ij} - q_{ij}) \\ y_j + h_j &\leq y_i + H(2 - p_{ij} - q_{ij}) \end{aligned} \qquad (2.3)$$

For example, if $p_{ij} = 0$ and $q_{ij} = 1$, equation $y_i + h_i \leq y_j$ must be satisfied, while the other Equations 2.3 will be insignificant due to the addition of the factor

$W(H)$. Thus, at least one of the Equations 2.2 will always be satisfied. The values of $W(H)$ can be set to $\sum w_i$ ($\sum h_i$) if no better values are known.

Module size constraint. Dismiss the assumption that w_i and h_i are known. Instead, a lower bound A_i from the area of each module M_i and the parameters for aspect ratios a_i, b_i are given:

$$w_i h_i \geq A_i$$
$$a_i \leq w_i/h_i \leq b_i \tag{2.4}$$

From the above equations, the range $w_i \in [w_{\min}, w_{\max}]$ and $h_i \in [h_{\min}, h_{\max}]$ are derived, where the equations can be satisfied. The values are given by:

$$w_{\min} = \sqrt{A_i a_i}$$
$$w_{\max} = \sqrt{A_i b_i}$$
$$h_{\min} = \sqrt{A_i/b_i} \tag{2.5}$$
$$h_{\max} = \sqrt{A_i/a_i}$$

The constraint $w_i h_i \geq A_i$ is a nonlinear equation and cannot be incorporated into our LP. We use a first-order approximation of the equation, that is, a straight line passing through $(w_{\min}, h_{\max})$ and $(w_{\max}, h_{\min})$ to approximate the equation:

$$h_i = \Delta_i w_i + c_i$$
$$\Delta_i = (h_{\max} - h_{\min})/(w_{\min} - w_{\max}) \tag{2.6}$$
$$c_i = h_{\max} - \Delta_i w_{\min}$$

The equations and their approximations are depicted in Figure 2.26. The linear equation can now be incorporated into Equations 2.2 giving:

$$x_i + w_i \leq x_j$$
$$x_j + w_j \leq x_i$$
$$y_i + \Delta_i + c_i \leq y_j \tag{2.7}$$
$$y_j + \Delta_j + c_j \leq y_i$$

Here x_i, y_i, and w_i are unknown. All the other variables ($\Delta_i, c_i, \Delta_j, c_j$) can be computed from A_i, a_i, and b_i, which are given. Remember that only one of

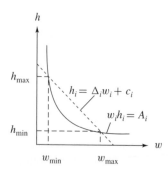

FIGURE 2.26
Area approximation using a straight line.

the above equations needs to be satisfied. Again, the selection variables, p_{ij}, q_{ij}, can be used to follow a formulation like Equations 2.3. When the solution for w_i is obtained, h_i can be computed from Equations 2.6.

Cost function. The function to be minimized is the total area of the floorplan $A = xy$. Again this is a nonlinear constraint. However, it is possible to fix the width W and minimize the height of the floorplan:

$$
\begin{aligned}
\min \quad & y \\
& x_i + w_i \leq W \\
& y \geq y_i + h_i
\end{aligned} \tag{2.8}
$$

Complete LP formulation. The complete mixed-integer LP formulation for the floorplanning problem is as follows:

$$
\begin{aligned}
\min \quad & y \\
& x_i, \ y_i, \ w_i \ \geq \ 0 \\
& p_{ij}, \ q_{ij} = 0 \ \text{or} \ 1 \\
& x_i + w_i \ \leq \ W \\
& y \ \geq \ y_i + h_i \\
& x_i + w_i \ \leq \ x_j + W(p_{ij} + q_{ij}) \\
& x_j + w_j \ \leq \ x_i + W(1 - p_{ij} + q_{ij}) \\
& y_i + h_i \ \leq \ y_j + H(1 + p_{ij} - q_{ij}) \\
& y_j + h_j \ \leq \ y_i + H(2 - p_{ij} - q_{ij})
\end{aligned} \tag{2.9}
$$

When h_i appears in the above equations, it is replaced by $h_i = \Delta_i w_i + c_i$. The unknown variables are x_i, y_i, w_i, p_{ij}, and q_{ij}. All other variables are known. Equations 2.9 are fed into an LP solver such as LINDO [314] to find a minimum cost solution for the unknown variables. A floorplan can be directly constructed from the solution of the LP. The procedure can be repeated for several values of W until a floorplan with a desirable cost and aspect ratio is obtained. In some cases, the *topology* (the relative location of all modules) of the floorplan is known. This means that p_{ij} and q_{ij} are known values and the number of unknown variables and equations can be reduced. The wire length and routing area can also be estimated and incorporated to create a set of sophisticated LP equations. Readers are referred to [348] for further details.

2.2.5 Discussion

One of the early approaches to floorplanning is recursive bipartition to construct a floorplan tree. The min-cut and other bipartitioning algorithms, for example, the Kernighan-Lin [181] and Fiduccia-Mattheyses [100] heuristics, are applicable. The idea is to minimize the number of nets crossing two halves of the bipartition. Four-way partition can also be used instead of bipartition. The main drawback of the approach is that it may result in routing congestion. The nets crossing the two halves may be congested in certain parts of the cutline, and there exists no simple method to control it.

Sliceable floorplans are widely used in practice because of their simplicity. Most problems related to sliceable floorplans have exact solutions without resorting to heuristics. The tree structure of sliceable floorplans facilitates algorithm design, and most sliceable floorplanning algorithms are inherently based on the slicing tree. The main disadvantage of sliceable floorplans is also their simplicity. The sliceable restriction may result in a less compact floorplan as compared to a nonsliceable floorplan. Also, the routing patterns of sliceable floorplans are more restricted.

Many floorplan-related problems for general nonsliceable floorplans have been shown to be intractable. Solutions of these types of floorplans have to resort to heuristic or stochastic methods. The solution quality is more difficult to control and sometimes unpredictable. In some cases, it is even difficult to estimate how far a solution is from the optimal solution.

Hierarchical floorplans have recently received a lot of attention in the research community. Some successful results have been reported for hierarchical floorplans of order 5 [376, 382]. There were also reports on using L-shaped modules in the hierarchical framework [390]. However, most work along this line is still in the research stage with very few reports on commercial floorplanning packages. Earlier studies on hierarchical floorplanning mostly focused on the top-down approach. One problem with the top-down approach is the difficulty in predicting the wiring cost. The bottom-up approach has received a lot of attention lately. This approach usually works well when applied simultaneously with later design phases, such as channel ordering and global routing. There are also approaches which combine the top-down and bottom-up strategies (also called the meet-in-the-middle approach).

The rectangular dual approach to floorplanning is a new area in floorplanning concepts. Most of the work is still in the early stages of research. Many properties of the dual graphs are not well understood. Research has focused on the quantitative aspects of the problem, such as existence, characterization, and identification. Little work has been done on the qualitative aspect of the problem, such as sizing. Most algorithms reported were fairly complicated and not suitable for wide-spread application. To date, no commercial floorplanning package is based on the rectangular dual graph approach. Combining global routing and floorplanning is not difficult because the dual graph is known. No such effort has been reported yet. However, this approach is the most promising, since it is explicitly driven by the connectivity requirements of the modules.

The floorplanning algorithm using simulated annealing is easy to implement and has been successfully implemented in many commercial design automation tools. The major drawback of simulated annealing is that it requires substantial computation resources. It also requires many parameters to control the behavior of the algorithm, for example, temperature and negative move probability. The setting of the parameters is usually based on empirical observations. Fine-tuning the parameters requires many trial runs and is a time-consuming process. Often, no single set of parameters is suitable for all the inputs and it may require repeated adjustments to obtain a good solution. Another problem is that in the simulated annealing algorithm it is possible to be close to an optimal solution but not quite there. If the parameters are changed slightly, a completely different solution may be found. It also makes it hard for engineering changes. One of the attractive features of simulated annealing is that the solution quality has room for improvement as CPU speed increases. Suppose there is a CPU with 10 times the throughput. The parameters can be set for a slower annealing rate to obtain a better solution within the same time. Such advantages cannot be attained with nonstatistical approaches without major changes in their software programs.

The genetic algorithm paradigm has also been applied to the floorplanning problem. Like simulated annealing, the algorithm has many characteristics which require empirical adjustments, and the results are usually unpredictable. However, some experimental reports have confirmed the applicability of the algorithm [63, 65].

The linear programming (LP) approach also requires substantial computational resources. Fortunately, linear programming is a rather mature subject. Many LP solver packages are widely available, and sometimes special-purpose hardware is used. Some LP solvers are able to exploit special structures of the problem to speed up the computation. However, such exploitation of the LP structure has not yet been reported for floorplanning problems. Generally, the LP technique cannot handle large input data. For moderate input data, the number of LP equations can be very large.

Studies on a higher-order floorplan, which allows more shapes than rectangle, have been very scarce. Many floorplanning algorithms are not immediately extendable to handle more general shapes. Relaxing the shape constraint introduces more variables into each module. There are more degrees of freedom in the solution space, and in general this should lead to better results.

2.3   PLACEMENT

The input to the placement problem is a set of modules and a net list, where each module is fixed. That is, each module has fixed shape and fixed terminal locations. The net list provides connection information among modules. The goal is to find the best position for each module on the chip (i.e., the plane domain) according to the appropriate cost functions. Typically, a subset of modules has pre-assigned positions, for example, I/O pads.

Placement algorithms can be divided into two major classes, iterative improvement and constructive placement. In iterative improvement, algorithms start with an initial placement and repeatedly modify it in search of a better solution, where a better solution is quantified as a cost reduction. In constructive placement, a good placement is constructed in a global sense. In this section, prevalent cost functions are introduced and three classes of algorithms are discussed. The first is the force-directed method, which has both iterative improvement and constructive placement techniques. The second class is based on simulated annealing and is classified with the iterative improvement algorithms. The others, partitioning placement and resistive network techniques, are classified as constructive placement algorithms. Last, the assignment problem and linear placement, which are subproblems of the general placement problems, will be discussed.

2.3.1 Cost Function

The main goal in the placement problem is to minimize the area. This parameter is very difficult to estimate; thus, alternative cost functions are employed. There are two prevalent cost functions:

- Wire length-based cost functions, $\mathcal{L}$.
- Cut-based cost functions, $\mathcal{K}$.

Since at the placement stage a final routing is not available, the exact wire length cannot be calculated. Designers usually use the following simple method, or a variation of it, to estimate the length. For each net, find the minimal rectangle which includes the net (called the minimal enclosing rectangle) [316], and use half of its perimeter ($p/2$) as the estimated length of that net. This is called the semiperimeter method. Thus, the total estimated length is:

$$\mathcal{L} = \sum_{N_i \in \mathcal{N}} \frac{p_i}{2}, \tag{2.10}$$

where p_i is the perimeter of net N_i. In practice, $\mathcal{L}$ is a good estimate of the final length because in a circuit most nets have only two or three terminals, and each net is routed using the shortest (Steiner) tree between its terminals, as shown in Figure 2.27. In most circuits, a small wire length indicates a small area.

In cut-based cost functions, take a vertical (or a horizontal) cutline and determine the number of nets crossing the cutline. This number $\mathcal{K}$ dictates the cost. Thus, a placement resulting in a small cost is obtained. This task is repeated in a recursive manner.

The two cost functions can be combined, in conjunction with a scaling parameter λ.

$$\text{cost} = \lambda \mathcal{K} + (1 - \lambda)\mathcal{L} \tag{2.11}$$

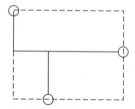

FIGURE 2.27
A two-terminal net and a three-terminal net with their minimal enclosing rectangles.

A trivial algorithm for the placement problem is as follows. Start with an arbitrary placement and repeatedly perform module exchanges. An exchange is accepted if it results in a cost decrease or in a small cost increase (if a cost decrease is not possible). The latter is to avoid getting trapped in a local minimum. This simple-minded algorithm works well when the number of modules is small (less than 15). For larger problems, more sophisticated techniques are required, as described in the following subsections.

2.3.2 Force-directed Methods

Modules that are highly interconnected need to be placed close to each other. One might say there is a force pulling these modules toward each other. Thus, the number of connection between two modules is related to a force attracting them toward each other. The interaction between two modules M_i and M_j can be expressed as

$$F_{ij} = -c_{ij}d_{ij}$$

where c_{ij} is a weighted sum of the nets between the two modules and d_{ij} is a vector directed from the center of M_i to the center of M_j. The magnitude of this force is the distance (in Euclidean or Manhattan geometry) between the center of the modules. For example, set $|d_{ij}| = |x_i - x_j| + |y_i - y_j|$, where (x_i, y_i) is the center of M_i, and similarly, (x_j, y_j) is the center of M_j. (Note that the above equation for F_{ij} is the expression for the force between two points connected by a spring with constant c_{ij}.) An optimal placement is defined as the one that minimizes the sum of the force vectors acting on the modules. Force-directed algorithms have been proposed in [9, 151, 266]. Next constructive and iterative-improvement approaches that solve the placement problem using a force-directed formulation will be described.

CONSTRUCTIVE FORCE-DIRECTED ALGORITHMS. First, an initial placement is constructed by placing the modules so that they are in equilibrium with respect to the forces acting on them. That is, find a placement so that the vector sum of the forces acting on each module is zero. A solution to this problem can be obtained by solving a nonlinear system of equations as follows. Let M_0 be a module with its final position denoted by (x_0, y_0). The set of modules connected to M_0 are denoted by $\{M_1, \ldots, M_s\}$, where M_j has the final position (x_j, y_j), for $1 \leq j \leq s$. The x-component of the set of forces acting on M_0 is set to zero, that is, $\sum_j c_{0j} D_{0j}^x = 0$, where D_{0j}^x is the magnitude of the x-component of the vector d_{0j} from (x_j, y_j) to (x_0, y_0). The magnitude of d_{ij} is the distance (in some metrics, e.g., the L_1 metric) between (x_0, y_0) and (x_j, y_j). As mentioned earlier, c_{0j} is a weighted sum of the nets between M_0 and M_j. Similarly, $\sum_j c_{0j} D_{0j}^y = 0$, where D_{0j}^y is the magnitude of the y-component of the vector d_{0j} from (x_j, y_j) to (x_0, y_0).

If there are no modules with predetermined positions, then a trivial solution is obtained by placing the center of all modules at an arbitrary point (x, y). This solution is not feasible, since in a legal placement the overlap of modules is not allowed. Indeed, overlap constraints cannot be expressed in the above formulation. However, if some modules have pre-assigned positions (some (x_j, y_j) have fixed values), a more reasonable (i.e., less overlapping) initial solution is obtained. In general, even when a large number of modules have pre-assigned positions, some overlap remains.

The notion of repulsive forces was introduced in [338]. The main idea is to introduce a repulsive force between two modules that are not connected. The force can be modeled as either a constant force or a force that is inversely proportional to the distance between the two modules. Repulsive forces decrease the amount of overlap. However, some overlap remains that is later removed by modifying the initial placement as little as possible. (A technique for removing overlaps is elaborated on in a later section on linear placement and relaxation algorithms.)

ITERATIVE FORCE-DIRECTED ALGORITHMS. The force-directed method as just described does not produce an effective final placement. Instead, it needs further improvement to produce a good solution. The solution can be improved by an iterative force-directed algorithm. Such algorithms can be (and have been) used to improve solutions obtained by other placement algorithms.

Iterative force-directed algorithms proceed in the following manner. They start with an initial position of the modules. A module with the maximum total force acting on it (i.e., a module that is in a bad place) is identified. Denote this module by M. There is an ideal location (x, y) for M. At this location, there is no force (sum) acting on M. If placing (the center of) M at (x, y) results in no overlap, then M is placed there. Otherwise, there are various strategies. One is to place M at (x, y) and move the module currently at (x, y) to where M was previously placed (assuming all the modules are similar in size). Another strategy is to place M at (x, y), cause overlap, and then remove the resulting overlaps at a later stage. The above procedure is repeated for other badly placed modules, one by one. Note that this process may continue indefinitely. To circumvent this, each module can be displaced a constant finite number of times, typically once or twice.

Let F_i be the total force acting on module M_i. One objective function that is used is to find a placement minimizing the total force $\mathcal{F} = \sum F_i$. Start with an initial placement and a given $\mathcal{F}$. For every pair of modules calculate, if they are exchanged, the resulting total force. The two modules with the best total force are selected, and their position is switched. These two modules are *locked*, that is, their position cannot change any more. This step is repeated until all the modules are locked. The best intermediate solution is identified as the best placement.

The previous discussion applies to layout environments where modules have the same sizes, for example, gate arrays. If the modules have arbitrary sizes, then there are other strategies, for example, switching the location of modules that have the same size, or, switching one larger module with several smaller modules.

2.3.3 Placement by Simulated Annealing

The basic procedure in simulated annealing is to start with an initial placement and accept all perturbations or moves which result in a reduction in cost [323]. Moves that result in a cost increase are accepted, with the probability of accepting such a move decreasing with the increase in cost and also decreasing in later stages of the algorithm. A parameter T, called the temperature, is used to control the acceptance probability of the cost-increasing moves.

TimberWolf [320] is the most wildly used and successful placement package based on simulated annealing. There are different versions of TimberWolf for placing standard cells [318, 320, 321] and for the floorplanning problem [319]. The basic algorithm for standard cell placement will be described here. Here the main part of TimberWolf, where the cells are placed using simulated annealing so as to minimize the estimated wire length, is described. A formal description of the simulated annealing algorithm was given in Section 2.2.3.

There are two methods for generating new configurations from the current configuration. Either a cell is chosen randomly and placed in a random location on the chip, or two cells are selected randomly and interchanged. The performance of the algorithm was observed to depend upon r, the ratio of displacements to interchanges. Experimentally, r is chosen between 3 and 8.

A temperature-dependent range limiter is used to limit the distance over which a cell can move. Initially, the span of the range limiter is twice the span of the chip. In other words, there is no effective range limiter for the high temperature range. The span decreases logarithmically with the temperature

$$L_{WV}(T) = L_{WV}(T_1)\frac{\log T}{\log T_1}$$

$$L_{WH}(T) = L_{WH}(T_1)\frac{\log T}{\log T_1},$$

where T is the current temperature, T_1 is the initial temperature, $L_{WV}(T_1)$ and $L_{WH}(T_1)$ are the initial values of the vertical and horizontal window spans $L_{WV}(T)$ and $L_{WH}(T)$, respectively.

The cost function is the sum of three components: the wire length cost C_1, the module overlap penalty C_2, and the row length control penalty C_3. The wire length cost C_1 is estimated using the semiperimeter method, with weighting of critical nets and independent weighting of horizontal and vertical wiring spans for each net:

$$C_1 = \sum_{\text{nets } i}[x(i)W_H(i) + y(i)W_V(i)],$$

where $x(i)$ and $y(i)$ are the vertical and horizontal spans of the net i's bounding rectangle, and $W_H(i)$ and $W_V(i)$ are the weights of the horizontal and vertical wiring spans. When critical nets are assigned a higher weight, the annealing algorithm will try to place the cells interconnected by critical nets close to each other. Independent horizontal and vertical weights give the user the flexibility to prefer connections in one direction over the other.

The module overlap penalty C_2 is parabolic in the amount of overlap

$$C_2 = W_2 \sum_{i \neq j} [O(i, j)]^2,$$

where $O(i, j)$ is the overlap between the ith and jth module, and W_2 is the penalty weight. Although cell overlap is not allowed in the final placement and has to be removed by shifting the cells, it is time consuming to remove all overlaps after every proposed move. This is why most algorithms allow overlap during the annealing process but penalize it. The penalty function depends on the amount of overlap.

The parameter C_3 controls the length of the rows and is a function of the difference between the actual row length and the desired row length. The penalty is given by:

$$C_3 = W_3 \sum_{\text{rows}} |L_R - \widehat{L}_R|,$$

where L_R is the actual row length, $\widehat{L}_R$ is the desired row length, and W_3 is the penalty weight.

The acceptance probability is given by $\exp(-\Delta C/T)$, where ΔC is the cost increase and T is the current temperature. When the cost increases, or when the temperature decreases, the acceptance probability gets closer to zero. Thus, the acceptance probability $\exp(-\Delta C/T)$ less than random$(0, 1)$ (a random number between 0 and 1) is high when ΔC is small and when T is large.

At each temperature, a fixed number of moves per cell is allowed. This number is specified by the user. The higher the maximum number of moves, the better the results obtained. However, the computation time increases rapidly. There is a recommended number of moves per cell as a function of the problem size in [323]. For example, for a 200-cell and 3000-cell circuit, 100 and 700 moves per cell are recommended, respectively.

The annealing process starts at a very high temperature, for example, $T_1 = 4{,}000{,}000$, to accept most of the moves. The cooling schedule is represented by

$$T_{i+1} = \alpha(T)T_i,$$

where $\alpha(T)$ is the cooling rate parameter and is determined experimentally. In the high and low temperature ranges, the temperature is reduced rapidly (e.g., $\alpha(T) \approx 0.8$). However, in the medium temperature range, the temperature is reduced slowly (e.g., $\alpha(T) \approx 0.95$). The algorithm is terminated when T is very small, for example, when $T < 0.1$.

2.3.4 Partitioning Placement

Partitioning placement, or min-cut placement, is based on the intuition that densely connected subcircuits should be placed closely. It repeatedly divides the input circuit into subcircuits such that the number of nets cut by the partition is minimized. Meanwhile, the chip area is partitioned alternately in the horizontal and vertical directions, and each subcircuit is assigned to one partitioned area. The core procedure is a variant of the Kernighan-Lin algorithm or the Fiduccia-Mattheyses algorithm for partitioning (see Section 2.1).

Consider a set of horizontal and vertical cutlines. Breuer [28, 29] proved that the problem of finding a placement minimizing the number of nets cut by this set of cutlines is equivalent to the problem of finding a placement minimizing the semiperimeter wire length. However, it is algorithmically difficult to minimize the cost functions. Instead, a sequential cutline cost function is used. The objective is to minimize the number of nets crossing one cutline. Certainly, a globally optimal solution will not be obtained in this manner.

A block-oriented min-cut placement algorithm, proposed by Breuer, proceeds as follows. Use a min-cut partitioning algorithm to partition the circuit into two subcircuits. According to the area of the modules of these two subcircuits, the position of the cutline on the chip is decided. Each subcircuit is placed on one side of the cutline. Recursively, perform the same procedure on each subcircuit until each subcircuit contains only one module.

Dunlop and Kernighan [88] consider not only the internal nets of the subcircuit but also the nets connected to external modules at higher levels of the hierarchy. An example is shown in Figure 2.28a, where the entire circuit is divided into two sections.

If a module is connected to an external terminal on the right, the module should preferably be assigned to the right side of the chip. Thus, the modules shown in Figure 2.28a are in the right region. The method by Dunlop and Kernighan is called the *terminal propagation technique* and will be illustrated using Figure 2.28b, c. Consider the net connecting modules 1, 2, and 3 in block A. This net is connected to other modules in blocks B and C. In Figure 2.28c the modules in block B are represented by a "dummy" module $P1$ on the boundary of block A. Similarly, the modules in block C are represented by a "dummy" module $P2$ on the boundary of block A. After partitioning, the net-cut would be minimized if modules 1, 2, and 3 were placed in the bottom half of A. Note that the partitioning procedure has to be done in a breadth-first manner in order to perform terminal propagation.

2.3.5 Module Placement Based on Resistive Network

An effective technique for module placement was proposed in [50]. This is a constructive placement method that employs resistive networks as a working domain. The cost function used is the sum of the squares of wire lengths (to make the transformation to the network domain straightforward). The algorithm includes the following procedures: optimization, scaling, relaxation, partitioning, and assignment. The method is efficient because it takes advantage of the net-list sparsity. The algorithm runs in $O(n^{1.4} \log n)$ time, where n is the number of modules.

Let the n modules be placed at positions (x_i, y_i), $i = 1, 2, \ldots, n$. These locations could be, for example, those of the centers of the modules. Let c_{ij}

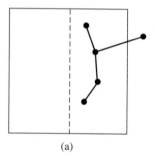

(a)

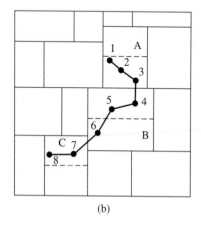

(b)

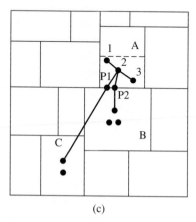

(c)

FIGURE 2.28
(a) Modules connected to an external terminal. (b) A net connecting modules in different blocks.
(c) Modules in blocks B and C are replaced by dummy modules P1 and P2. Each point denotes a
module or an external terminal.

denote the number of wires (connectivity) between module i and module j. The
cost function is given by:

$$\Phi(x, y) = \frac{1}{2} \sum_{i,j=1}^{n} c_{ij}[(x_i - x_j)^2 + (y_i - y_j)^2].\tag{2.12}$$

The cost function can be rewritten as follows [124]:

$$\Phi(x, y) = x^T Bx + y^T By,\tag{2.13}$$

where

$$B = D - C\tag{2.14}$$

is an $n \times n$ symmetric matrix, $C = [c_{ij}]$ is the connectivity matrix, and D is a
diagonal matrix whose ith element d_{ii} is equal to $\sum_{j=1}^{n} c_{ij}$. For the optimization
problem, only the one-dimensional problem needs to be considered because of the
symmetry between x and y in Equation 2.12.

   B in Equation 2.14 is of the same form as the indefinite admittance ma-
trixof an n-terminal linear passive resistive network in circuit theory. Model the

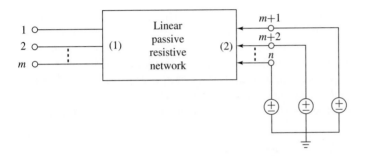

FIGURE 2.29
An n-terminal linear, passive resistive network whose first m nodes are floating and the remaining $(n - m)$ nodes are connected to voltage sources.

x-coordinate of module i, x_i with a node voltage v_i at node i. The reference coordinate $x = 0$ is thus the datum (or zero) voltage. The term $-c_{ij}$ in Equation 2.14 is, then, the mutual admittance between node i and node j, and $d_{ii} = \sum_{j=1}^{n} c_{ij}$ is the self-admittance at node i.

The power dissipation in the resistive network is given by:

$$P = v^T Y_n v, \tag{2.15}$$

where v is an n-vector representing the node voltage vector and Y_n is the indefinite admittance matrix, which is symmetric. Thus the cost function of the placement problem becomes the power dissipation in the linear passive resistive network. It is known that the current will distribute itself in such a way that the power dissipation is minimum [82]. Therefore, the techniques developed for problems in circuit theory can be used to solve the modules placement problem.

Consider the n-terminal resistive network shown in Figure 2.29. The first m nodes are floating, and their voltages are denoted by an m-vector v_1. The remaining $(n - m)$ nodes are connected to voltage sources denoted by an $(n - m)$-vector v_2. Thus the coordinates of the n modules are represented by an n-vector

$$v = \begin{bmatrix} v_1 \\ v_2 \end{bmatrix},$$

where the coordinates of the fixed modules are specified by v_2, and the coordinates of the movable modules, to be determined, are represented by v_1. The network equations are:

$$0 = y_{11}v_1 + y_{12}v_2, \tag{2.16}$$

$$i_2 = y_{21}v_1 + y_{22}v_2, \tag{2.17}$$

where y_{11}, $y_{12} = y_{21}^T$, and y_{22} are the short-circuit admittance submatrices of the indefinite admittance matrix Y_n. Equation 2.16 gives:

$$v_1 = -y_{11}^{-1} y_{12} v_2, \tag{2.18}$$

which gives the solution of the movable modules in terms of the fixed modules and the admittance submatrices.

Movable modules must be located on slots. That is, the voltage vector v_1 must represent a set of legal values. Let the prescribed slots in terms of the permutation vector be $p = [p_1, p_2, \ldots, p_m]^T$, where p_i is the ith legal value and m is the total number of movable modules. Note that the order of p_i is not important because it appears in summation form. Let $v_1 = [x_1, x_2, \ldots, x_m]^T$, where x_i denotes the coordinate of module i or the voltage at node i. The following set of equations represents the constraints on the modules which are required to be on slots:

$$\sum_{i=1}^{m} x_i = \sum_{i=1}^{m} p_i$$

$$\sum_{i=1}^{m} x_i^2 = \sum_{i=1}^{m} p_i^2$$

$$\vdots$$

$$\sum_{i=1}^{m} x_i^m = \sum_{i=1}^{m} p_i^m \qquad (2.19)$$

The first equation can be rewritten as

$$I^T v_1 = I^T p \equiv d, \qquad (2.20)$$

where I is a unit vector and d is a constant which is equal to the sum of the m legal values.

Designing an exact algorithm for solving this problem is very time-consuming and not practical. The following heuristic was proposed in [50]. The main idea is to solve a simple optimization problem using the linear resistive network analogy repeatedly, and, in the meantime, the movable modules are assigned to slots. Node voltages and module coordinates are interchangeable according to the context.

From Equations 2.15, 2.16, and 2.17, the goal is to minimize the power dissipation

$$P = v^T Y_n v = \begin{bmatrix} v_1^T, v_2^T \end{bmatrix} \begin{bmatrix} y_{11} & y_{12} \\ y_{21} & y_{22} \end{bmatrix} \begin{bmatrix} v_1 \\ v_2 \end{bmatrix} \qquad (2.21)$$

$$= v_1^T y_{11} v_1 + 2 v_1^T y_{12} v_2 + v_2^T y_{22} v_2 \qquad (2.22)$$

subject to the set of constraint equations in Equations 2.19. It is only practical to use the first linear constraint to get the initial solution. The solution is given by the well-known Kuhn-Tucker conditions [193]:

$$v_1 = y_{11}^{-1}[-y_{12} v_2 + i_1], \qquad (2.23)$$

where

$$i_1 \equiv \frac{d + I^T y_{11}^{-1} y_{12} v_2}{I^T y_{11}^{-1} I} I. \qquad (2.24)$$

The first term in Equation 2.23 is the solution when there is no constraint on slots. The second term arises because of the constraint on slots.

The result can be improved by considering the deviation of the solution of Equation 2.23 under the constraint of Equation 2.20,

$$\mathbf{I}^T \delta v_1 = 0, \tag{2.25}$$

where δv_1 denotes a deviation of v_1. Then the power dissipation is increased by

$$\Delta P = \delta v_1^T y_{11} \delta v_1 \tag{2.26}$$

because of Equations 2.22, 2.23, and 2.25. Furthermore, from a theorem of Gerschgorin [372], it is known that the eigenvalues of y_{11} are not larger than $2y_M$, where y_M is the largest diagonal element in y_{11}. This gives

$$\Delta P = \delta v_1^T y_{11} \delta v_1 \le ||y_{11}|| \, ||\delta v_1||^2 \le 2y_M \sum_{i=1}^{m} \delta v_i^2. \tag{2.27}$$

Next, add the second-order slot constraint, and try to minimize the deviation of the minimum power. This procedure is called *scaling*.

Again, assume that in the region there are k modules, and the legal values are given by the permutation vector $[p_1, p_2, \ldots, p_k]$. Let $[x_{01}, x_{02}, \ldots, x_{0k}]$ denote the solution obtained from optimization with the linear constraint, and let $[x_{n1}, x_{n2}, \ldots, x_{nk}]$ denote the new solution after scaling. Thus the problem is to minimize

$$\sum_{i=1}^{k} (x_{ni} - x_{0i})^2 \tag{2.28}$$

under the constraints

$$\sum_{i=1}^{k} x_{ni} = \sum_{i=1}^{k} p_i \tag{2.29}$$

and

$$\sum_{i=1}^{k} x_{ni}^2 = \sum_{i=1}^{k} p_i^2. \tag{2.30}$$

The solution is also given by the Kuhn-Tucker conditions, that is, for $i = 1, 2, \ldots, k$,

$$x_{ni} = \frac{x_{0i} - c_0}{a_0} a_n + c_n, \tag{2.31}$$

where c_n, a_n, c_0, and a_0 are functions of k, p_i, and x_{0i} (see [50] for details).

The relaxation step calls for repeated use of scaling and optimization over subregions to be specified by designers. When a subregion is considered, modules outside the subregion are kept fixed. It is proposed that subregions be chosen by first starting from one end of the region, then from the other end, and, finally, from the middle.

The overall procedure is as follows. After the initial optimization over the entire region, three steps of scaling and optimization over subregions are carried out. Then the region is evenly partitioned into two subregions and scaling is performed once more on both of them. The result gives two partitioned subregions along with their associated modules. The process is repeated for each subregion; that is, optimization, relaxation, and partitioning are performed independently for each subregion until each subregion contains only one module.

2.3.6 Regular Placement: Assignment Problem

The assignment problem is a special class of the placement problem. Here, possible positions of the modules are predetermined. Such positions are called target cells (or targets, for simplicity). The problem is to assign each module to a target. In gate arrays, each target stands for a real gate, and we use them to implement circuit gates. In other words, the gates of the circuit are assigned to gates of a gate array. The assignment problem is typically solved in two steps, relaxed placement, and removing overlaps.

In the relaxed placement phase, the positions of modules are determined using a cost function. In this step, overlap of modules is allowed; that is, two or more modules may be assigned to the same target. A subset of modules, for example, I/O pads, have preassigned locations (otherwise the problem becomes trivial). All overlaps will be removed in subsequent stages. Here, the linear programming method is applied to the relaxed placement problem.

For a net N_i, define three variables,

$X_{l,i}$: the leftmost position of net N_i,

$X_{r,i}$: the rightmost position of net N_i, and

X_v : possible location of module M_v, where module M_v is connected to net N_i.

According to the definition of $X_{l,i}$ and $X_{r,i}$,

$$X_l \leq X_{l,i} \leq X_v \leq X_{r,i} \leq X_r, \qquad (2.32)$$

where X_l and X_r are the values of the left and right boundaries of the chip, respectively. For fixed modules, $X_{v1}, X_{v2}, ..., X_{vk}$,

$$X_{vi} = X_i. \qquad (2.33)$$

Designers can add a weight $W(i)$ to each net N_i. In this case the goal is to minimize

$$\sum_{N_i \in \mathcal{N}} W(i)(X_{r,i} - X_{l,i}). \qquad (2.34)$$

By Equations 2.32, 2.33, and 2.34 (and the three symmetric equations for the y-coordinate), the assignment problem is formulated as a LP and existing LP packages (called LP solvers) can be used to find a relaxed assignment of the modules.

At the end of the relaxed placement phase, the solution may have overlapping modules. Therefore, all overlaps should be removed. Assign to each circuit module M_i a cost C_{ij}, if it is assigned to a target H_j. Typically, C_{ij} indicates the distance from the original position of M_i (obtained in the relaxed placement phase) to the target H_i. Define a complete bipartite graph with modules as one set of vertices, and targets as the other set of vertices. The weight of the edges is dictated by costs C_{ij}. Now, the problem of removing overlaps becomes a minimal weighted matching problem because each module is to be assigned to exactly one target. The classical solution to the matching problem runs in $O(n^3)$ [272], where n is the number of modules.

Figure 2.30 shows an example. First, an assignment is obtained in the relaxed placement phase (Figure 2.30a). There is a cost assigned to each module-target pair (Figure 2.30b). This cost is assigned by the designer. For example, $C_{a1} = 0$ indicates that it is good (i.e., costs nothing) to assign module a to target 1, and $C_{a4} = 100$ indicates that it is very bad (i.e., costs a lot) to assign module a

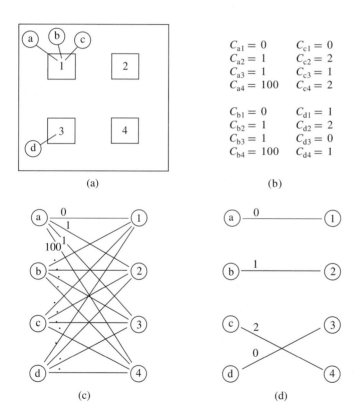

(a)

$$C_{a1} = 0 \qquad C_{c1} = 0$$
$$C_{a2} = 1 \qquad C_{c2} = 2$$
$$C_{a3} = 1 \qquad C_{c3} = 1$$
$$C_{a4} = 100 \qquad C_{c4} = 2$$

$$C_{b1} = 0 \qquad C_{d1} = 1$$
$$C_{b2} = 1 \qquad C_{d2} = 2$$
$$C_{b3} = 1 \qquad C_{d3} = 0$$
$$C_{b4} = 100 \qquad C_{d4} = 1$$

(b)

(c)

(d)

FIGURE 2.30
(a) An assignment obtained by solving the relaxed placement problem; (b) cost of assigning modules to targets; (c) a complete bipartite weighted graph; (d) a min-weight perfect matching. The values on edges are the corresponding costs.

to target 4. Thus a complete bipartite weighted graph is formed (Figure 2.30c). Finally, targets are matched to appropriate modules as dictated by a min-cost perfect matching (Figure 2.30d).

A summary of the algorithm follows.

1. Generate equations for the variables of the nets and modules.
2. Use a linear programming subroutine to solve the equations and obtain a relaxed placement solution.
3. Assign a cost C_{ij} if module M_i is assigned to a target H_j for each module and target; a complete bipartite weighted graph is formed.
4. Use a minimal weighted matching algorithm to solve the perfect (i.e., one in which all vertices are matched) bipartite weighted graph.

2.3.7 Linear Placement

Linear placement is a one-dimensional placement problem. It can be used as a heuristic or as an initial solution for the two-dimensional placement problems, or it can be employed as a cell generator (see Chapter 6). In particular, after a good linear placement is constructed according to an appropriate cost function (e.g., total length, min-cut), it can be modified to obtain an effective placement for standard cells. However, optimal linear arrangement problems under both cost functions (cut-based and length-based) are NP-hard [113].

One heuristic method for solving the linear placement problem is as follows. First, a clustering tree is created by combining highly connected nodes. Then the nodes are placed according to the leaf sequence (i.e., postordering of the leaves). In order to improve the result, certain operations are performed on the inner nodes (e.g., permuting children to get a better placement), as shown in Figure 2.31. There, node X has 3 children. There are 3! different arrangements. The relative positions of modules d and e are fixed, and so are the relative positions of modules a and b. Note that while tracing the tree, either in a top-down or bottom-up manner, the best local permutation is found by evaluating the corresponding cost. Once it is found, it will not be changed in later stages. Figure 2.31c shows a possible initial placement for a standard cell. It is a "snake"-type folding of the linear placement.

Such an algorithm will spend at most $d!$ time at each internal node of the tree, where d is the maximal degree of the tree. There are at most n inner nodes, and the cost evaluation for each permutation takes $O(n)$ time. Therefore, the total time complexity is $O(n^2 d!)$. If $d = n$, the time complexity is $O(n^2 n!)$; a good solution is obtained, sacrificing time. If $d = 2$, the time complexity is $O(n^2)$; it takes less time, but does not explore many possibilities. There is a natural trade-off between the quality of the final solution and the time complexity of the algorithm.

An advantage of linear placement is that cost evaluation is easy. In fact, for some problems (e.g., when the interconnections among the modules form a tree), an optimal linear placement can be obtained in polynomial time. The main

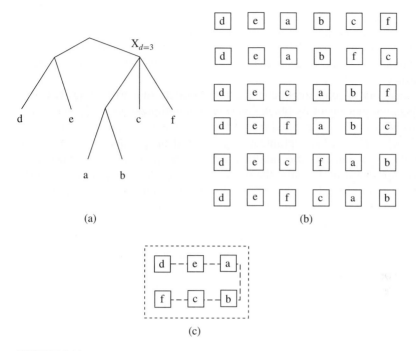

FIGURE 2.31

An example of linear placement. (a) A clustering tree; (b) possible placements corresponding to inner node X; (c) an application to standard cell.

disadvantage of the technique is that, in most situations, it is only an approximation of the two-dimensional reality. Linear placement and cell generation will be discussed in more detail in Chapter 6.

2.3.8 Discussion

Simulated annealing, genetic algorithms [322], neural networks [341, 403], and a special class of force-directed methods [61] are among the iterative improvement techniques in general use. Genetic algorithms work by emulating the natural process of evolution as a means of progressing toward the optimum. Most of them obtain near-optimal results, but spend enormous amounts of time. Iterative improvement algorithms obtain good results but are time-consuming.

One of the recent methodologies for the placement problem is based on neural networks [176]. The module placement is formulated as a binary-valued optimization problem. The binary formulations allow these complex problems to be solved using specialized hardware or efficient mathematical software. The placement problem has been formulated as a hierarchical quadratic binary optimization problem, which is a form that can be solved using a new neural network model, an artificial neural network (a network of analog devices called *neurons* is used to solve the problem). The classic network that has been employed is the

Hopfield network. To deal with the placement problem (which involves a quadratic cost function), [176] proposed a modified neural network. The modified network consists of two interconnected Hopfield networks. The main disadvantage of the neural network-based algorithms is that they must be implemented in hardware to be fully utilized.

The category of constructive placement consists of partitioning, resistive network analogy, placement by block Gauss-Seidel optimization [359], and force-directed methods [8] Force-directed method, resistive network analogy, and placement by block Gauss-Seidel optimization are based on physical law analogies. They do not consume much time, but the results are not as good as those achieved by iterative improvement. Usually, the results of the constructive placement methods are used as the initial placements in the iterative improvement algorithms. However, when there is an enormous number of cells, as in a sea-of-gates layout, the constructive placement algorithms are more practical [359].

In [334], there is a comparison between linear and quadratic cost functions. The linear cost function used in a GordianL placement tool achieves results with up to 20% less area than the quadratic cost function of the original Gordian procedure. The reduction by GordianL of the net length for nets connecting only two or three modules is mainly responsible for the improvement. Other ideas have also been proposed, [132, 367].

There are other heuristic approaches [310, 380] for the assignment problem that are fairly fast and are used to avoid the high time complexity of the linear programming approach. Other cost functions for removing overlaps have also been proposed [21, 269].

Recently, timing constraints and delay optimization in the placement problem have been investigated [135, 165]. These issues will be discussed in the next chapter.

EXERCISES

2.1. Consider a hypergraph H, where each hyperedge interconnects at most three vertices. We model each hyperedge of degree-3 with three edges of weight $\frac{1}{2}$, on the same set of vertices, to obtain a weighted graph G. Prove that an optimal balanced partitioning of G corresponds to an optimal balanced partitioning of H. Prove this cannot be done if each edge of H interconnects at most four vertices (i.e., give a counter example).

2.2. Design an algorithm for transforming a hypergraph into a graph. Discuss effectiveness of your technique as it applies to the bipartition problem.

2.3. Give an example of an unweighted graph for which the Kernighan-Lin algorithm does not produce an optimal solution. How bad (as a function of n, where n is the number of vertices in the graph) can the ratio C_{kl}/C_{opt}, where C_{kl} is the cost obtained by the Kernighan-Lin algorithm and C_{opt} is the optimal cost?

2.4. Consider a path graph $v_1, \ldots, v_n$. That is, v_i is connected to v_{i+1}, for $1 \leq i \leq n - 1$. Apply the Kernighan-Lin algorithm to this graph. As the initial partition, let v_a, for all odd values of a be in one set, and v_b, for all even values of b, be in the other set.

2.5. Consider a complete binary tree with n nodes. Apply the Kernighan-Lin algorithm to this graph. As the initial partition, let v_a, for all internal vertices, be in one set and v_b, for all leaves, be in the other set.

2.6. Consider a graph with n vertices and maximum degree k. Design an algorithm for partitioning the graph into g groups such that the number of vertices in each group is at most s, and the number of edges connected to each group is at most b. Analyze the quality of your algorithm for different values of k, g, and b. For what values is your algorithm optimal?

2.7. Consider a circuit whose adjacency graph is a complete binary tree with seven nodes. Find an initial placement of the modules, using a constructive force-directed algorithm, in a 3×3 gate array environment. Write a set of nonlinear equations and solve them to find an initial placement of the modules. In your formulation, place the branches of the tree on the four corner modules.

2.8. Design a cost function for the general building-block placement problem which considers the wire length, estimated area, module overlap, and aspect ratio of the entire layout.

2.9. Prove that there is a one-to-one correspondence between a sliceable floorplan and a normalized Polish expression.

2.10. Given a Polish expression corresponding to a given slicing floorplan, show that the expression 12—3— ... —n— can be reached, and vice versa, using OP1, OP2, and OP3.

2.11. Consider two lists, $A = \{(x_1, y_1) \ldots (x_n, y_n)\}$ and $B = \{(a_1, b_1) \ldots (a_m, b_m)\}$, with $x_i \leq x_{i+1}$, $y_i \geq y_{i+1}$, $a_i \leq a_{i+1}$, and $b_i \geq b_{i+1}$. Combine A and B by considering each element (x_i, y_i) of A and each element (a_j, b_j) of B to produce an element of a list C: $(x_i + a_j, max(y_i, b_j))$. Thus, C has $m \times n$ elements. If there are two elements (c_i, d_i) and (c_j, d_j) in C with $c_i \leq c_j$ and $d_i < d_j$ then delete (c_j, d_j) from C. Prove that the resulting list C has at most a linear function of m and n elements. Find that linear function.

2.12. Find an optimal implementation of modules $M_1 \ldots M_8$ for sizing each of the following sliceable floorplans:

     Floor 1:   1 2 V 3 4 V H 5 6 V 7 8 V H V
     Floor 2:   1 2 V 3 H 4 V 5 H 6 V 7 H 8 V

     M_1: $4 \times 33 \times 43 \times 62 \times 81 \times 9$
     M_2: $4 \times 55 \times 4$
     M_3: $4 \times 43 \times 55 \times 3$
     M_4: 3×5
     M_5: $5 \times 66 \times 5$
     M_6: $2 \times 64 \times 4$
     M_7: 5×5
     M_8: $1 \times 52 \times 43 \times 35 \times 1$

2.13. Solve the following generalization of the slicing floorplan sizing problem. Given a slicing tree corresponding to a set of modules, each module has a set of implementations and each implementation is specified by three integers (w, h, p). As before, w and h, respectively, represent the width and the height of the implementation, and p represents the the power consumption of the implementation.

Design an algorithm that finds an implementation of the modules that minimizes $A + \lambda P$, where A is the area of the slicing floorplan, P is the power consumption of the floorplan (being the sum of the power consumption of each module), and λ is a user specified constant. Analyze the time complexity of your algorithm.

2.14. Design an algorithm for the linear arrangement (or placement in one row) problem. How good is your solution? Solve the same problem if the input graph is known to be a (not necessarily balanced) binary tree.

COMPUTER EXERCISES

2.1. Implement the Kernighan-Lin algorithm for a hypergraph. Our goal is to find a balanced partition with minimum cost.

Input format. Each input starts with the weight of a hyperedge followed by the vertices interconnected by it. Specification of the hyperedges are separated by commas.

> 3 1 4,   (* there is a hyperedge of weight 3 connecting vertices 1 and 4 *)
> 2 1 4 2,
> 6 2 3 5, 1 4 5

Output format. The output format is shown in Figure CE2.1. There should be options for:

> Not showing names of the vertices,
> Not showing hyperedges of weight less than k (k is specified by the user)
> Color coordinating the edges (e.g., red for large weights, blue for average).
> Any other options that help us learn about the algorithm.

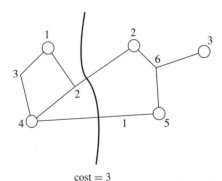

cost = 3

FIGURE CE2.1
Output format, a partition.

2.2. Implement a multiway version of the Kernighan-Lin algorithm. The vertices are to be partitioned into g groups, each with size at most s and number of edges connected to each group at most b (this is a bounded capacity, bounded I/O multiway partitioning problem).

Input format. The first line of the input consists of three integers:

g b s

The rest of the input is the same as CE 2.1.

2.3. Consider a weighted hypergraph H. Implement an algorithm that transforms it into a weighted graph G (see Section 2.1.1). Implement the Kernighan-Lin algorithm to partition G into V_1 and V_2. Find the cost C_G of the partition (V_1, V_2) in H. Now apply the Kernighan-Lin algorithm directly to H. Let C_H denote the cost of this partition. Compare values of C_G and C_H for different numbers of vertices, hyperedges, and various weight distribution. Report your conclusion.

2.4. Implement the aatio-cut algorithm. Use the input and output formats of CE 2.1. Implement another version of the algorithm that consists of phases 1 and 3 only (do not include any iterative shifting). Experiment with both versions using graphs with different numbers of nodes, hyperedges, and distributions of weight. Report your conclusion.

2.5. Implement a vertex-weighted version of the Kernighan-Lin algorithm on a weighted hypergraph. As before, each hyperedge is assigned a weight. In addition, each vertex is also assigned a weight. A balanced partition is one in which the sum of the weights of the vertices in one set is (approximately) the same as the sum of the weights of the vertices in the other set. Do you expect the algorithm to perform better when the vertex weights are almost the same or when they are very different? How can you verify your claim experimentally?

Input format. The first set of data contains the names of the vertices along with their weights. Specifications of two vertices are separated by a comma. After the weights of the vertices, there is a semicolon, followed by the weight of each hyperedge along with the vertices it connects. This part is the same as shown in CE 2.1.

1 3, 2 4, 3 1, 4 2, 5 6;   (* vertex 1 has weight 3, etc. *)
3 1 4,   (* there is a hyper edge of weight 3 connecting vertices 1 and 4 *)
2 1 4 2,
6 2 3 5, 1 4 5

Output format. The output format is the same as Figure CE2.1. You should have an option that displays the weight of the vertices.

2.6. Consider a module adjacency graph G. Design and implement a greedy algorithm that finds a rectangular dual F corresponding to G. Since your algorithm is a greedy one (and probably is not optimal), then F^d (the adjacency graph of F) is not, in general, equal to G. Try to minimize the difference Δ between G and F^d. Design an algorithm assuming:

Case 1) Δ is the number of edges that are in G and not in F^d,
Case 2) Δ is the number of edges that are in G and not in F^d, plus the number of edges that are in F^d and not in G.

Input format. The input is specified by a set of edges separated by commas, the same format used in the previous computer exercise. For example, see the graph coloring problem of CE 1.7.

Output format. The output format is shown in Figure 2.10 (show the adjacency graph on the left and the floorplan on the right). Highlight (with a different color) the edges whose adjacency is not satisfied. Show (with a different color) edges corresponding to adjacencies that were not in G and have been created by your algorithm.

2.7. Solve the previous problem assuming edges of G have weights. Δ is defined as the sum of the weight of the edges that are in G and not in F^d.

Input format. The input is given by a set of edges. Each edge is specified by three integers: the two vertices it connects and its weight. Specifications of the edges is separated by commas. For example,

2 3 1, 3 1 6

indicates that there are two edges. The first one connects vertex 2 to vertex 3 and its weight is 1. The second edge connects vertex 3 to vertex 1 and its weight is 6.

Output format. The output format is as shown in Figure 2.10.

2.8. Solve the previous exercise (input is a weighted graph) employing a bottom-up hierarchical algorithm. Now solve the problem using a top-down hierarchical algorithm. Compare the two results. Is one algorithm better in all cases or does it depend on the input structure? Tabulate your results and discuss.

2.9. Consider a set of modules $\mathcal{M} = \{M_1, \ldots M_n\}$. The size of M_i is (w_i, h_i). Use simulated annealing to find a minimum-area slicing floorplan of $\mathcal{M}$. Size and orientation of each module is fixed.

Input format.

2 2, 2 2,
2 1, 2 3

The example corresponds to Figure 2.22.

Output format. The output is the final floorplan (see Figure 2.22, the last floorplan).

2.10. Consider a set of modules $\mathcal{M}$. Each module has a set of implementations. Generate a random slicing tree T corresponding to $\mathcal{M}$. Then implement the sizing procedure (discussed in Section 2.2.4) to find an optimal implementaion of the modules as dictated by T. Run your algorithm several times to generate different slicing trees T. Output the best three implementations.

Input format. The input is a set of implementations for each module. Implementations of two modules are separated by a period, and two implementations of the same module are separated by a comma. For example:

4 1, 3 2, 2 3. 5 2, 6 1

indicates that one modules has three implementations (4 1, 3 2, 2 3) and another module has two implementations (5 2, 6 1).

Output format. The output should show the floorplans (with wasted area shaded) and the corresponding trees. As an example, see Figure 2.19b.

2.11. Consider a set of modules. Each module has a set of terminals on the upper side of its horizontal edge. Find a linear placement with small density. Draw the modules and nets, and report the density of your solution.

Input format.

```
3                    (* number of modules *)
M1 6, 3 1, 5 4;      (* module 1 occupies 6 grid points, at the 3rd grid
                        point there is a terminal of net 1 and at the 5th
                        grid point there is a terminal of net 4 *)

M2 4, 2 1;
M3 7, 3 1, 2 4;
```

Output format. The output format is shown in Figure CE 2.11. Show all nets and their routing.

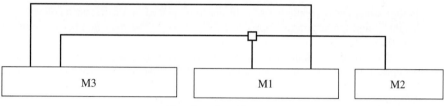

No. of tracks = 2

FIGURE CE2.11
Output format, a linear placement.

2.12. Consider a set of modules in a gate array environment. Find a placement with minimum cost. The cost of a net is the smallest rectangle enclosing all terminals of a net (the distance between two adjacent modules in the same row is 1). The cost of a solution is the sum of the costs of the nets. Start with a random placement of these modules. Implement an iterative force-directed algorithm that improves the initial placement. Next start with a better initial placement (not a random one) and apply the same iterative force-directed algorithm to it. Which one performs better?

Input format.

```
3             (* number of modules *)
M1 2 3        (* module 1 is connected to modules 2 and 3 *)
M2 1
M3 1
```

Output format. The output format is a gate array placement as shown in Figure CE 2.12. Show all nets and write the total length of your placement.

total cost = 2   **FIGURE CE2.12**
Output format, a gate array placement.

2.13. Solve the previous computer exercise employing a greedy algorithm. How can you verify that your solution is good?

2.14. Given a set of standard cells and a net list, implement an efficient algorithm for placing all modules. The goal is to minimize the sum of bounding boxes of all nets. The number of rows and the width of each row are given numbers. The separation between two rows is given. Note that in each cell the same terminal appears on both the top side and the bottom side of the cell, and there are no terminals on the left and the right side of the cell.

Input format.

3 7 9 200
1 3 0 4, . . . ,
0 2 3 1 0 0

In the format above, a number of cells are to be placed in 3 rows, each cell row has height 7, and the separation between two rows is 9. The total number of columns are 200.

Two of the modules are shown, the first and the last one. The first module occupies four columns: net 1 has a terminal at the first column, net 3 at the second column, the third column is empty, and net 4 has a terminal at the fourth column. Note that two modules should be placed at least one unit apart.

Output format. Your output can be similar to the one shown in Figure CE 2.14. Show a placement (containing the name of the modules) and a connection of all nets (as straight lines). Indicate the total length of your result.

length of half-perimeter $= 9 + 7 + 9 + 1$

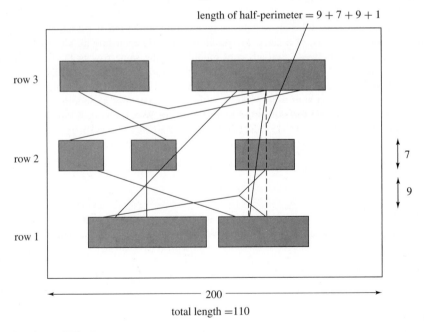

total length $= 110$

FIGURE CE2.14
Output format, placement of standard cells.

CHAPTER
3

THE TOP-DOWN APPROACH: ROUTING

This chapter focuses on the routing problem, in particular solving global and detailed routing problems while minimizing the chip area and total wire length. In the subsequent chapters routing problems are considered in relation to other performance measures such as signal delays, number of vias, and power consumption.

In the discussion here of fundamental problems of circuit layout, several techniques for each problem will be described in order to familiarize the reader with more than one algorithm in this area. At the end of each section, other known techniques will be outlined, along with their advantages and disadvantages.

3.1 FUNDAMENTALS

Before considering the (global and detailed) routing problem, several fundamental concepts, maze running, line searching, and Steiner trees, will be covered. These concepts have been used in various routing algorithms.

3.1.1 Maze Running

Maze running was studied in the 1960s in connection with the problem of finding a shortest path in a geometric domain. A classic algorithm for finding a shortest path between two terminals (or points) on a grid with obstacles was proposed in [200]. The idea is to start from one of the terminals, called the source terminal, and then label all grid points adjacent to the source with 1 until we reach a terminal

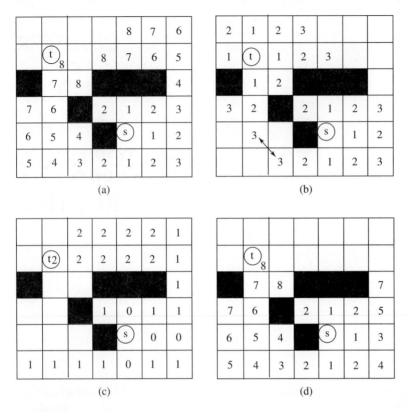

FIGURE 3.1
An example demonstrating Lee's algorithm from source s to sink t. (a) Minimum length path;
(b) bidirectional search; (c) minimum bend path; (d) minimum weight path.

called the sink terminal. The label of a point indicates its distance from the source.
Any unlabeled grid point p that is adjacent to a grid point with label i is assigned
label $i + 1$. All label i's are assigned first before assigning any label $i + 1$. Note
that two points are called adjacent only if they are either horizontally or vertically
adjacent; diagonally adjacent points are not considered to be adjacent, because
routing problems deal with rectilinear distances. The task is repeated until the other
terminal of the net is reached. See Figure 3.1a. The just described procedure is
called the maze-running algorithm, Lee's algorithm, or the Lee-Moore algorithm.

> **Theorem 3.1.** The maze-running algorithm finds a shortest route if the net has two
> terminals.
>
> *Proof.* Inductively, assume
>
> **H1.** The length of the shortest path from s to every grid point labeled i is i
> units. Furthermore,
> **H2.** No other grid point that has label j should have label i.

This is certainly true for $i = 1$. The induction is obvious since all grid points that are labeled $i + 1$ are at distance one from a grid point that has label i; thus they are at a distance at most $i + 1$ from s (based on H1). Furthermore (based on H2) they cannot have label i or less.

Several extensions of maze running have been studied. The reason for extending the maze running techniques, as opposed to inventing new tools, is that the approach is simple and easy to understand, and its performance is predictable.

OPTIMIZATION OF MEMORY USAGE. A major drawback of maze-running approaches is the huge amount of memory used to label the grid points in the process. Attempts have been made to circumvent this difficulty. One solution is to use an encoding scheme where a grid just points to its neighbors instead of storing the actual distance. Therefore, at each grid point $O(1)$ bits are stored instead of $O(\log n)$ bits, where n is the number of grid points. There are other effective memory optimization schemes that are primarily for speeding up the process. One of the most effective approaches is based on bidirectional search as discussed below.

BIDIRECTIONAL SEARCH. An effective approach to speeding up the maze-running approach and, as a result, optimizing the memory usage, is to perform a *bidirectional search*. The algorithm starts from the source and labels all adjacent grid points with 1. Then, it starts from the sink and labels all adjacent grid points with 1. Then all grid points adjacent to a grid point with label 1 are labeled 2. In general, at stage i, all grid points that are still unlabeled and are adjacent to grid points with the label $i - 1$ are labeled with i. This task is repeated until the search from the source reaches the search from the sink at stage j: if they are reaching diagonally, as in Figure 3.1b, then the length of the shortest path is $2j + 1$. Otherwise, if they are reaching horizontally or vertically, the length of the path is $2j$.

A true saving in time and memory results because in the traditional technique, it is necessary to search a (rectilinear) circle with diameter l, where l is the length of the shortest path between the source and the sink, thus, the number of grid points visited is approximately $l^2/2$ (assuming there are no obstacles). In the new bidirectional search scheme, the search is in two circles of diameter $l/2$ each. In this case, the number of visited grid points is about $2(l/2)^2/2$, or $l^2/4$.

MINIMUM COST PATHS. Previously, the goal was to minimize the length of the path between the source and the sink, that is, minimize the distance in L_1 metric. However, there are also other objectives.

To minimize the number of bends in the path, proceed as follows. All grid points that are reachable with zero bends from the source are labeled zero. Note that these grid points are either adjacent to the source or are adjacent to a grid point with label zero. All grid points that are reachable from a grid point with label zero with one bend are labeled with one. In general, in stage i, all grid points that are reachable from a grid point with label $i - 1$ with one bend are labeled i. Note that for each grid point with label i, it is also necessary to store the direction of the path (if there are more than one path, all directions—at most 4—need to be stored) that connects the source to that grid point with i bends. An example is shown in Figure 3.1c.

It is also possible to obtain a minimum cost path, where the cost of a path is defined by the user. For example, with reference to Figure 3.1d, a path could be found that minimizes the use of the right boundary. To accomplish this, assign every connection that uses the right boundary a cost of 2. Then, proceed with the traditional maze-running algorithm. Except for grid points that are on the right boundary, skip assigning labels every other time. That is, if the grid point is adjacent to a grid point with label i, do not assign a label to it until stage $i + 2$.

MULTILAYER ROUTING. Multilayer routing can be achieved with the classic maze-running algorithm. The main difference is that the maze is now a three-dimensional maze (or a three-dimensional grid). The labeling proceeds as before. To minimize the number of layer changes, assign a higher cost to traversing the grid in the third dimension.

MULTITERMINAL ROUTING. This procedure involves first interconnecting two terminals, as before. Then, starting from a third terminal, label the points until the path between the first two terminals is reached. This task is repeated for all unconnected terminals. The resulting solution may be suboptimal.

3.1.2 Line Searching

There are two classes of search algorithms. The first one is the grid search, where the environment is the entire grid, as in the Lee-Moore search techniques. In grid searching it is very easy to construct the search space; however, the time and space complexities are too high. The second class of search techniques, first proposed by Hightower [136], is called *line searching*. The environment is a track graph that is obtained by extending the boundary of modules. A track graph is much smaller than the corresponding grid-graph, and, thus, computations can be done more efficiently on it. The original version of the line-searching algorithm is as follows.

The algorithm starts from both points to be connected and passes a horizontal and a vertical line through both points. These lines are called *probes* (for that reason, the technique is sometimes called *line probing*). Specifically, the lines generating from the source are called *source probes*, and the lines starting from the sink are called *sink probes*. These probes are called first-level probes. If the sink probes and the source probes meet, then a path between the source and the sink is found. Otherwise, either the source probes or the sink probes (or both)

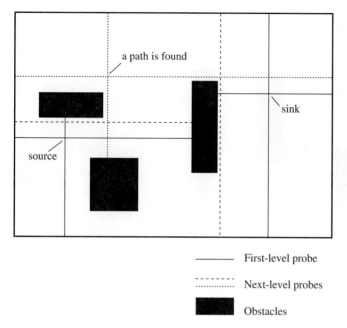

————	First-level probe
--------	Next-level probes
▉	Obstacles

FIGURE 3.2
An example demonstrating the line-searching algorithm.

have intersected an obstacle. From these intersections, pass a line perpendicular to the previous probe. Probes constructed after the first-level probes are called next-level probes. This task is repeated until at least one source probe meets at least one sink probe; a path between the source and the sink has been found (see Figure 3.2).

Although, in the worst case, it might take a long time to find a path, on average a path is found quickly. In particular, the time it takes to find a path between the source and the sink is at most equal to the number of obstacles in the routing environment (see Exercise 3.3 to understand why). In contrast, in the maze-running approach, the time is proportional to the shortest path (number of grid points) between the source and the sink.

A more effective way of dealing with line searching is to form a *track graph*. A track graph is obtained by extending the horizontal and vertical sides of each obstacle until another obstacle is reached, in addition to passing a horizontal and a vertical line from the source and the sink. The track graph corresponding to the problem in Figure 3.2 is shown in Figure 3.3. The idea is to search for a shortest path between the source and the sink in the track graph instead of the entire grid. It is known (and easy to show) that the shortest path between the source and the sink is a path in the track graph. Therefore, a shortest path algorithm such as Dijkstra's algorithm can be employed to find a path of minimum length between the source-sink pair.

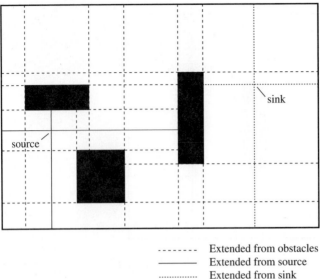

────── Extended from obstacles
─────── Extended from source
·············· Extended from sink

FIGURE 3.3
An example of a track graph.

3.1.3   Steiner Trees

A tree interconnecting a set $P = \{p_1, \ldots, p_n\}$ of specified points in the rectilinear plane and some arbitrary points is called a *(rectilinear) Steiner tree* of P. A Steiner tree with minimum total cost (e.g., length), is called a *Steiner minimal tree* (SMT). The problem with constructing an SMT has been given much attention. The general SMT problem is NP-hard [112]; only restricted classes of the problem can be solved in polynomial time. Here, all distances are measured in the L_1 metric.[1] SMT algorithms, exact and heuristic, are reviewed below in different models and environments. In particular, the following classes of problems are of interest.

- **Minimum length Steiner trees.** In exact and approximation algorithms for the SMT problem, the goal is to minimize the sum of the length of the edges of the tree.

- **Weighted Steiner trees.** Given a plane partitioned into a collection of weighted regions, an edge with length ℓ in a region with weight w has cost ℓw. The goal is to minimize the total cost of the tree.

- **Steiner trees with arbitrary orientations.** Rectilinear geometry is generalized to allow lines in other (fixed) directions, for example, $+45°$ and $-45°$ lines in addition to horizontal and vertical lines.

---

[1]In the L_1 metric, the distance between two points $P_1 = (x_1, y_1)$ and $P_2 = (x_2, y_2)$ is $|x_1 - x_2| + |y_1 - y_2|$.

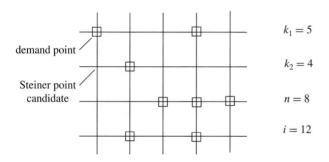

FIGURE 3.4
Candidate Steiner points.

MINIMUM LENGTH STEINER TREES. This section describes the classic version of the SMT problem, which consists of a set of points in a two-dimensional plane where all directions are rectilinear and distances are measured in the L_1 metric. As the general problem is NP-complete, first, subclasses for which exact algorithms are known are reviewed. Then, the general problem and recent heuristics are considered.

Exact Steiner trees. Several classes of SMT problems have been solved optimally. An exact algorithm for solving an arbitrary instance of SMT is considered first. Clearly this algorithm runs in exponential time. After that, other instances that can be efficiently solved are described.

Consider the point set P involving n points. By passing a vertical and a horizontal line through each point of P, a $k_1 \times k_2$ grid $(k_1, k_2 \le n)$ $G = (V, E)$, is obtained, where V is the set of grid points and E is the set of grid edges. It is called the Hanan grid. See Figure 3.4. It has been proven [130] that there exists an SMT of Q where all the Steiner points are a subset of vertices of G. In an SMT of Q, there are between zero and $n - 2$ Steiner vertices. If it is known that there are i Steiner points in an SMT of P, then an optimal tree can be found by enumerating all the Steiner trees with i Steiner points and n demand points. There are $\binom{k_1 k_2 - n}{i}$ ways to select the i points. For each selection, find a *minimum spanning tree* (MST) of $n + i$ points. Thus, the number of candidate Steiner trees is

$$f_n = \sum_{i=0}^{n-2} \binom{k_1 k_2 - n}{i}.$$

In the above equation, $k_1 k_2 - n$, instead of $k_1 k_2$ is used since a Steiner point and a demand point may not coincide. For each choice of Steiner points $O((n + i) \log(n + i)) = O(n \log n)$ time is spent to obtain an MST of $n + i$ points. Thus, an optimal Steiner tree, which is obtained for at least one selection of the i points, is constructed in $O(f_n n \log n)$ time. This is a straightforward approach and, by itself, not effective. Heuristic improvements of the above exhaustive search have

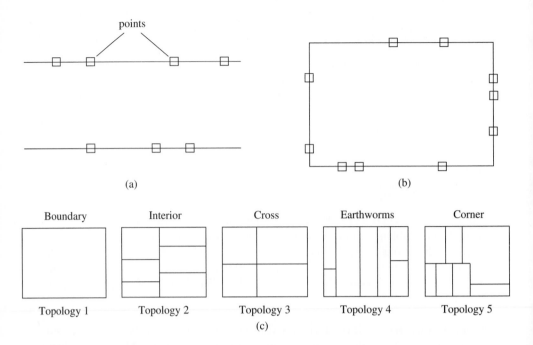

FIGURE 3.5
Channel and switchbox Steiner trees. (a) A channel; (b) a switchbox; (c) topologies of an SMT.

been considered; these algorithms have been designed in a graph environment and have not been investigated in the rectilinear plane. These algorithms still ensure an SMT (SMT by definition means optimal length); however, they try to reduce the time complexity. Various heuristics for improving the time complexity (e.g., to $O(n^2 3^n)$ time) have been proposed. This class of algorithms is applicable only when the number of points is small.

There are other instances that can be "efficiently" solved. Consider an instance of an SMT where points are on two parallel lines called a *channel*, as shown in Figure 3.5a. This problem can be solved employing a dynamic programming technique (i.e., a left-to-right scan of the points) in $O(n)$ time [3]. A generalization of this problem, when the points are on the boundary of a rectangle, also called a *switchbox*, was studied in [2, 66]. An instance is shown in Figure 3.5b. The key idea for designing an efficient algorithm for the switchbox problem is to establish that the number of topologies of an SMT is bounded by a small constant. The part of a Steiner tree that is inside the switchbox is called an *interior segment*. Otherwise, a segment is a *boundary segment*. With reference to Figure 3.5c, consider the following topologies (in Topology 1 all segments are boundary segments and in Topologies 2–5 some of the segments are interior segments and some are boundary segments). Only interior segments are shown in the figure (symmetric cases are omitted).

Topology 1. *Boundary:* In this case, the entire Steiner tree is on the boundary of the switchbox.

Topology 2. *Interior:* There is one vertical line and a set of interior horizontal line segments are connected to it. These segments are alternatively connected to the left and right. There is also a symmetric case where there is one horizontal line segment and a collection of alternating vertical line segments are connected to it.

Topology 3. *Cross:* There are two interior segments, a horizontal line segment and a vertical line segment.

Topology 4. *Earthworms:* There is a set of vertical line segments. There are at most two horizontal segments that connect the first and the last vertical line segments to the boundary. There is also a symmetric case, where there is a set of horizontal line segments and at most two vertical line segments connected to it.

Topology 5. *Corner:* There is a corner path consisting of one vertical and one horizontal line segment. A horizontal line segment connects this path to the right boundary and a set of alternating vertical line segments connect the horizontal part of the path to the top and the bottom boundaries. There are also symmetric cases.

Theorem 3.2. [2, 66] There exists an SMT interconnecting the points of a switchbox having Topologies 1–5.

It was shown that it takes linear time to construct an optimal Steiner tree for each of the topologies described above. Thus, the entire algorithm runs in linear time.

Heuristic Steiner trees. A number of heuristics for Steiner tree construction have been proposed. Most heuristics start with a minimum spanning tree (MST) and modify it to obtain shorter trees. This is because it has been shown that the length of an MST over the length of an SMT is less than $\frac{3}{2}$ [156]. Thus, an MST is a good approximation of an SMT. A brief description of some SMT algorithms are presented below.

- **Iterated One-Steiner [172].** In each *round*, the algorithm adds one Steiner point. Among all possible Steiner points, the one that minimizes the cost will be selected. The cost of a Steiner tree is the length of an MST of the demand points and previously selected Steiner points. While iterating on the Steiner points, if the cost of a round is not improved over the previous round, the algorithm terminates. It improves over an MST by about 10.9% for $n = 40$.

- **L-shaped and Z-shaped optimal embedding [142].** This technique finds a separable MST, that is, an MST with the bounding box of every two nonadjacent edges being nonintersecting. Then, allowing either an L-shaped layout or a Z-shaped layout of the edges, an optimal Steiner tree is constructed. For $n = 100$, an L-shape improves about 9.7% over an MST and a Z-shape improves about 10.2%.

- **Neighboring Steinerization [375].** This is also an algorithm that improves the Steiner tree by local modification. The idea is to lay out a set of pairwise nonintersecting edges at each step. For $n < 35$, it improves about 8.5% over an MST.

- **Hierarchical construction [308].** The idea is to construct an MST and a binary decomposition thereof to balance the tree. Edges are laid out in a hierarchical (bottom-up) fashion. One extension of this work is to construct an optimal tree of subsets of points, where each subset contains k points, k being a design parameter. Considering $n = 100$, for $k = 2$ about 8.5% improvement over an MST is offered. For $k = 4$ this improvement is 10% and for $k = 8$ it is about 10.3%.

- **Prim-based [291].** This technique starts with one point and at each step adds the closest point to it, as in Prim's MST algorithm [281]. The overlapping of various edges produces Steiner points (this is also the case for the following two algorithms). This technique offers about 9.0% improvement for $n \leq 10$ and only about 5.3% for $n = 100$.

- **Kruskal-based.** Following the MST algorithm of Kruskal [192], at each step the smallest edge is added to the previously constructed tree. This technique offers about 8.5% improvement for $n = 100$.

- **Delaunay triangulation-based [337].** As all MST edges belong to Delaunay triangulation of the input point set, a tree is constructed by considering the edges of this triangulation. It shows about 8.6% on the average over an MST.

- **Greedy embedding [291].** A number of heuristics have been proposed based on local modification of a Steiner tree to reduce its length. For $n = 500$, the best of these algorithms improves about 9.5% over an MST.

It would be interesting to design a super-algorithm that effectively combines the best features of each algorithm.

Many of the above-mentioned algorithms have been used in global routers. An SMT is constructed that guides the routing of each net. However, as obstacles (modules) are not considered, the constructed SMT needs to be modified locally.

WEIGHTED STEINER TREES. Consider a tessellation of the plane into a collection of weighted regions. A *minimum-weight Steiner tree* is a Steiner tree with minimum total weight, where an edge with length ℓ in a region with weight w has total weight ℓw. The dotted path interconnecting the two points labeled p_i in Figure 3.6 corresponds to minimizing a total weight path.

Consider a tessellation (or dissection) of the plane into a collection $\mathcal{R} = \{R_1, \ldots, R_m\}$ of regions. Region R_i is assigned a weight w_i. Consider a path $\mathcal{P}$ connecting two points p_a and p_b. $\mathcal{P}$ passes through different regions.

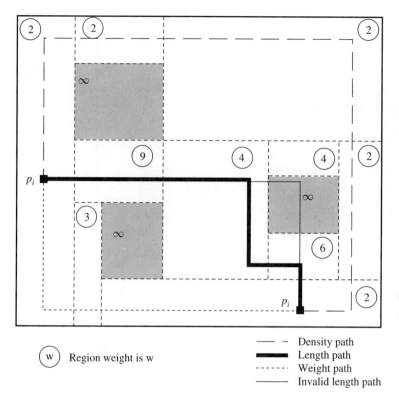

FIGURE 3.6
Connecting two points in weighted regions.

Convention 3.1. If some part $\mathcal{P}_t$ of $\mathcal{P}$ is on the boundary between two regions R_i and R_j and $w_i \leq w_j$, then it is said that $\mathcal{P}_t$ passes through R_i.

Let ℓ_i denote the length of $\mathcal{P}$ in region R_i; that is, $\ell_i = |\mathcal{P} \cap R_i|$. The weight of $\mathcal{P}$ is $W(\mathcal{P}) = \sum \ell_i w_i$.

Given a point set P in $\mathcal{R}$, obtain a minimum-weight Steiner tree interconnecting points of P, considering only rectilinear paths. The problem is referred to as the weighted rectilinear Steiner tree (WRST) problem. Certainly, the WRST problem is NP-complete as a restricted class of it—when $w_i = 1$ for all i—is the traditional minimum-length Steiner tree problem. Here, an effective approach is proposed to construct a WRST. It is also shown that the weight of constructed trees is at most two times the weight of the corresponding minimum spanning tree; furthermore, the bound of two is tight in the worst case (i.e., there are examples with the weight of constructed minimum spanning trees two times the weight of optimal WRST).

First find a minimum-weight path between two points p_a and p_b in $\mathcal{R}$. Construct a graph G, called the *search graph* of $\mathcal{R}$. This graph is obtained by extending boundaries of each region until they reach either the external boundary

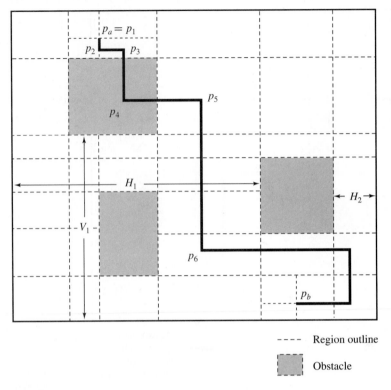

FIGURE 3.7
Finding a path in weighted regions.

or an obstacle (i.e., a region with infinite weight). The intersection of these edges defines the vertices of G and and the lines interconnecting them define the edges. The solid line in Figure 3.7 show the search graph of the regions in Figure 3.6. Given a point set Q, obtain the search graph G_Q in $(\mathcal{R}, Q)$ by passing a horizontal and a vertical line segment through each point in Q until each segment hits an edge of G. The intersection of the new lines with edges of G along with the original vertices of G define the vertices of G_Q. The lines connecting adjacent vertices form the edges thereof. The union of dashed and dotted lines in Figure 3.7 correspond to $G_{\{p_a, p_b\}}$.

Consider two adjacent horizontal lines in G. The maximal rectangle, bounded either by the boundary or by an obstacle, defined by the two lines is called *an elementary horizontal rectangle* (H_1 and H_2 in Figure 3.7). An *elementary vertical rectangle* (V_1 in Figure 3.7) is defined similarly. Two points of P that belong to the same elementary horizontal rectangle are called *horizontally dependent*. Similarly, two points of P that belong to the same elementary vertical rectangle are called *vertically dependent*. Otherwise, two points are *independent*. The following lemma was proved in [55]. The idea is to inductively show that each bend (i.e., $p_2, p_3, \ldots$ in Figure 3.7) can be shifted to a neighboring grid point.

Lemma 3.1. There is a minimum-weight path between two independent points p_a and p_b that is a path in $G_{\{p_a, p_b\}}$.

Graph G_P has e edges. The upper bound for e is $m^2 + n$, where $m = |\mathcal{R}|$ and $n = |P|$. However, in circuit layout applications (e.g., in gatearrays) $e = m + n$. To obtain a minimum-weight path between two independent points p_a and p_b, it is sufficient to find a minimum-weight path in G in $O(e \log e)$ time. A weighted MST $T_P = (P, L)$ of P is a minimum-weight tree interconnecting all points. Assuming the points of P are pairwise independent, an MST of P can be constructed by finding an MST among the points in G_P, for G_P contains all shortest paths among the points.

In general, a shortest path between two horizontally (or vertically) dependent points is not in G_P. Consider an elementary horizontal region containing a subset P_i of points with $n_i = |P_i|$ points. The shortest path between two points of P_i, p_a and p_b, is any x-y monotone[1] path between the two points. An MST of P_i is denoted by T_i. T_i is obtained as follows. Consider a complete graph $\mathcal{G}_i$ with vertices corresponding to the points of P_i and the weight of an edge corresponding to its distance in the elementary horizontal region. The weights are calculated as follows. Let p_j be a point in P_i. Pass a horizontal line ℓ through p_j and a vertical line from each point p_k (in P_i) to ℓ. The intersection of a point p_k with ℓ is denoted by p_k^*. In $O(n_i)$ time, using a simple scan, find the distance between p_j and p_k^*, for all k. In an additional $O(n_i)$ time, find the distance between each p_k and p_k^*. Repeat this task for all points to find weight of all edges of $\mathcal{G}_i$ in $O(n_i^2)$ time. From this an MST T_i of points in P_i, that is, an MST of $\mathcal{G}_i$ in $O(n_i^2)$ time, is obtained employing Prim's algorithm.

Certainly, all the points of G_P are not independent. It takes $\sum n_i^2$ time to find all MSTs. This is bounded by $O(n^2)$. Note that n is the multiplicity of a net and, in practice, is bounded by a small constant. Again n^2 is a pessimistic bound; although it may happen in the worst case, for example, when all the points belong to the same elementary region. Specifically, when points are uniformly distributed and the number of points is equal to or less than the number of regions in G, then $\sum n_i^2 = O(1)$.

A graph G^*, being the union of G_P and the edges of the T_i's, is produced. Based on Lemma 3.1 and the fact that all possible edges are being considered, conclude that an MST of G^* is indeed an MST in the environment $(\mathcal{R}, P)$. G^* has e^* edges and thus an MST of it can be found in $O(e^* \log e^*)$ time, where e^* is bounded by $n + e = O(e)$. The total time to find an MST is, then, $O(n^2 + e \log e)$, where $e = m^2 + n$. Based on the well-known technique of doubling, relating SMT length to the length of a Hamiltonian circuit, the following lemma can be verified.

Lemma 3.2. $\ell_{\text{MST}} \leq 2\ell_{\text{WRST}}^*$, and there exists an instance $(\mathcal{R}, P)$ with $\ell_{\text{MST}} = 2\ell_{\text{WRST}}^*$.

---

[1] An x-y monotone path is one that is cut by any vertical or horizontal line at most once.

The lemma suggests the following algorithm for constructing a WRST. Start with an MST T_P of P and modify it to obtain a WRST. If the weight of the resulting WRST is less than the weight of the original MST, then, by virtue of the previously stated lemma, it is at most twice the optimal weight. That is, the MST T_P will be used as a "skeleton" of the constructed WRST to ensure a good worst-case performance. Each edge of T_P is laid out, one by one, using a shortest path realizing that edge. A formal description of the algorithm and a theorem follow.

> *procedure* LAYOUT-WRST ($\mathcal{R}$, P)
>   *begin-1*
> $T_P :=$ An MST of P;
>   *for* $j = 1$ *to* $n - 1$ *do*
>   *begin-2*
>   *for* $i = 1$ *to* k *do*
>   *begin-3*
> $L_i :=$ MERGE (sub1, sub2, $PATH_i(e_j)$);
>   CLEANUP ($\mathcal{L}_j$);
>   (* CLEANUP removes repeated edges to obtain a tree *);
>   *end-3*;
>   SAVE the minimum-weight tree;
>   *end-2*;
>   *end-1*.

Theorem 3.3. LAYOUT-WRST finds a two-approximation to a WRST of an environment $(\mathcal{R}, P)$ in $O(n^2 + e \log e)$ time, where $n = |P|$, $m = |\mathcal{R}|$, and $e \le m^2 + n$.

ARBITRARY ORIENTATIONS. The previous sections have considered only rectilinear geometry. The most commonly employed geometric environments are Euclidean space and rectilinear space, where distances are measured in L_2-metric and L_1-metric, respectively. In the former geometry, infinitely many orientations are allowed; but, in the latter, only horizontal and vertical orientations are permitted. To fill the gap between Euclidean geometry and rectilinear geometry (in the x-y plane) and in order to better understand practical environments (e.g., $45°$ environments) the *(uniform)* λ-*geometry* is used. This allows orientations making angles $i\pi/\lambda$ with the x-axis, for all i, where λ is an integer, where either λ divides 360 or 360 divides λ, and $\lambda \ge 2$. In the λ-metric the distance between two points is measured using directions allowed in λ-geometry. Note that rectilinear geometry is the two-geometry and Euclidean geometry is the ∞-geometry (see Figure 3.8).

Consider two points p_1 and p_2 in the x-y plane. To find the distance between p_1 and p_2 in λ-geometry, for a fixed λ, a set of line segments $\ell_1, \ldots, \ell_m$ from p_1 to p_2 (where the intersection of the line segments form a path) must be obtained, where each line segment has a legal direction in λ-geometry. If the length of ℓ_i is denoted by d_i and $D = \sum_i d_i$, then a set of line segments with minimum D measures the distance between p_1 and p_2, and D is the distance between p_1 and p_2 (see Figure 3.9a). It is easy to show that there exist two line segments ℓ_1 and

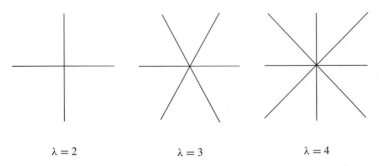

$\lambda = 2$ $\lambda = 3$ $\lambda = 4$

FIGURE 3.8
Examples of λ-geometry.

ℓ_2 with this property. Find the two lines having legal directions in λ-geometry, making the minimum possible angle with the line $\overline{p_1 p_2}$. The intersection of the two lines defines two line segments that are ℓ_1 and ℓ_2 (see Figure 3.9b).

It was shown that the weight of an MST is at most $\frac{3}{2}$ the weight of an SMT in two-geometry and at most $\frac{2}{\sqrt{3}}$ the weight of an SMT in ∞-geometry [86]. Employing the $\frac{3}{2}$ bound establishes a bound on the weight of the produced Steiner trees. A similar relation needs to be established in λ-geometry, for each value of λ. Let W^λ_{type} denote the weight of a type tree (or type edge) in λ-geometry.

Consider two points p_1 and p_2 in the x-y plane (see Figure 3.10a). Let ℓ_1 and ℓ_2 be the line segments measuring the distance between p_1 and p_2 in λ-geometry and x_1 and x_2 denote the lengths of ℓ_1 and ℓ_2, respectively. The angle between $\overline{p_1 p_2}$ and ℓ_1 is denoted by α_1. Similarly, the angle between $\overline{p_1 p_2}$ and ℓ_2 is denoted by α_2. Denoting x_{12} as the ∞-geometry (or Euclidean) distance between p_1 and

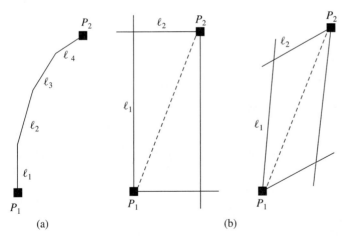

(a)   (b)

FIGURE 3.9
Distances in λ-geometry. (a) Measuring a distance; (b) distances in specific λ-geometries.

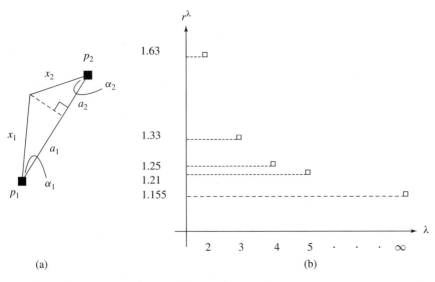

(a)   (b)

FIGURE 3.10
Specific values of r^λ for various λ values.

p_2 then $(x_1+x_2)/x_{12}$ is maximized, over all possible points p_1 and p_2 with a fixed distance x_{12} in ∞-geometry, when $\alpha_1 = \alpha_2 = \pi/(2\lambda)$, and thus, $a_1 = a_2$. (This fraction is the maximum ratio of the distance in λ-geometry and the distance in ∞-geometry between two arbitrary points.) Therefore,

$$\cos\left(\frac{\pi}{2\lambda}\right) W^\lambda_{p_1 p_2} \le W^\infty_{p_1 p_2}$$

and

$$\cos\left(\frac{\pi}{2\lambda}\right) W^\lambda_{\text{MST}} \le W^\infty_{\text{MST}}.$$

As shown in [86], $W^\infty_{\text{MST}} \le \sigma W^\infty_{\text{SMT}}$, where $\sigma = \frac{2}{\sqrt{3}}$ (this result was conjectured in 1970 and was recently proved). Since $W^\infty_{\text{SMT}} \le W^\lambda_{\text{SMT}}$ (by definition),

$$r^\lambda = \frac{W^\lambda_{\text{MST}}}{W^\lambda_{\text{SMT}}} \le \frac{2}{\sqrt{3}\cos(\frac{\pi}{2\lambda})}.$$

A plot of an upper bound on r^λ (by the inequality, given above) versus λ appears in Figure 3.10b. Thus, in λ-geometry, any heuristic Steiner tree that has a weight less than the weight of an MST on the same point set has total length (at most) r^λ-times the optimal length. Previously, $r^2 = \frac{3}{2}$ (i.e., a better value than that shown in Figure 3.10b) and $r^\infty = \frac{2}{\sqrt{3}}$ have been obtained. However, for other values of λ, there were no bounds known on r^λ, except a trivial bound of 2.

A theorem can be established regarding the location of Steiner points that is similar to Hanan's but is a counter-intuitive generalization of it. Before that the following properties of an SMT T_1 in λ-geometry interconnecting n demand points are needed. Assume distances of edges in T_1 are measured in the λ-metric; however the direction of edges in T_1 may not be legal directions in λ-geometry. The following properties have been established in [308]:

Prop 1. T_1 can be replaced by another SMT T_2 with the direction of edges legal in λ-geometry.

Prop 2. All Steiner points of T_2 have degree three or four and there are at most $n - 2$ Steiner points.

Prop 3. The line segments connected to a Steiner point divide the angle 2π as evenly as possible.

A generalization of the LAYOUT-WRST algorithm can be employed to effectively construct a Steiner tree in λ-geometry. The traditional lemma of Hanan [134] does not directly generalize to λ-geometry. Construct a graph G_λ^* as follows. First, G_λ^0 consists of demand points as vertices and has no edges. Assume graph G_λ^i has been obtained, $i \geq 0$. G_λ^{i+1} is constructed by passing lines through each vertex v in G_λ^i in all λ-directions if that line does not already pass through v. The set of vertices created, as the intersections of new lines with new lines or the intersections of new lines with previous lines, is the set of vertices of G_λ^{i+1}. Edges (i.e., line segments) between the vertices are edges of G_λ^{i+1}. The graph $G_\lambda^{\lfloor n/2 \rfloor}$ is denoted by G_λ^*. The following theorem was proved in [311].

Theorem 3.4. There exists an SMT with vertices being a subset of G_λ^*.

Applying previous heuristics designed for rectilinear geometry to λ-geometry remains. As noted above, such an extension may not be trivial. A lot of work needs to be done in this context.

3.2 GLOBAL ROUTING

The purpose of a global router is to decompose a large routing problem into small and manageable subproblems (detailed routing). This decomposition is carried out by finding a rough path for each net (i.e., sequence of subregions it passes through) in order to reduce chip size, shorten wire length, and evenly distribute the congestion over routing area. The definition of subregion depends on whether the global router is performed after floorplanning or placement.

In the case that only a floorplan is given, the dual graph G of the floorplan is used for global routing (see Chapter 2 for the definition of dual graph G). First, each edge e' is assigned a weight $w(e')$ that is equal to the capacity of the corresponding boundary (in the floorplan) and a value $l(e')$ which is the edge length. All the edges of the dual graph are weighted and the lengths are based on the Manhattan or Euclidean distances between the centers of the rectangular tiles of the floorplan. Then, a global routing for a two-terminal net N' with its terminals in rectangles r_1 and r_2 is specified as the path connecting vertices v_1 and

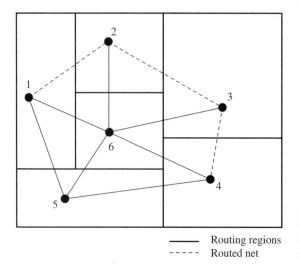

FIGURE 3.11
The global routing graph of a given floorplan. The dotted lines show one routed net.

―――― Routing regions
- - - - Routed net

v_2 in G (where v_i is the vertex corresponding to rectangle r_i). See Figure 3.11. Note that in such an approach, the result of global routing can help in determining the exact position of each module in the corresponding rectangle.

When a placement is given, the routing region (i.e., the region of the plane external to the modules) is partitioned into simpler regions. These subregions are typically rectangular in shape. There are several approaches for the partitioning problem with different objectives in mind. For example, one approach is to minimize the total perimeter of the regions [292]. However, in practice, any reasonable partitioning algorithm works well. A technique similar to that used in the floorplanning process can also be applied to define the routing graph. Now, each $v \in V$ corresponds to a channel, and two vertices are connected only when the two corresponding channels are adjacent. The weight of the edge is the capacity of the common boundary of the two channels. Figure 3.12 shows a placement of three modules. The routing region has been partitioned and the routing graph drawn.

Formally, in global routing of multiterminal nets there are a set $\eta = \{N_1, \ldots, N_n\}$ of multiterminal nets. The *layout environment* is a two-dimensional tessellation of the plane. For simplicity, assume a rectangular tessellation (i.e., all regions are rectangles). However, the algorithms to be described work for any rectilinear tessellation. Each k-terminal net N is specified by a k-tuple $(R_1, \ldots R_k)$, where R_i, $1 \leq i \leq k$, are the regions containing terminals of N. R_i are not necessarily distinct (a net may have more than one terminal in a region). In a global routing, for each net, a sequence of regions through which it passes is specified.

The following concepts are demonstrated in Figure 3.13. In a global routing (i.e., output of a global router), let $d(i, j)$ denote the number of nets crossing the border of regions R_i and R_j. Let $c(i, j)$ denote the *capacity* of the border of regions R_i and R_j (i.e., the maximum number of nets that can cross this border). An instance of global routing is specified by a collection of regions, a set η of

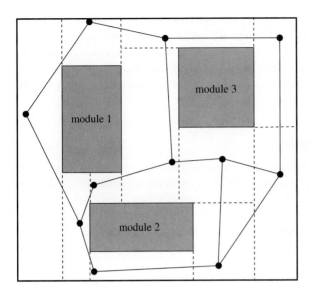

FIGURE 3.12
The global routing graph given the placement of three modules.

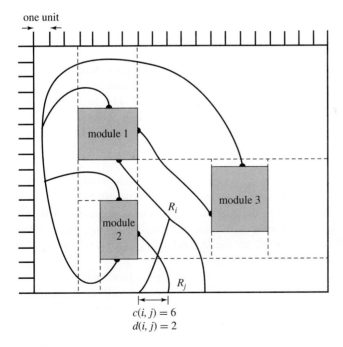

$$c(i, j) = 6$$
$$d(i, j) = 2$$

FIGURE 3.13
An instance of global routing.

multiterminal nets, and a capacity function C. A k-terminal net is said to have *multiplicity* k. Multiplicity of η is the maximum number of terminals per net, over all nets in η. The global routing problem (GRP) of an arbitrary instance (η, C) is to find a global routing with $d(i, j) \leq c(i, j)$ for every two adjacent regions R_i and R_j.

In the next few subsections several fundamental approaches to global routing will be discussed. At the end of these sections, a summary of several other known techniques will be provided.

3.2.1   Sequential Approaches

In sequential approaches to global routing, nets are routed sequentially; that is, one at a time. First an ordering of the nets is obtained. Nets are normally ordered based on their bounding-box length, the number of terminals they have, and criticality (the notion of timing critical nets will be discussed in the next chapter: these are nets that need to be made as short as possible due to timing requirements). In the second step, each net is routed (i.e., a Steiner tree is obtained) as dictated by the ordering.

A classic algorithm for finding a shortest path between two points is Lee's maze running algorithm, as described in the previous section. Most known sequential routing techniques are variations of Lee's algorithm.

The problem faced in global routing is solvable in polynomial time provided there are no restrictions on the traffic through the regions. The objective of the problem would then only be that of minimizing the lengths of each net's route. Since the routes are confined to the routing graph mentioned earlier, simply solve the problem using the minimum spanning tree approach. A popular modification of the Prim's Algorithm for a minimum spanning tree is presented here.

Generally, there are two classes of search algorithms. The first one is the grid search, where the environment is the entire grid, as in Lee-Moore type search techniques. In this case, it is very easy to construct the search space; however, the time and space complexities are too great. The second class of search techniques is called line searching. The environment is a track graph that is obtained by extending the boundary of modules. A track graph is much smaller than the corresponding grid-graph; thus computation can be done more efficiently on it.

Although a maze router finds a Steiner tree interconnecting a set of terminals dictated by a net, there are more direct approaches for finding a Steiner tree. Most recent sequential global routers formulate the second step (the first step being the ordering of the nets) as a Steiner tree problem.

Global routers have made an attempt to find a Steiner tree minimizing wire lengths and traffic through the regions. As achieving both objectives seemed difficult, many routers have employed a minimum-length Steiner tree [10, 258, 355], where length is a purely geometrical concept and traffic has been heuristically minimized as the second objective.

The following technique simultaneously minimizes traffic through the regions and wire lengths [54]. Consider Figure 3.6, where the weight of a region depends on its area, the number of nets going through it, and the number of terminals it contains. Certainly, the weights of regions corresponding to the modules are set to infinity (∞) to indicate that no nets can go through them. Recall that a minimum-weight Steiner tree is a Steiner tree with minimum total weight, where an edge with length ℓ_i in a region R_i with weight w_i has total weight $\ell_i w_i$. With reference to Figure 3.6, if the two terminals labeled i are to be connected, then a minimum-length Steiner tree is not appropriate as it goes through critical regions, including the modules. Such a path needs to be modified so it does not go through any modules. After the modification (shown by a thick line in Figure 3.6) the path still goes through critical regions (e.g., the region with weight 9 in Figure 3.6), and the optimality of its length may be violated. A Steiner tree that only tries to avoid crowded regions, regardless of its length, is not suitable by itself, for it is excessively long. However, a minimum-weight Steiner tree is an effective connection. The notion of weighted Steiner trees was introduced to overcome difficulties that arise when dealing with length or density alone.

An algorithm is chosen for finding a Steiner tree, called LAYOUT-WRST. The nets are routed one by one, employing LAYOUT-WRST. Constants λ^j are introduced to balance weight and length in a WRST. At the jth pass, we will obtain a WRST of net N with minimum $\sum_i \lambda_i^j w_i^j \ell_i^j$ for all nets, where w_i^j is the weight of region R_i that N passes through and ℓ_i^j is the length of N in R_i. The value λ_i^j is selected so that $\lambda_i^j w_i^j$ approaches 1 as j increases.

After several iterations, λ_i^j is selected to make $\lambda_i^j w_i^j$ equal to 1, and thus, only ℓ_i^j is important when calculating the total weight of a WRST. At the jth pass of the algorithm, if the total weight of routing of a net N is less than the total weight of routing of N from pass $j - 1$ then it is accepted. Otherwise, the routing is rejected. A formal description of the proposed approach follows.

procedure ROUTING $(\mathcal{R}, P)$
begin-1
    W := initial weight function of $\mathcal{R}$;
    for $i = 1$ to n *do*
    $W(N_i) = \infty$; (* initialize weight of all nets *);
    Pass j;
    for $i = 1$ to n *do*
    begin-2
        N_i := current net;
        temp := LAYOUT-WRST(N_i); (* minimum $\sum(\lambda_i^j w_i^j)\ell_i^j$ of N_i *);
        if temp $\leq W(N_i)$ *then*
        begin-3
            N_i := temp; (* accepting routing of N_i *);
            update W;
        end-3;
    end-2;
end-1.

This algorithm considers the weight and length of a WRST at the same time. As j increases, length is given more attention. Since the weights are updated after each routing, each net may generate a different layout at each iteration. In practice one selects $\lambda_i^j = (1/j + 1/w_i^j)$ and the total number of iterations (maximum value of j) as equal to 20.

The sequential techniques are very efficient at finding routes for nets as they employ well-known shortest path algorithms. But, rerouting of nets in case of conflict is purely heuristic.

A general technique used in connection with routing algorithms (both global routing and detailed routing, and, in particular, sequential algorithms) uses the notion of *rip up and reroute*. The main concept is to remove some of the routings previously done in order to make room for new nets to be routed. And hopefully, the nets just ripped up can be rerouted. This technique is used in a large number of industrial routers and is an effective one.

3.2.2 Hierarchical Approaches

Hierarchical routing algorithms use the hierarchy on the routing graph to decompose a large routing problem into subproblems of manageable size. The subproblems are solved independently and their solutions are combined to form the solution of the original problem.

There are two methods for hierarchical routing: top-down and bottom-up. Both methods utilize the hierarchy imposed by the cut tree of the routing graph G. Figure 3.14 shows a routing graph and one possible cut tree. Each interior node in the cut tree represents a primitive global routing problem. Then, each subproblem is solved optimally by translating it into an integer programming problem. The composition of partial solutions is done by techniques such as integer programming or sequential routing [235].

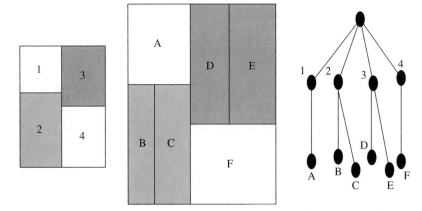

FIGURE 3.14
A routing graph and a cut tree of it.

Let the root of T be at level 1, the leaves of T are at level h, where h is the height of T. Top-down methods use the primitive routing problems that refine the routing step by step. They traverse the cut tree T top-down, level by level. At level i, the floorplan patterns corresponding to nodes larger than i are deleted. This gives rise to the routing problem instance I_i, also referred to as the *level-i abstraction* of the problem instance. Then, a solution is obtained for each (updated) routing graph that is associated with nodes at level i. Each such solution is then combined with solutions at level $i - 1$; this step refines the routing to cover one more level of the cut tree. There are various approaches to combine the solutions. The final solution is obtained when i is equal to h. A formal description of the top-down approach follows.

> *procedure* TOP_DOWN_ROUTING
> *begin-1*
>   compute solution to the level-1 abstraction of the problem; (* I_1*);
>   *for* $i = 2$ to d *do*
>   *begin-2*
>   *for* all nodes n at level $i - 1$ *do*
>   compute solution R_n of the routing problem I_n;
>   combine the solutions R_n for all nodes n
>   and the solution R_{i-1} into solution R_i;
>   *end-2*;
> *end-1.*

The solutions to all the problem instances R_n are obtained using an integer programming formulation. Combining the solutions of one level into that of the next level is a crucial step in the algorithm and is done using various techniques.

- *Sequencing of nets* is done by considering one net at a time in the combining process. The nets are ordered according to heuristics. The minimum Steiner tree for every net in R_n and R_{i-1} is obtained one by one, according to the ordering. The length of an edge in the minimum Steiner tree of the net is 0 if it exists in the routing of the net in R_n or in R_{i-1}. Also, as the loads on the edges increase, their lengths are increased to reduce the traffic through them.

- *Integer programming* [38] is an algorithm for routing graphs that are grid-graphs. The hierarchy is defined either by horizontal or by vertical lines. Take two adjacent rows (or columns) and decompose each net into subnets that can be realized. Thus a binary tree is obtained and minimum Steiner tree algorithms can be used to solve the problem. Alternatively the problem can be solved by a divide-and-conquer approach. Here, the routing problem is decomposed into a sequence of problems whose routing graph is a cycle of length 4. These decomposed problems can be solved exactly, using integer programming.

A bottom-up method was proposed in [157] that uses the binary cut tree. In the first phase, the routing problem associated with each branch in T is solved by integer programming. Then, the partial routings are combined by processing internal tree nodes in a bottom-up manner. Note that each node corresponds to a slice in the floorplan. Thus, each net that runs through the cutline must be interconnected (while maintaining the capacity constraints) when the results of two nodes originating from the same node are combined. The disadvantage of such an approach is that it does not have a global picture until the late stages of the process.

> *procedure* BOTTOM_UP_ROUTING
> *begin-1*
>   compute solutions to the level-k abstraction of the problem;
>   (∗ k is the highest level in the tree ∗);
>   *for* $i = k$ to 1 *do*
>   *begin-2*
>   *for* all nodes n at level $i - 1$ *do*
>   *begin-3*
>   compute solution R_n of the routing problem I_n
>   by combining the solution to the children of node n;
>   *end-3*;
>   *end-2*;
> *end-1.*

The time complexity of hierarchical approaches depends on the type of procedure used in each step. For example, one that uses integer programming is slow. Also, the number of levels and the number of solutions combined at each level affect the time complexity.

A more recent ILP-based algorithm was proposed in [211]. The proposed method formulates the problem as an ILP and considers various performance issues in the formulation. The main idea is to consider a set of patterns for the global routing of each net. Deciding which pattern to use for a net is left to an ILP solver.

3.2.3   Multicommodity Flow-based Techniques

The global routing problem can be formulated as a multicommodity flow problem in a graph where nodes represent collections of terminals to be wired, and the edges correspond to the channels through which the wires run. Though the multicommodity flow problem has been proven to be NP-complete, the proposed algorithm uses effective shortest path techniques and has produced solutions quite efficiently. This approach works only for two-terminal nets but could be extended to multiterminal nets by decomposition of a multiterminal net into several two-terminal nets. (The conservation of flow fails when dealing with multiterminal nets directly.) This approach helps in considering multilayer routing and nonrec-

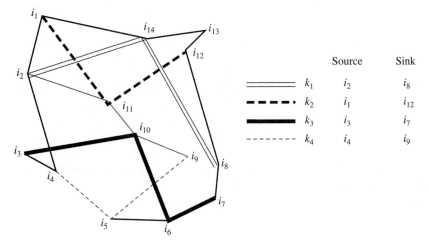

		Source	Sink
════════	k_1	i_2	i_8
▪ ▪ ▪ ▪	k_2	i_1	i_{12}
▬▬▬▬	k_3	i_3	i_7
- - - - - -	k_4	i_4	i_9

FIGURE 3.15
Multicommodity flow approach to global routing.

tilinear structures. Figure 3.15 shows a routing graph on which four commodities k_1 through k_4 are routed. The sources and sinks are shown.

Multicommodity flow formulation. A graph $G = (I, J)$ is obtained as mentioned earlier. The graph is an undirected graph with a set of commodities $K = \{1, 2, \ldots, t\}$ defined on the set of nodes $I = \{1, 2, \ldots, m\}$. Only two-terminal nets are considered here and hence there is only one sink and one source for each commodity. It has to choose the edges from the set $J = \{1, 2, \ldots, n\}$ along which the commodities can be transported. The edges have capacities c_j corresponding to the capacities of the channels. p_j is the shipment cost along the edge j. a_{ik} is the amount of commodity k produced or consumed at the node i, and $a_{ik} = 0 \pm 1$. ω_{ij} is the node-edge incidence matrix,[1] $\omega_{ij} = 0, \pm 1$. x_{jk} is the flow of commodity k by arc j. Then the problem is formulated as

$$\min \sum_{k=1}^{t} \sum_{j=1}^{n} p_j |x_{jk}|$$

subject to

$$\sum_{j=1}^{n} \omega_{ij} x_{jk} = a_{ik} \quad \forall i, k; \quad \sum_{k=1}^{t} |x_{jk}| \le c_j \quad \forall i, j; \quad x_{jk} = 0, \pm 1 \quad \forall j, k$$

Multicommodity flow algorithm. [40, 325] give an algorithm for global routing based on multicommodity flow formulation. Each step is described below.

---

[1] In this matrix, rows correspond to vertices and columns correspond to edges. If an edge is incident to a node, then the corresponding entry is 1; otherwise, it is zero.

Initial step. Solve the above problem with constraints on the capacities removed. When there are no constraints on the capacities of edges, the problem can be separated into one of solving the problem for each commodity k independently.

$$\min \sum_j p_j |x_{jk}|$$

$$\sum_j \omega_{ij} x_{jk} = a_{ik} \forall i, k; \, x_{jk} = 0, \pm 1 \forall j, k$$

This problem is similar to the shortest path problem and the familiar Dijkstra's algorithm can be used for each commodity.

General step. Let r be the iteration number. x_{jk}^{r-1} is the solution after $r-1$ iterations.

Step 1. Find a maximal level of overflow M_r over the graph G after $r-1$ iterations

$$M_r = \max_j \left(\sum_k |x_{jk}^{r-1}| - c_j \right)$$

If $M_r \le 0$, then stop; otherwise continue.

Step 2. Find a set of arcs J_r^0 at the maximum and one below maximum level of overflow

$$J_r^0 = \left\{ j \in J : \left\{ \sum_k |x_{jk}^{r-1}| - c_j + 1 \ge M_r \right\} \right\}$$

and a set of arcs $J_r \in J_r^0$, so that

$$\sum_k |x_{jk}^{r-1}| - cj = M_r.$$

Step 3. Assign new costs to the arcs of the graph G.

$$p_j^r = \infty \text{ for } j \in J_r^0, \, p_j^r = p_j^{r-1} \text{ for } j \notin J_r^0.$$

Step 4. Define a set of connections K_r^0, so that

$$K_r^0 \{ k_r \in K, \, j \in J_r^0 : \{ |x_{jk}^{r-1}| = 1 \} \};$$

and subset $K_r \in K_r^0$, so that

$$K_r = \{ k_r \in K_r^0, \, j \in J_r : \{ |x_{jk}^{r-1}| = 1 \} \}.$$

Step 5. Solve shortest path problems for all $k \in K_r$. Let x_{jk*}^r denote the solutions of these problems. If a solution of the shortest path problem satisfies condition $\min \sum_j p_j^r |x_{jk}^{r-1}| \ne \infty$ as minimum for one $k \in K_r$, then go to Step 6. Otherwise, go to Step 9.

Step 6. Find connections $k^0 \in K_r$ such that

$$\sum_j (p_{j0}^r |x_{jk}^r| - p_j^{r-1} |x_{jk}^{r-1}|) = \min_{k* \in K_r} \sum_j (p_j^r |x_{jk*}^r| - p_j^{r-1} |x_{jk*}^{r-1}|).$$

Step 7. Assign $x_{jk}^r = x_{jk^0}^r \forall j$ for $k = k^0$; $x_{jk}^r = x_{jk}^{r-1} \forall j$ for $k \neq k^0$.

Step 8. Increase the iteration counter $r = r + 1$ and go to Step 1.

Step 9. Check condition

$$x_{pk}^{r-1} \times x_{qk}^{r-1} = 0 \forall p, q \in J_r^0 (p \neq q), \forall k \in K_r^0.$$

The above multicommodity flow algorithm is very efficient in practice, although the proof of efficiency cannot be given since the original problem is NP-complete and the algorithm uses an escape procedure whenever it fails to produce a feasible solution.

3.2.4 Randomized Routing

This section discusses the approach of Raghavan et al. [283], based on the integer linear program formulation of the global routing problem. The idea is to omit the integral constraint of the integer program and solve the new linear relaxation problem. The next step is to obtain integer solutions which are close to the optimal solution.

The integer linear program formulation for global routing is usually a 0,1-integer program (i.e., the solutions are either 0 or 1). Let R be such a problem. Then the following algorithm gives a near-optimal result.

Step 1. Obtain a solution to the problem R with the integral constraint removed. Let $x = \alpha$ be such a solution.

Step 2. Use α_i, the solution to the variable x_i, as a probability. Choose values for x_i by tossing a biased coin that yields a 1 with probability α_i and a 0 with probability $1 - \alpha_i$. The coin tossing is done for all x_i independently of one another.

Step 3. Repeat Step 2 depending on the time allocation for the problem to generate another solution.

Step 4. Choose the best feasible solution.

Step 2 is based on Bernoulli's trials. The expected value of the variable X_i is α_i. That is,

$$E[X_i] = \alpha_i.$$

The objective function is

$$\min \ c(x),$$

where

$$c(x) = \sum_{i=1}^{n} c_i x_i$$

and c_i are real values in the interval $[0,1]$. If ψ is the cost obtained after a single evaluation of Step 2, then

$$\psi = \sum_{i=1}^{n} c_i X_i.$$

The expected value of ψ is

$$E[\psi] = E\left[\sum_{i=1}^{n} c_i X_i\right] = \sum_{i=1}^{n} c_i E[X_i] = \sum_{i=1}^{n} c_i \alpha_i.$$

Thus the expected value of ψ is the optimal cost of the linear relaxation of the problem. It has been found that the probability that the outcome of ψ being close to $E[\psi]$ is very high. Furthermore, the possibility of getting very near optimal solutions is increased by doing Step 2 a number of times and selecting the best solution obtained.

Randomized routing is an improvement in efficiency over integer programming because the linear relaxation of the integer program is solved here. The time spent on the algorithm can be varied depending on the accuracy of the solution required.

3.2.5 One-Step Approach

This approach has been discussed in [179] and involves the decomposition of the chip area by horizontal and vertical lines into an $n \times n$ matrix. The tiles so formed have one or more terminals depending on the restrictions of the algorithm used. The terminals are marked by the corresponding net numbers, and terminals with the same number are to be connected as a net. [179] gives a good solution to the global routing problem P with single-turn routing.

Let $w(R)$ denote the maximum number of wires passing from one cell into an adjacent one in a global routing R. Thus, an optimal global routing is one with minimum $w(R)$ over all global routings for the given problem P. $w(P)$ denotes density (width) of an optimal routing R that solves P. $w(n)$ is the maximum density (width) of any problem instance defined on an $n \times n$ array. p is the maximum number of nets that can be connected to any of the terminals.

Over all subsquares of the $n \times n$ array, let cut(P) denote the maximum number of nets which must cross the border of the squares, divided by the perimeter of that square. It is easy to see that cut(P) is a lower bound on $w(P)$. Divide the chip into squares whose sides have length $\lambda = \text{cut}(P)/p$. Route these squares independently, in an arbitrary one-turn manner with width at most $O(\text{cut}(P))$. Route nets that must leave a square arbitrarily to a point on the perimeter of that square. Then proceed through n/λ levels of bottom-up recursion, at each level pasting together four squares from the previous level in a 2×2 pattern, and using at most $O(\text{cut}(P))$ additional width to route all nets that leave the newly constructed square to the perimeter of that square.

Figure 3.16 shows a 2×2 pattern which has to be routed. If nets in part A are connected to only those in part D and nets in part C to those in part B, then

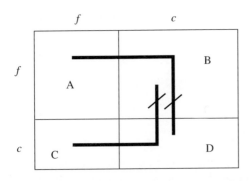

FIGURE 3.16
One-step approach in a 2×2 array.

we have a maximum $O(\text{cut}(P))$ additional nets crossing through the perimeter of any of the squares. This can be seen in Figure 3.16, where the thicker lines show the routings from A to D and from B to C. Thus we have a polynomial time algorithm to produce a global route such that $w(R) \le O(w(P). \log(pn/w(P)))$ for any problem instance P.

The one-step approach is extremely fast, but not very effective by itself. It needs to be combined with effective heuristics to be of practical use.

3.2.6 Integer Linear Programming

Let each net be defined in terms of its multiplicity (the number of terminals). If a net $n \in N$ has a multiplicity of k_n then the net n is defined as a set $n = \{(n, 1), (n, 2), \ldots, (n, k_n)\}$. T_n^j, $j \le I_n$, is a route available to the net n. Each net n is labeled with a cost factor $w(n) \ge 0$.

$x_{n,j}$ is a variable for each net n and route T_n^j.

$$x_{n,j} = \begin{cases} 1 & \text{if net } n \text{ uses the route } T_n^j \\ 0 & \text{otherwise} \end{cases}$$

A routing R is defined by assignment to the variables $x_{n,j}$. The load on the edge e is given by

$$U(x, e) = \sum_{(n,j)|e \in T_n^j} \sum_n w(n) x_{n,j}.$$

E is the set of edges in the routing graph.

$$W(x) = \sum_{e \in E} l(e) U(x, e)$$

where $l(e)$ is the length of the edge e.

There are two basic versions that formulate the problem further. The first (the constrained version) considers capacity of the edges. The second (the unconstrained version) ignores the capacities and solves a relaxed version of the problem.

CONSTRAINED GLOBAL ROUTING. The constraints of the integer program for this problem are:

$$x_{n,j} \in \{0, 1\} \quad \text{for all nets } n \text{ and all } j,$$

$$\sum_{j=1}^{j=I_n} x_{n,j} \leq 1 \quad \text{for all nets } n, \text{ and}$$

$$U(x, e) \leq c(e) \quad \text{for all edges } e \in E.$$

The first two constraints ensure that at most one admissible route is chosen for each net. The third constraint implements the capacity constraints on all edges. Here, the goal is to minimize the wire length and increase the number of nets routed at the same time. Thus, the cost function to be minimized is a linear combination of these two terms.

$$c = \lambda \sum_{n \in N} w(n) \left(1 - \sum_{j=1}^{j=I_n} x_{n,j} \right) + W(x)$$

UNCONSTRAINED GLOBAL ROUTING. The capacity constraint is eliminated here. It is also ensured that all nets are routed. So,

$$\sum_j x_{n,j} = 1.$$

Let x_L denote the maximum load on any edge; then

$$U(e)/c(e) \leq x_L,$$

for all $e \in E$. The cost function to be minimized is

$$c = \lambda x_L + W(x).$$

The general version of the integer linear programming problem is NP-complete. The solutions to the above problem tend to be slow.

3.2.7   Discussion

Several major approaches to global routing have been considered in this chapter.

- **Sequential approach.** This technique does not consider the interdependence of nets and hence is not global in nature. This approach is order dependent.
- **Hierarchical approach.** This approach could either be bottom-up or top-down. In either case, the approach is order independent but does not deal with the requirements of uniformity of wiring on the chip. In the bottom-up approach there are effective rerouting schemes based on good heuristics.
- **Multicommodity flow approach.** This approach is more universal in that it not only considers rectilinear but also nonrectilinear routes. A number of generalizations could be added to the global routing graph. The only limitation to this approach is that the original problem is NP-complete.

- **One-step approach.** This is useful only for a specific set of problems. In general, it must be combined with other heuristics to be of practical use.

- **Integer programming approach.** This has a major disadvantage of being extremely slow. It is used in other approaches, for example, in the hierarchical approach, in solving parts of the bigger problem.

Apart from these approaches there have been attempts to use the simulated annealing formulation to solve the global routing problem [206, 373]. Every net has a set of possible routes. In the first step of the generic algorithm using simulated annealing, this set is obtained. Next, each net is given a route at random, and the capacity constraints are checked for each channel. If the capacity constraints are satisfied, a feasible solution has been obtained. If not, then any arbitrary net is selected from among those that are overloading some channel, and a new route is selected from the set of admissible routes for the net. The objective here is to minimize the total excess length used by the nets over the total length used by all the nets if they had been routed independently using the shortest-path techniques. Simulated annealing gives a good model to combine all the problems related to routing, like pin assignment, placement, and global routing itself, into one single problem; but the algorithm need not be time efficient and developing the cooling equations might be an involved task.

3.3 DETAILED ROUTING

The classic technique for solving the detailed routing problem is the Lee-Moore maze-running algorithm, described earlier in this chapter. This algorithm can be used as a global router (as described), or as a detailed router (once the global routing is known). It is also possible to bypass the global routing stage and use the Lee-Moore technique to perform area routing, that is, start the detailed routing stage in the entire routing region. It is not practical to proceed in this manner for larger problems. Application of the Lee-Moore algorithm to detailed or area routing is exactly the same as in global routing, except that the routing environment is a fine grid instead of a coarse grid or a general graph.

The two-stage routing method, global routing followed by detailed routing, is a powerful (and commonly used) technique for realizing interconnections in VLSI circuits. In its global stage, the method first partitions the routing region into a collection of disjoint rectilinear subregions. Typically, the routing region is decomposed into a collection of rectangles. It then determines a sequence of subregions to be used to interconnect each net. All nets crossing a given boundary of a routing region are called *floating terminals*. Once the subregion is routed, and thus positions of all crossings become fixed, the crossings (or floating terminals) become *fixed terminals* for subsequent regions. There are normally two kinds of rectilinear subregions, channels and switchboxes. *Channels* refer to routing regions having two parallel rows of fixed terminals. *Switchboxes* are generalizations of channels that allow fixed terminals on all four sides of the region. In the detailed routing stage, channel routers and switchbox routers are used to complete the connections.

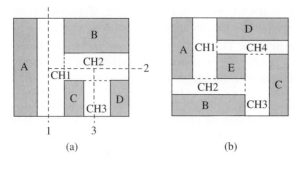

FIGURE 3.17
(a) A slicing placement topology.
(b) A nonslicing placement topology.

Consider an example of a placement shown in Figure 3.17. Depending on the placement or floorplanning result, the routing region is classified either as a slicing or a nonslicing structure. These terms were defined earlier from a floorplanning/placement point of view; in the context of routing, *slicing* means that horizontal or vertical lines are used recursively to divide the chip into two parts. Figure 3.17a illustrates a slicing placement/floorplanning topology. Line 1 separates block A from the rest. Slicing is applied again on the remaining structure. Now, line 2 separates block B from the rest. Finally, line 3 separates blocks C and D. The slicing structure refers to those structures in which slicing can continue to be applied until every block is separated. The slicing scheme gives the reverse order of channels to be routed. Channel 3 must be routed first, then channel 2 and finally channel 1. If the topology of placement yields a slicing structure, then the detailed routing can be completed by means of channel routers. Figure 3.17b illustrates a nonslicing placement topology. The placement topology contains channels with cyclic constraints. CH1 must be routed before CH2, CH2 before CH3, CH3 before CH4, and completing the cycle, CH4 before CH1. A channel router cannot be used to complete this routing, so a special router called a switchbox router is needed (thus, e.g., CH1 becomes a switchbox). Channel routers are usually easier to implement and faster to run than the more general two-dimensional (e.g., switchbox) routers because the problem under consideration is inherently simpler.

Switchboxes are predominately encountered in building-block style routing. Channels are widely used in other design styles (e.g., standard cells, gate arrays, sea-of-gates, and FPGAs). In latter cases, the interconnection problem can be transformed into a collection of one-dimensional routing problems. Several researchers have proved that the general channel-routing problem is an NP-complete problem in various models ([304, 351]).

Recall that the traditional model of detailed routing is the two-layer Manhattan model with reserved layers, where horizontal wires are routed in one layer and vertical wires are routed in the other layer. For integrated circuits, the horizontal segments are typically realized in metal while the vertical segments are realized in polysilicon. In order to interconnect a horizontal and a vertical segment, a contact (via) must be placed at the intersection points. Consequently, two layout styles are typically adopted to perform detailed interconnections. In the *reserved layer model*, vertical wires are placed in one layer and horizontal wires in another layer. In the *free-hand layer model*, both vertical and horizontal wires can run in both layers. Generally, the reserved model produces more vias than the free-hand model and the free-hand model generally requires less area (and fewer layers in multilayer domains) than the reserved model to connect all nets. In the reserved model a post-processing can be done to minimize the number of vias.

Here only the Manhattan reserved model is discussed in detail. The remaining models will be examined in the discussion section.

3.3.1   Channel Routing

A channel, also called a two-shore channel, is a routing region bounded by two parallel boundaries. For a horizontal channel, fixed terminals are located on the upper and lower boundaries and floating terminals are allowed on the left and right (open) ends. The left end may also contain fixed terminals. The *channel routing problem* (CRP) is to route a specified net list between two rows of terminals across a two-layer channel. The task of a channel router is to route all the nets successfully in the minimum possible area. Each row in a channel is called a track. The number of tracks is called the *channel width w* of the channel.

When the channel length is fixed, the area goal is to minimize the channel width, the *channel width minimization problem*. The classic channel routing problem is as follows. Given a collection of nets $\eta = \{N_1, \ldots, N_n\}$, connect them while keeping the channel width minimum. The channel routing problem with channel width minimization has been an area of extensive research in the 1970s and early 1980s. One reason is that it is the most convenient way of solving the detailed routing problems and therefore can be applied to many layout methodologies. In standard-cell layout, the channel width minimization problem formalizes the detailed routing problem between each pair of adjacent routing rows of standard cells. In an incremental approach to detailed routing in a slicing placement/floorplanning structure, the detailed routing processes the slicing tree in a bottom-up manner. Each internal tree node defines an instance of the channel width minimization problem.

This section describes channel width minimization problems for the Manhattan routing model. In order to ensure that two distinct nets are not shorted, it is necessary to ensure that no vertical segments overlap and, similarly, no horizontal segments overlap. That is

- The input consists of two rows of terminals, one on the upper boundary and the other on the lower boundary of a channel. TOP = $t(1), t(2), \ldots, t(n)$ and BOT = $b(1), b(2), \ldots, b(n)$. $t(i)$ is the terminal located at position i on the top row. $b(i)$ is similarly defined.
- The output consists of Steiner trees with no vertical/horizontal overlaps and no bending of two wires at a grid point (called knock-knees).
- The goal is to minimize the number of tracks.

Two fundamental graphs (see Figure 3.18) are associated with a channel and are important for routing. Every column i, such that $t(i)$ and $b(i)$ are not zeros,

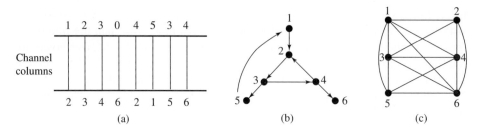

FIGURE 3.18
(a) A netlist upper terminal with label i should be connected to the lower terminal with label i;
(b) vertical constraints graph VG; (c) horizontal constraints graph HG.

introduces a directed edge from node $t(i)$ to node $b(i)$ in the *vertical constraints graph* VG (zeros represent empty positions). Each vertex of VG corresponds to a terminal. Thus VG is a directed graph associated with the channel. The *horizontal constraints graph* HG is constructed as follows. With every net i there is an interval $I(i)$, where the left point of $I(i)$ is the minimum column number k such that $t(k)$ or $b(k)$ equals i, and the right point of $I(i)$ is the maximum such column number. HG is the *intersection graph* with the set of intervals as a vertex set, and an edge connects two vertices labeled with $I(i)$ and $I(j)$ if there exists a vertical line crossing two intervals $I(i)$ and $I(j)$.

Define the *local density* d_i as the number of nets that must cross the vertical line between columns i and $i+1$. The local density can be found by scanning the channel from left to right. The local density is zero initially, $d_0 = 0$. If the current column i has a left terminal of a net on one of two boundaries, $d_i = d_{i-1} + 1$. If it has a right terminal of a net, $d_i = d_{i-1} - 1$. Otherwise, $d_i = d_{i-1}$.

The *global density* d of HG, or clique number, is the maximal number of intervals crossing a vertical line, and presents a lower bound on channel width.

> **Lemma 3.3.** In an arbitrary instance of the channel-routing problem (in the two-layer Manhattan model), $t \geq d = \omega(HG)$, where d is the maximum density of the problem, ω denotes clique number of HG, and t is the minimal number of horizontal tracks required.
>
> **Proof.** Consider a column i with $d_i = d$. There are exactly d nets that cross (or pass through) column i. Since each of these nets need at least one track (in the two-layer Manhattan model, the overlap of wires is not allowed), then at least d tracks are needed at column i. That is, at least d tracks are needed to route the entire channel.

To formalize another type of lower bound, consider one extreme instance of CRP, the *shift-right-1* shown in Figure 3.19. Note that the global density of every instance of shift-right-1 is one. If the left-edge algorithm (to be described in the next subsection) is applied, $t = n$. However, if a detour is allowed as shown in Figures 3.19b, c, the channel density can be significantly reduced. Here s is an interval such that a detour is allowed for every net positioned at the abscissa with multiples of s. In general, when such detours are allowed, $t \leq s - 1 + 2n/s$. Note that t is minimized when $s = \sqrt{n}$. Therefore, $t = O(\sqrt{n})$.

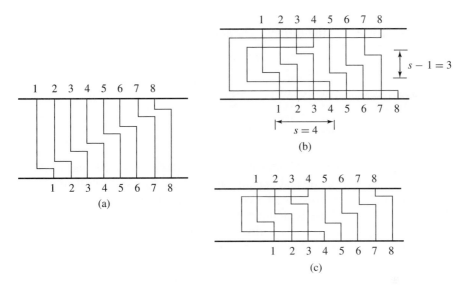

FIGURE 3.19
(a) An instance of shift-right-1. (b) Its solution using $t = s - 1 + 2n/s = 7$. (c) Its solution using minimum tracks, $t = 5$.

Lemma 3.4. [30] In an instant of the shift-right-1 problem with n nets, $t = \Omega(\sqrt{n})$.

Proof. An informal proof of the lemma is given. In order to route net i in track j, the net $i - 1$ should "go out" (of the way) in a track lower than j. For example, in Figure 3.19c, to route net 5 in track 2, net 4 goes out in track 1. Note that at most two nets can go out in track 1. And if net i goes out in track j then nets $i + 1 \ldots k$ can be routed in $k - i$ tracks. Thus if nets $n_1, n_2, \ldots, n_s$ go out the rest of the nets can be routed in max $(n_2 - n_1 - 1, \ldots, n_s - n_{s-1} - 1)$. To route the nets that go out two tracks are needed for every two nets, that is, total of s tracks. It is easy to see the total number of tracks $t = s + $ max $(n_2 - n_1 - 1, \ldots, n_s - n_{s-1} - 1)$ is minimized when nets go out evenly. Then the total number of required tracks is $t = \Omega(\sqrt{n})$.

The just described notion was formalized and called *flux*. The concept of flux provides an alternative lower bound for the channel width. Suppose that instead of making vertical cuts in the channel, we make a horizontal cut on one side of the channel (top or bottom) which isolates a set of contiguous columns. A net is split by a horizontal cut if it contains terminals both within the cut and outside of it. The flux of a channel is defined as the largest integer f for which there exists a horizontal cut of size (number of columns) $2f^2$ which splits at least $2f^2 - f$ nontrivial nets. [12] proved that every channel with density d and flux f requires channel width of at least max(f, d) tracks. Flux is usually one or two for practical problems. In the shift-right-1 problem, $f = \Omega(\sqrt{n})$. A technique for reducing flux was proposed in [197].

Several fundamental approaches to the channel routing problem will be described in detail: the classical left-edge algorithm and its extensions, the greedy

TABLE 3.1
Router comparison for the difficult channel

Router	Tracks	Vias	Wire length
LEA	31	290	6526
Greedy channel router	20	403	5381
Hierarchical router	19	336	5023
YACR-II	19	287	5020

rule-based channel router, the hierarchical channel router based on linear integer programming, and the divide-and-conquer approaches. Deutsch's difficult example published in [83] (an instance of a channel consisting of 174 columns, $d = 19$ and $f = 3$) has become the benchmark by which channel routers are measured. Table 3.1 gives a summary of the various channel routers to be discussed for this example (arranged in order of decreasing wire length).

LEFT-EDGE ALGORITHMS. The first channel-routing algorithm was proposed by Hashimoto and Stevens [134]. Their algorithm, also known as the *left-edge algorithm* (LEA), used a top-down row-by-row approach. If a top terminal and a bottom terminal have the same abscissa and they are to be connected to distinct nets, the horizontal segments of the net connected to the top terminal must be placed above the horizontal segment of the net connected to the bottom terminal; otherwise, the vertical segments would partially overlap. On each horizontal track, the leftmost net which has no ancestors in the vertical constraint graph is placed first. The process then continues until all the nets are embedded. If two terminals of different nets are in the same column (if a vertical constraint exists), the horizontal segments connected to the top terminal must be placed above the horizontal segment connected to the bottom terminal; otherwise the vertical segments would partially overlap. This algorithm always yields a solution with the minimum possible number of tracks, provided that there are no vertical constraints. The problem is analogous to the *job scheduling problem* that aims to minimize the number of processors. The job scheduling problem is much easier than the CRP because no vertical constraints are in the problem.

Lemma 3.5. The left-edge algorithm solves an arbitrary instance of the channel-routing problem with no vertical constraint using d tracks, where d is the maximum density of the problem.

Proof. Consider the subproblem I_i which is defined as the set of nets assigned to tracks above track $i - 1$, $i \geq 1$. Prove, by induction on i, that removal of nets assigned to the ith track decreases the maximum density of I_i by 1. By induction, the maximum density of $I - i$ is $d - i$. Consider an arbitrary column c with maximum density in I_i. Let D denote the set of nets in I_i contributing to this maximum density. If one of the nets in D is in track i (and since c was chosen arbitrarily) the result is established. Now assume that none of the nets in D is in track i. Let I^* be the rightmost net in track i that is to the left of column c. The only reason that none of the nets in D was assigned to track i is that they all overlap with I^*. This is a contradiction, since if all the nets in D overlap with I^*, then the maximum density of I_i is (at least) $d - i + 1$. Thus there is at least one net from each maximum density cut of I_i in track i. By induction, removing the nets in the track i reduces the density of the remaining problem by 1, $i > 1$. Thus, after d tracks, the maximum density is equal to zero; that is, all nets have been placed in some track.

Next, connect every terminal to the appropriate track. Since the problem has no vertical constraints no additional tracks are necessary.

Yet another channel router (YACR, [302]) operates under the assumption that vertical tracks can be added whenever required within a channel. Experimental results showed improvement in both the overall routing area and the number of vias over classic routers. An approach to handle the vertical constraints was introduced in YACR-II [303], the follow-up version of YACR that is very fast and practical. YACR-II allows the addition of horizontal tracks and the introduction of horizontal jogs on the vertical layer which may result in wire overlap. To describe the algorithm, *trunks* are defined as the horizontal wire segments placed in tracks and *branches* are the vertical wires connecting trunks to the top and bottom of the channel. This router employs a two-phase approach. A modified LEA that allows cycles in vertical constraint graphs (in other words, vertical constraint conflict exists in the problem instances) is followed by a branch layer routing phase that can find more complex routings than just strictly vertical segments. That is, after completing the track-assignment phase using LEA, the branch layer assignments are placed for all columns that do not violate vertical constraints. YACR-II then attempts to connect all of the remaining trunk layer segments using maze-routing techniques that are not based on Lee's [200] algorithm but use a pattern router. While a Lee-based algorithm could be used, the authors observed that many of the conflicts can be resolved by simple connections.

A pattern router seeks solutions that match a set of particular topologies. These patterns avoid the overhead of a complete maze search and give better control over the number of vias used. YACR-II used three patterns: maze 1, maze 2, and maze 3. The key to the maze 1 pattern is to resolve the vertical constraint conflict by connecting constrained branches using horizontal jogs. Figure 3.20 shows an example. The other two patterns attempt to dogleg [83] branch connections using the existing space on the trunk layer (allowing overlaps). If all three patterns fail, YACR-II adds an additional track. To break cycles in a vertical constraint graph (if all of the maze patterns have failed to resolve a constraint that is

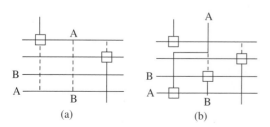

FIGURE 3.20
Resolving constraints using YACR-II.
(a) Vertical constraint. (b) Resolution
with maze I pattern.

part of a cycle), a new column is added. This process continues until all nets are connected.

GREEDY CHANNEL ROUTER. While the above-mentioned channel routers are all based on row-by-row routing, the *greedy channel router* ([294]) scans the channel in a left-to-right, column by column manner, completing the wiring within a given column before proceeding to the next. It always finds a solution, usually using no more than one track above the required channel density. Note that extra columns to the right of the channel might be used; however, in practice the number of used extra columns is small (e.g., 0, 1, or 2). This algorithm is the result of a great deal of experimentation and evaluation of variations on the basic idea of scanning down the channel from left to right and routing all nets as one proceeds. In each column, the router tries to optimize the utilization of wiring tracks in a greedy fashion using the following steps.

1. Make feasible connections to any terminals at the top and bottom of the column. Bring the nets safely to the first track which is either empty or contains the same net.
2. Free up as many tracks as possible. Make vertical jogs to collapse nets that currently occupy more than one track.
3. Shrink the range of tracks occupied by nets still occupying more than one track. Add doglegs to reduce the range of split nets by bringing the nets into empty tracks so that collapsing these nets later will be easier.
4. Introduce a jog to move the net to an empty track close to the boundary of its target terminal. This tends to maximize the amount of useful vertical wiring so that it reduces the upcoming column's congestion.
5. Add a new track. If a terminal can not be connected up in step 1 because the channel is full, then the router widens the channel by inserting a new track between existing tracks.
6. Extend processing to the next column. When the processing of the current column is complete, the router extends the wiring into the next column and repeats the same procedure.

The router usually starts with the number of tracks equal to channel density. However the initial channel width can be used as a controlling parameter in the algorithm. Another useful controlling parameter to reduce the number of vias is minimum jog length, the lower bound for the dogleg length allowed in phases 2, 3 and 4.

Because the greedy channel router makes fewer assumptions about the topology of the connections, it is more flexible in the placement of doglegs than the LEAs. Another nice feature of the greedy router is that its control structure is very flexible and robust; it is easy to make variations in the heuristics employed to achieve special effects or to rearrange priorities. The algorithm takes about 10 seconds on the difficult example and completes the whole channel in 20 tracks.

The greedy router often results in fewer tracks than a left-edge solution but uses more vias and more wire bends in the solution. Also, the algorithm lacks the ability to control the global behavior.

HIERARCHICAL ROUTING. Another approach designed to handle large-scale routing problems is hierarchical decision-making that is applied at each level of the hierarchy to consider all nets at once. The divide-and-conquer approach to hierarchical routing is attractive, as it provides a problem of reduced-size at each level of hierarchy.

Two schemes have been proposed, the top-down and the bottom-up approaches. In the top-down scheme, as one proceeds downwards, one obtains a better focus on all connections of each level of hierarchy. As one reaches the last step, a clear view of the complete problem is obtained [38]. The idea of the bottom-up recursion is to cut the chip into square cells (called global cells) small enough to be handled easily and then paste the adjacent square cells successively [154]. In both schemes, the local decision made at each step of hierarchy may result in an inability to find a final solution. However, the hierarchical approach works extremely fast and is invaluable in solving many instances practically.

One of the top-down schemes proposed in [37, 38] is as follows. The proposed top-down approach (see Figure 3.21) starts from the top with a 2×2 super cell representing the whole chip (Figure 3.21b). The 2×2 structure is routed first. The next level of hierarchy is considered in the horizontal direction first (Figure 3.21c). The vertical hierarchy is next (Figure 3.21e) and the necessary connections across the boundary are made. The problem can be naturally reduced to the problem of global wiring within a $2 \times N$ grid as shown in Figure 3.21f. This technique is not commonly used.

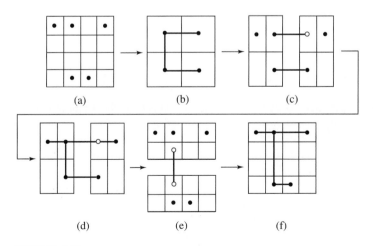

FIGURE 3.21
Illustration of the top-down hierarchical routing approach.

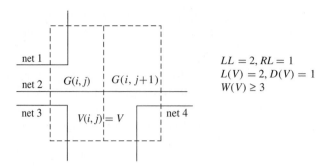

$LL = 2, RL = 1$
$L(V) = 2, D(V) = 1$
$W(V) \geq 3$

FIGURE 3.22
Legal routing.

A DIVIDE-AND-CONQUER APPROACH. Here another approach to the channel routing problem is presented, applying divide-and-conquer technique based on the restricted hierarchical $(2 \times N)$ wiring. The divide-and-conquer router was the first to automatically complete the Deutsch's difficult channel in 19 tracks.

A route R is said to cross a boundary B if the pair of cells forming this boundary are crossed by R (e.g., Net 2 in Figure 3.22). A route R turns within a cell G if it crosses two nonopposite boundaries of G (e.g., nets 1, 3, 4 in Figure 3.22). Suppose a routing is specified. Consider an arbitrary vertical boundary $V(i, j)$. Let LL (left load) and RL (right load) denote the number of routes not crossing $V(i, j)$ but turning within the cell $G(i, j)$ $(G(i, j + 1))$. The load of the vertical boundary $V(i, j)$ is defined as $L(V) = \max(LL, RL)$. For any boundary B let $D(B)$ $(W(B))$ denote the number of routes crossing B (boundary capacity of B). Routing R is called legal if and only if

(i) for any boundary B: $D(B) \leq W(B)$;
(ii) for any vertical boundary V: $D(V) + L(V) \leq W(V)$.

If all boundary capacities are equal to 1, the legal routing corresponds to the traditional Manhattan channel routing, because condition (ii) rules out knock-knees.

The proposed approach to channel routing is based on the reduction to a generalized problem for a $(2 \times N)$ grid. The reduction is performed on every level of the top-down hierarchy as follows. First partition the grid into two parts, $(\lceil m/2 \rceil \times N)$ and $(\lfloor m/2 \rfloor \times N)$ subgrids. Assuming that the routing within this $(2 \times N)$ grid is obtained, now partition each of the horizontal strips of the grid into two parts, thus generating two $(2 \times N)$ subproblems (second level of hierarchy). The global routes from the previous level define terminal positions for wiring of new $(2 \times N)$ subproblems. Figure 3.23 illustrates these steps.

Proceed with the resulting $(2 \times N)$ routing in this manner until a single cell solution is reached. At this level the routing will be completed. If the $(2 \times N)$ router fails to obtain a legal routing, that is, a routing satisfying (i) and (ii),

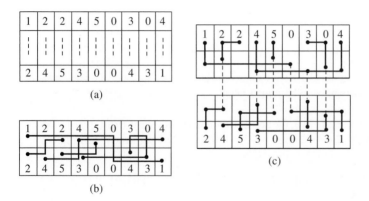

FIGURE 3.23
An example of restricted hierarchical $2 \times N$ routing. (a) An instance; (b) first partition; (c) second partition.

then condition (ii) is relaxed. In this case, all vertical capacities of boundaries in the same horizontal strip are increased by one. The increase is equivalent to the addition of a new horizontal track to the original problem.

The problem of the $(2 \times N)$ (also called TBN) routing is defined as follows.

- The input consists of two rows of terminals on the $(2 \times N)$ grid. EL is a $(2 \times N)$ Boolean matrix; if $EL(i, j) = \text{TRUE}$, then the corresponding cell is a terminal cell. HC is an n element vector which stores the costs of crossing the horizontal boundaries; $HC(j)$ indicates the cost associated with the number of wires crossing the boundary between $G(1, j)$ and $G(2, j)$. Similarly, $VC(i, j)$ is a $2 \times (N - 1)$ matrix that stores the costs of vertical boundary crossings; $VC(i, j)$ indicates the cost associated with the number of wires crossing the boundary between $G(i, j)$ and $G(i, j + 1)$, $i = 1, 2$.
- The output consists of Steiner trees with no vertical/horizontal overlaps and no knock-knees.
- The goal is to minimize the number of tracks; that is, to find a minimal cost tree interconnecting a set of elements located on a $(2 \times N)$ grid.

The solution to the TBN is an extension of dynamic programming procedure described in [3] which finds a Steiner tree that interconnects a set of elements located on a $(2 \times N)$ grid. Instead of the minimum length tree, in a TBN the minimum cost tree in which its cost is associated with the number of nets crossing a boundary is found.

Definition 3.1.

- Let $T^1(k)$ denote the minimal cost tree which interconnects the following set of cells: $\{G(i, j) : (j \leq k) \& (EL(i, j) = \text{TRUE})\} \cup \{G(1, k)\}$.
- Let $T^2(k)$ denote the minimal cost tree which interconnects the following set of cells: $\{G(i, j) : (j \leq k) \& (EL(i, j) = \text{TRUE})\} \cup \{G(2, k)\}$.

- Let $T^3(k)$ denote the minimal cost tree which interconnects the following set of cells: $\{G(i, j) : (j \le k)\&(EL(i, j) = \text{TRUE})\} \cup \{G(1, k); G(2, k)\}$.
- Let $T^4(k)$ denote the minimal cost forest, consisting of two different trees T^* and T^{**}: $T^*(T^{**})$ uses cell $G(1, k)$ $(G(2, k))$. The set $\{G(i, j) : (j \le k)\&(EL(i, j) = \text{TRUE})\}$ is interconnected by either one of them, meaning that the trees have to be joined later.

The trees $T^i(k + 1)$ $(i = 1, 2, 3, 4)$ are computed recursively from $T^i(k)$. Denote by FIRST and LAST the abscissas of the leftmost and rightmost terminal cells; then

$$\text{FIRST} = \min\{k : EL(1, k) \cup EL(2, k) = \text{TRUE}\},$$

$$\text{LAST} = \max\{k : EL(1, k) \cup EL(2, k) = \text{TRUE}\}.$$

Trees $T^i(k)$ for $k \le \text{FIRST}$ are computed trivially and serve as a basis for recursion. In fact, for $k \le \text{FIRST}$, $T^l(k)(l = 1, 2)$ consists of a single vertex, $G(l, k)$, $T^4(k)$ consists of the disjoint pair of vertices: $G(1, k)$ and $G(2, k)$. However, $T^3(k)$ is obviously a path $G(1, k), \ldots, G(1, s), G(2, s), \ldots, G(2, k)$ where $1 \le s \le k$ and

$$HC(s) + \sum_{j=s}^{k-1}(VC(1, j) + VC(2, j))$$

is minimal. In general, the cost $HC(k)$ might be high and detouring might result in cheaper routes (see Figure 3.24).

Suppose that $\text{FIRST} \le k \le \text{LAST}$ and the trees $T^i(k)$, $i = 1, 2, 3, 4$ are constructed. To construct $T^j(k+1)$, enumerate all possible extensions from $T^i(k)$, $i = 1, 2, 3, 4$ and select the cheapest one. For example, to construct $T^1(k + 1)$, select the cheapest among $T^i(k)$, $i = 1, 2, 3, 4$. When the value $k = \text{LAST}$ is reached, our trees are covering all terminal cells. Consider the case where the costs $HC(s)$ for $s \le \text{LAST}$ are so high that it is cheaper to take a right side detour. This case is similar to the FIRST case (see Figure 3.24).

Lemma 3.6. The proposed divide-and-conquer algorithm runs in $O(Nn \log m)$ time, where N is the number of nets, n is the length of the channel, and m is the width of the channel.

$H(s) \qquad H(k)$

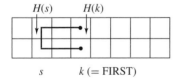

$s \qquad k\ (= \text{FIRST})$

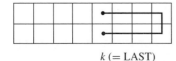

$k\ (= \text{LAST})$

FIGURE 3.24
Detouring.

Proof. Note that at each stage of recursion a constant number of comparisons is made, and the total computation time is $O(n)$. The time spent for routing a single net at any level of hierarchy is proportional to the total length of horizontal segments of this net. It is observed that the total length of a net at any level of hierarchy is bounded by $O(n)$. Since there are N nets, and $\log_2 m$ levels of hierarchy, the upper bound for algorithm complexity can be expressed as $O(Nn \log m)$. This running time is for the worst case, when almost all nets are expanded through the whole channel. On the average, the algorithm runs faster.

3.3.2 Switchbox Routing

The routing region with fixed terminals on four sides is called a switchbox. Switchboxes are predominately encountered in building block-style routing. The goal of a switchbox router is to interconnect all terminals belonging to the same net with minimum total length and vias.

A powerful technique for handling large-scale routing problems which avoid net ordering is a hierarchical method. The hierarchical method developed in [38] can be used both for ordinary channels and switchboxes. It is based on the divide-and-conquer paradigm and leads to excellent results for channel routing (but fails to complete the Burstein's difficult switchbox). The hierarchical switchbox router, due to its high speed and global view of the problem, can quickly generate solutions to large problems. However, the local decision made at each step of hierarchy may result in an inability to find a final solution. It may also serve as a generator of good starting solutions for other algorithms (e.g., rip up and reroute algorithms). Another switchbox router is WEAVER [170], which uses a knowledge-based expert (e.g., on constraint propagation, wire length and congestion). WEAVER produces excellent routing results, but takes an enormous amount of time even for simple problems. However, both of them are conceptually difficult due to their inherent limitations. Two routing schemes that are effective while being simpler than the algorithms proposed in [37, 169] are BEAVER and greedy switchbox routers.

BEAVER: CAUTIOUS STEP FORWARD. A heuristic router, BEAVER [64], has produced excellent results. The algorithm consists of three successive parts: corner routing, line-sweep routing, and thread routing. All three subrouters are given a priority queue of nets to route. A priority queue is used to determine the order that nets are routed and the prioritized control of individual track and column usage to prevent routing conflicts.

The *corner router* connects terminals that form a corner connection. Such a connection is formed by two terminals if

- they belong to same net;
- they lie on adjacent sides of the switchbox;
- there are no terminals belonging to the net that lie between them on the adjacent sides.

Corner routing is preferred for two reasons; its connections tend heuristically to be part of the minimal rectilinear Steiner trees for the nets, and it is fast. If a

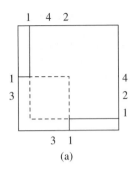
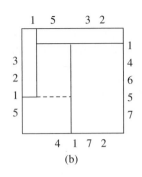

FIGURE 3.25
(a) An overlap cycle; (b) a four-terminal cycle.

net has terminals on either two or three sides of the switchbox, no additional analysis takes place. However, the corner ordering is performed for four-corner nets since there are two forms of cycles, overlap cycles and four-terminal cycles. An overlap cycle occurs when a pair of opposite corners overlap, as illustrated in Figure 3.25a. Net 1's lower corner connection blocks net 2 and 4, and its upper right corner connection does not block any nets, so the upper right corner connection is preferred. The corner router prefers one corner route with the least impact on the routability of others. A four-terminal cycle occurs when a four-terminal net has its terminal positioned at four sides, as shown in Figure 3.25b. For this case, the order is determined such that the corner is routed to maximize the routability of subsequent nets. The ordering criteria are similar to that of the overlap cycle case: the corner with the most potential blocks is preferred least. Net 1's lower left corner can block net 5, so it is placed last in the ordering (note that none of the other three corner routes have any potential blocks, thus their order is insignificant).

The *line sweep router* is an adaptation of the computational geometry technique of plane sweeping. Here a minimal length connection that joins two of the current net's disjoint subnets is heuristically sought. The line sweep priority queue is initialized with the unrealized nets. The line sweep router considers only a wire with a single bend, a single straight-line wire, a dogleg connection with a unit length cross-piece, and three wires arranged as either a horseshoe or a stairstep. As shown respectively from left to right in Figure 3.26, a dogleg connection with a unit-length cross-piece differs from a stairstep connection.

In addition, BEAVER only introduces a via or assigns a wire segment to a layer if it is immediately necessary. The router uses a plane sweep technique [17]. This technique involves scanning a geometric plane with a single line and

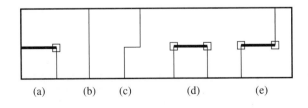

(a)   (b)   (c)   (d)   (e)

FIGURE 3.26
Prototype line sweep connections: (a) single bend; (b) straight-line wire; (c) dogleg connection; (d) horseshoe; (e) stairstep.

observing its interconnections in the space. It is used in finding straight-line connections that join two disjoint subnets of the net in question. Here, the switchbox is swept both on a row-by-row and a column-by-column basis. BEAVER prefers minimal length connections. If there is more than one minimal length connection, BEAVER performs several checks on the possible connections to determine the preferred one. The first check determines if one of the possible connections is more likely to use fewer vias than the others. If more than one connection meets this criterion, a second check is performed to determine which connection is likely to have the least impact on the remaining unfinished nets.

The *thread router* is a maze-type router that does not restrict its search for a connection to any preferential form. This router is a maze-type router [200, 251] that seeks minimal length connections to realize the remaining unconnected nets. Since the thread router does not restrict its connection to any preferential forms, it will find a connection for a net if one exists. When processing the current net i, the goal is to find a minimal length connection that joins i's currently smallest disjoint subnet s to some other disjoint subnet of net i, where the size of the subnet is measured in terms of wire length. By selecting s for maze expansion over net i's other subnets there is a two-fold benefit, the maze expansion is heuristically minimized since it reduces the number of starting points, and the number of possible destination points for the maze expansion is maximized. In addition, an important feature of track control is employed to prevent routing conflicts. The router tends to give a more global view and serves as a guide to the heuristic routing process.

A formal description of the algorithm follows.

> *Algorithm* BEAVER
>> *begin-1*
>>> initialize control information;
>>> initialize corner-pq;
>>> corner route;
>>> *if-2* there are unrealized nets *then*
>>>> initialize linesweep-pq;
>>>> line sweep route;
>>>> *if-3* there are unrealized nets *then*
>>>>> relax control constraints;
>>>>> reinitialize linesweep-pq;
>>>>> line sweep route;
>>>>> *if-4* there are unrealized nets *then*
>>>>>> initialize thread-pq;
>>>>>> thread route;
>>>>> *end if-4*;
>>>> *end if-3*;
>>> *end if-2*;
>>> perform layer assignment;
>> *end-1*.

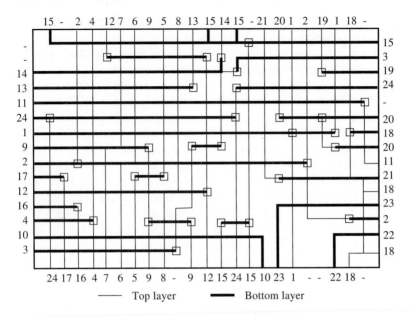

FIGURE 3.27
BEAVER's solution to the difficult switchbox.

Figure 3.27 shows BEAVER's solution to the difficult switchbox. All connections were completed by the corner and line sweep routers except for net 3's two terminals, which were thread-routed. Note that BEAVER allows both knock-knees (e.g., between net 14 and net 3) and overlaps (e.g., between net 9 and net 7).

There are six other switchbox routers that can solve the difficult switchbox (see Table 3.2 for comparison): DETOUR [125], GSR [234] (discussed below), BBLSR [240], MDR [147], WEAVER [170], and SWR [178].

GREEDY SWITCHBOX ROUTERS. A class of popular switchbox routing heuristics such as [147, 178, 234] uses a descendant of the greedy channel router [294].

TABLE 3.2
Router comparison for the difficult switchbox

Router	Number of vias	Length	Extra tracks	Prewired nets	Run time (sec)
DETOUR	67	564	0	1	?
GSR	58	577	1	0	1
MDR	63	567	0	0	?
BBLSR	59	560	0	0	5
WEAVER	41	531	0	0	1508
SWR	65	543	0	0	3
BEAVER	35	547	0	0	1

These routers scan a switchbox from left to right, column by column (or from bottom to top, row by row) and take an action according to a set of prioritized rules at each column (or row) before proceeding to the next one. The generic greedy algorithm is outlined below.

> **Algorithm** Greedy switchbox
> *begin-1*
>   initialize the left side of the switchbox;
>   determine goal tracks;
>   ColumnCount = 1;
>   *while-2* ((goal tracks not reached) and (ColumnCount ≤ MaxCol)) *do*
>   greedy route column;
>   ColumnCount = ColumnCount + 1;
>   *end-2*;
>   perform layer assignment;
> *end-1*.

For example, in each step of greedy column routing, the following rules are executed in the order of their priority.

1. *Bring nets into column.* Each column may have the nets connected to its top or bottom terminals. Connections to these terminals are brought into the switchbox and pulled up to a track already assigned to the net or to a free track, whichever is closer to the terminal.

2. *Collapse split nets.* A net may enter a column on more than one track. Obviously, many nets running on several tracks passing through a column may lead to unroutable situations. Thus, higher priority is given to such split nets to be merged earlier, so that more empty tracks are available for future routing. Since one vertical connection may block another one, vertical wires which block the fewest are considered first. Usually, among those which block the fewest number of other connections, the ones with the longest vertical runs are chosen to be routed.

3. *Reduce the range of split nets.* It is undesirable for split nets to extend over many columns. Thus, in the routed column, vertical connections are made to reduce the range of tracks over which a net is split.

4. *Pull nets toward goal tracks.* Nets, not yet on their goal tracks (the tracks matching the terminals belonging to the same nets at the right side of the switchbox), are shifted to a set of free unblocked tracks closest to their goal tracks. In this process nets that occupy other nets' goal tracks might be shifted away from their goal tracks.

5. *Choose goal tracks.* A greedy router sweeping the switchbox area from the left has to match the terminal positions on the right side. The task becomes more difficult when nets have multiple terminals on the right side. In such a case a net has to be split to meet multiple goals.

Although these greedy heuristics are fast and simple, they are often ineffective for more complex routing problems. They may fail to route all nets because of the unidirectional plane sweeping. Their performance on the Burstein's difficult switchbox is shown in Table 3.2.

3.3.3   Discussion

Throughout this section the focus has been on the Manhattan model. There are other models that have been given attention. One of these models is based on knock-knees. In the knock-knee model, wires are free to be routed in both directions on each layer, but overlap is forbidden; however *knock-knees* (points where two wires bend at a grid point) are allowed (e.g., see [248, 279]). Channel routing is especially easy in the knock-knee model. If only two-terminal nets are involved, a greedy algorithm solves the channel width minimization problem [248] optimally. If multiterminal nets are involved the problem becomes NP-complete [304].

The other type of detailed routing model is planar routing [274]. In planar routing the interconnection topology of the nets is planar; that is, no net intersections are allowed. This model is very restrictive; many instances of the detailed routing problem are not routable without crossing wires. The planar routing model is used for such tasks as routing the chip inputs and outputs to the pads on the chip boundary or routing critical nets such as power/ground/clock nets on a single layer. Chapter 5 is devoted to a discussion of planar routing problems.

Remaining types of channel routing algorithms are as follows. The router proposed in [242] operates under the assumption that wires can be routed on both layers in both directions, but no wire overlap on the same layer is allowed. However, wires can overlap when they are on different layers; that is, they can run on top of each other on different layers. This model is called the free-hand model. [293] also considered a similar wiring model. Because wiring overlap is allowed, channel density d is no longer a lower bound for the channel width. In this case, all that can be trivially guaranteed as a lower bound is $d/2$. However, the paper presented an algorithm requiring $2d - 1$ tracks in the worst case for two-terminal nets.

Another overlap model [53, 109, 127] of detailed routing is to allow a prespecified length of overlaps between signal lines. Overlapping of nets is generally undesirable because of increased capacitive coupling between the nets. Nevertheless, a very short length of overlap is tolerable.

Segmented channel routing considers the more restricted case where the routing is constrained to use fixed wiring segments of predetermined lengths and locations within the channel. Refer to [118] for details.

Unlike the previous grid-based models, the gridless routing model uses a continuous plane as a routing domain and incorporates the geometrical constraints (e.g., design rules) of the fabrication process.

In printed circuit board wiring, more than two layers are provided. For this application, a multilayer model is required that represents each wiring layer explicitly. Assume that k layers are available and overlaps are allowed in both directions. [126] proved that $\lceil d/(k-1) \rceil$ is a lower bound on the minimum channel width. The best known algorithm [18] uses $d/(k-1) + O(\sqrt{d/k})$ tracks.

The assumption that wires can run on top of each other for arbitrary lengths is not suitable for high performance systems. The reason is that long stretches of overlapping wires may introduce cross-talk noises in chip technologies. Cross-talk noise is a parasitic coupling (i.e., mutual capacitances and inductances) phenomenon between neighboring signal lines. This noise becomes serious in a high-density wiring substrate with narrow line pitch. The closer the lines, or the higher they are from the ground plane, or the longer they are adjacent, the more coupling occurs. The coupling between the lines can be minimized by making sure that no two lines are laid out in parallel or next to each other for longer than a maximum length. Thus, to account for this problem, researchers have introduced more restrictive models for multilayer routing. Two of these multilayer models with restricted overlap are defined here.

The first one is the *unit-vertical overlap model*. Wire overlaps are not allowed in the horizontal direction. In the vertical direction, wires may overlap, but not for two consecutive units. In the unit-vertical overlap model, channel density d is still a lower bound on the channel width, since overlap is allowed in only vertical direction. [110] routed the multilayer channel using $3d/2 + O(\sqrt{d})$ tracks.

Note that, for both unit-vertical overlap and for unrestricted overlap, although the algorithms are near-optimal, in practice the number of tracks used exceeds the lower bound by a high amount. For this reason, these algorithms are not used in practice.

The second one is the *L/k-layer model*. Overlapping wires have to be L layers apart. Thus, the rationale of this model is to reduce cross-talk noise by separating wires so they are far apart. In the model, $\lceil d/\lceil k/L \rceil \rceil$ tracks are a trivial lower bound on the channel width. [25] presents an algorithm for the model that uses $\lceil (d+1)/\lceil k/L \rceil \rceil + 2$ tracks.

There is also a directional multilayer model. This model assigns preferred wiring direction, horizontal or vertical, to layers. In a k-layer model, each layer specified by either H or V that defines the wiring direction. However, although preferred routing directions have been used in many applications, the strict limitation on wiring direction is not necessary and many vias between layer pairs (due to multiple bends in a net) can be introduced. Refer to [95] for a heuristic for channel routing in that model.

Other paradigms for the channel routing problem have also been proposed. For example, [107] has proposed a neural net algorithm for the problem. Their algorithm has been applied to the four-layer channel routing problem, allowing two horizontal layers and two vertical layers. Their algorithm can run both sequentially and in parallel. The proposed technique is based on a three-dimensional artificial neural network model with a highly interconnected network of a set of simple processing elements. As mentioned before, neural net-based techniques are slower then most other techniques. They may produce good solutions to particular instances of the problem.

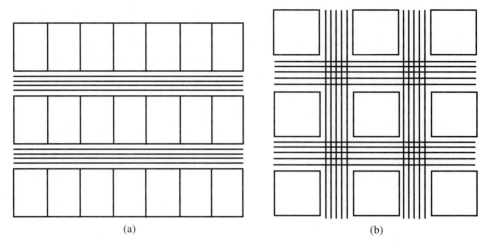

(a)   (b)

FIGURE 3.28
Two major classes of FPGA architectures. (a) Row-based FPGA; (b) array-based FPGA.

3.4 ROUTING IN FIELD-PROGRAMMABLE GATE ARRAYS

The field-programmable gate array (FPGA) [93, 394] is a relatively new technology for rapid design prototyping and implementation. An FPGA chip consists of prefabricated logic cells, wires, and switches. Each cell can implement any Boolean function of its inputs. The number of inputs to each cell is bounded by a constant K; typically K is 5 or 6. There is also a small number of flipflops (e.g., 1 or 2) in each logic cell to allow the implementation of sequential circuits. Two major cell architectures are currently available, lookup table- (LUT)-based and multiplexor- (MUX)-based. In a LUT-based cell architecture, each logic cell mainly consists of a K-input single-output programmable memory capable of implementing any Boolean function of K inputs. In a MUX-based cell architecture, multiplexors are used to implement arbitrary Boolean functions of up to K inputs. Cell terminals are connected to routing wires via programmable switches. There are also programmable switches to interconnect the wires to achieve the desired routing patterns.

Two major classes of commercial FPGA architecture are currently being used, row-based architecture and array-based architecture. The two classes of FPGA are depicted in Figure 3.28. The row-based architecture interleaves cells and routing wires in row fashion, like a standard cell layout style. The routing channels consist of horizontal wires segmented by programmable switches. Cells are organized in rows interspersed with column wires to allow the router to connect terminals on different rows. Alternatively, the array-based architecture uses a two-dimensional grid organization. Cells, routing channels and switches are uniformly laid out. Horizontal and vertical wires meet at programmable switchboxes where electrical connections can be made.

The foremost objective in the routing problem of FPGAs is to achieve 100% routing completion, because the routing resources are prefabricated and scarce. If a router cannot achieve 100% routing, then the previous stages in the automation process, namely, the cell placement and technology mapping steps, would have to be redone, which is time-consuming. To avoid such recurrence, research is being directed toward devising more efficient and more successful FPGA routers, and the following two issues have caught the attention of many researchers:

- Analysis of the routability of existing routing structures and design of better routing resources in order to increase the routability and probability of successful routing, and
- Design of routability-driven technology mapping.

The next two sections focus on FPGA routing problems for array-based and row-based FPGAs, respectively. Each section sets up the notation and formulates the problem for the corresponding FPGA architecture and then presents a routing algorithm for the formulated problem.

3.4.1 Array-based FPGAs

The routing architecture for array-based FPGAsis depicted in Figure 3.29. Wire segments run between logic cells (L blocks) in vertical and horizontal channels. Personalization of routing resources is done through the proper programming of the switches. Note that unlike the traditional gate arrays, in array-based FPGAs, all the routing resources (wire segments and switches) are prefabricated at fixed positions. This distinguishes the routing problems in array-based FPGAs from those in traditional gate arrays. The programmable switches are grouped inside connection (C) and switch (S) blocks. These switches are shown as dotted lines in Figure 3.29. The *flexibility* of a C block, F_c, is defined as the number of tracks a logic pin can connect to, and the flexibility of an S block, F_s, is defined as the number of outgoing tracks that an incoming track can connect to. Since there are significant resistive and capacitive attributes associated with the switches, it is desirable to have routing resources that allow the minimization of switches on the routed paths to improve performance. As a solution to this, *long wire segments* that span more than a single C block are frequently accommodated in modern FPGAs to allow connections with few switches along the routing paths to lower the unwanted parasitic effects. Note that such long wires cannot be subdivided into isolated routing segments. The channel densities resulting from global and detailed routing are denoted by W_g and W_d, respectively. All channels have the same number of prefabricated tracks W. The route of a net is called *simple* if all its terminals are connected to exactly one track, otherwise the route of the net is a dogleg route.

According to experimental studies in [32, 212, 295, 335, 360], a full flexible F_c and F_s of 3 yield a reasonable flexibility for routing resources. Studies in [146] show that an F_s value larger than 3 does not improve routability noticeably in the existence of full flexibility for blocks. However, lowering F_s to less than 3

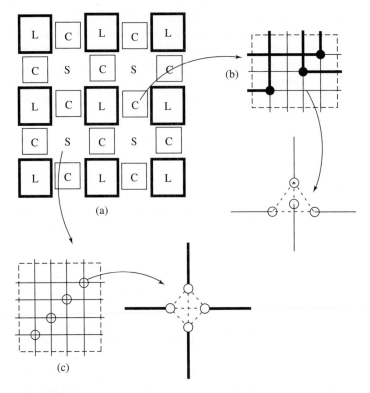

FIGURE 3.29
Routing model for array-based FPGA architectures. (a) L(ogic) block; (b) C(onnection) block; (c) S(witch) block.

causes sharp degradation of routability regardless of increases in F_c. The routing architecture for the rest of this section is therefore considered to have:

- full flexibility for C blocks;
- $F_s = 3$ for S blocks; further, all the S blocks are assumed to have the same topology, with switches along the diagonal as shown in Figure 3.29; and
- all wire segments as single-length segments that must be connected through the switches (the distance between centers of adjacent S blocks is considered to be one unit); in other words, long wire segments are not considered.

The routing problem for array-based FPGA architecture is formulated as follows.

> **Problem instance.** Given a 2D array of S blocks, a mapping of circuit elements into logic blocks, and a net list.

> **Problem objective.** Determine the track assignment for the nets such that all the connections are made, and W, the number of tracks which need to be available in each channel to guarantee routing completion, is minimized.

Note that the C blocks are assumed to have full flexibility. They do not appear in formulation of the problem. For reference purposes, the tracks in each horizontal channel are numbered from top to bottom, and the tracks in each vertical channel are numbered from right to left. The number assigned to a track is referred to as that track's id. The diagonal positions of switches in S blocks would then mean that a track in a horizontal channel can only connect to tracks with the same id in a vertical channel and vice versa. This is called a *diagonal S block* architecture.

It is easy to see that the routing problem in this architecture can be viewed as a 2D interval packing problem, with the wire segments of each net as intervals that have to be packed into horizontal and vertical tracks. In this view of the problem, the wire segments of a net must be connected in a two-dimensional fashion, passing through a sequence of S blocks. In a diagonal S block architecture a net can switch from channel to channel but must always remain on tracks with the same id. This will result in the formulation of the global-to-detailed routing stated above as a graph-coloring problem.

Problem instance. Graph $G = (V, E)$.

Problem objective. Assign colors to vertices of G such that total number of colors is minimized and two vertices are assigned different colors if they are connected by an edge.

The graph $G = (V, E)$ corresponding to a given instance of FPGA routing problem is defined as follows. Given a set of nets, each net is represented by a vertex. If two nets share at least one C block, the corresponding vertices are connected by an edge.

Figure 3.30 shows an instance of the routing problem and its corresponding graph coloring instance. The C blocks shared between more than one are shaded in this figure. This means that a minimal coloring of graph G corresponds to a solution of the global to detailed routing problem. The heuristics for graph coloring can therefore be used for this routing problem.

A drawback of the diagonal S block architecture is that the detailed routing problem in such architecture, as shown above, can be modeled as a graph coloring problem, which is an NP-hard problem [113]. A more general architecture is also discussed in [392] in which the S blocks are organized such that $F_s = 3$ and in each row and each column of the S blocks there is exactly one switching point. Each switching point can connect the incoming wire from any channel to one particular wire in each of the other three outgoing channels (note that the connected incoming and outgoing tracks do not necessarily have the same id). This means that the tracks with different id values can be connected through a switching point. An example of an S block at a switching point is shown in Figure 3.31.

A greedy bin-packing heuristic is proposed for the resulting routing problem in which each bin is a *track domain*, a set of tracks which are connected through switching points, and the objective is to pack the nets into the bins such that the

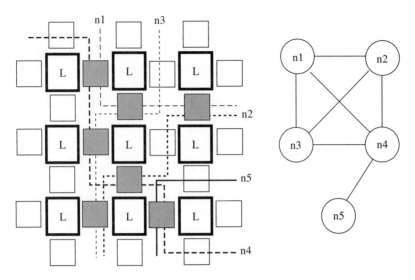

FIGURE 3.30
A routing instance and its corresponding graph-coloring instance.

minimum number of bins are used. Bin capacities are all the same and fixed, but depending on the capacity of the available routing resources, the object sizes may be extended. Some objects cannot share a bin. These objects are the nets that share pins located in at least one common C block. Such nets are called *confronting nets*.

A *confronting graph H* is formed, in which each vertex corresponds to a net, and two vertices are connected by an edge if their corresponding nets are confronting nets. The bin-packing heuristic is guided by the constructed confronting graph *H*. The nets are assigned to tracks, observing the constraints posed by the confronting graph. The proposed approach tries to route as many nets as possible in the currently processed track domain until it is full. The confronting graph *H* is gradually reduced after each track domain packing step. The routing is complete when the confronting graph is empty. The proposed algorithm runs in a series

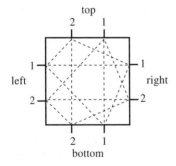

FIGURE 3.31
An example of an S block at a switching point.

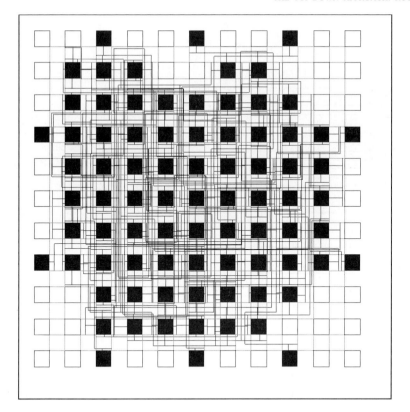

FIGURE 3.32
Final routing of a benchmark circuit (the name of the benchmark is *9symml*).

of passes for each track domain. In each pass, the nets to be routed are picked in order, giving higher priority to the longer nets because bin-packing heuristics perform better when larger objects are placed first [168].

To evaluate the efficiency of various algorithms, a set of examples based on real designs have been produced. These examples are called benchmarks and have been used by various algorithms' designers to evaluate the efficiency of their algorithms. Figure 3.32 shows the final routing of a benchmark example. Note that many of the cells are unused. This is a typical situation in FPGA designs because there are few fixed-sized FPGAs, and the size of the circuit to be implemented varies greatly.

3.4.2 Row-based FPGAs

The routing architecture for row-based FPGAs is depicted in Figure 3.33. The logic cells are laid out in rows separated by routing channels. The inputs and outputs of the cells each connect to a dedicated vertical segment. Programmable

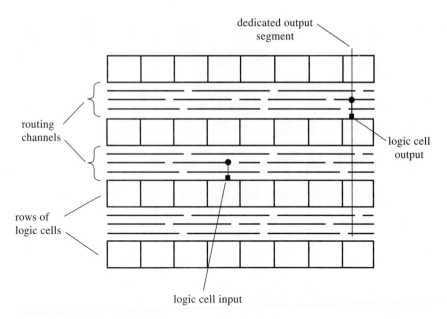

FIGURE 3.33
Routing model for row-based FPGAs.

switches are located at each crossing of vertical and horizontal segments and also between pairs of adjacent horizontal segments in the same track. By programming a switch, a low-resistance path is created between two crossing segments. The routing problem for row-based FPGAs is frequently referred to as *segmented channel routing* problem.

The segmented channel routing problem has been studied in [32, 36, 93, 121, 301, 406]. With complete freedom to configure the wiring offered in mask programming environment, the *left-edge algorithm* [134] will always find a routing solution using the number of tracks equal to the density of the connections. This is because there are no vertical constraints in the problems considered here (because there are dedicated vertical segments for the input and output pins on each logic cell). Achieving this complete freedom in an FPGA would require a programming switch at every cross point. Furthermore, switches would be needed between each pair of cross points along a routing track so that the track could be subdivided into segments of arbitrary length. This is undesirable because each switch occupies a considerable amount of area and, more importantly, switches incur significant resistance and capacitance to the routed nets, which would cause unacceptable delays through the routing. Another alternative would be to provide a number of continuous tracks long enough to accommodate all nets. This will use an excessive amount of area, and the capacitance due to unused portions of the tracks is added to the capacitance of routed nets, which is also undesirable.

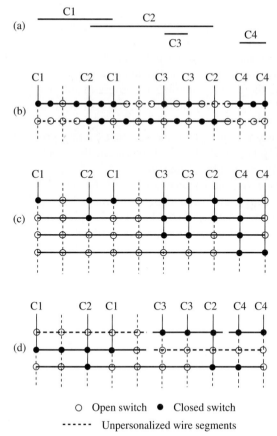

FIGURE 3.34

Examples of segmented channel routing. (a) Set of connections to be routed; (b) routing in a fully segmented channel; (c) routing in an unsegmented channel; (d) routing in a moderately segmented channel.

A segmented routing channel combines the characteristics of both approaches. Tracks are subdivided into segments of varying lengths. Figure 3.34 shows a routing problem in a fully segmented channel, an unsegmented channel, and a moderately segmented channel.

If there are no switches where segments in the same track meet, each connection must be routed in a single segment. The resulting routing problem is called the *one-segment routing problem*. Figure 3.34d depicts a segment routing example. If additional switches are available to allow segments that abut on the same track to be connected, the connections may span more than a single segment. This flexibility, however, greatly increases the complexity of the routing problem.

A *k-segment routing* for a set of connections in a segmented channel is an assignment of each connection to a track in which no segment is occupied by more than one connection, and also each connection occupies at most *k* segments.

[36] proposes a segmentation model with different length segments in which channels are partitioned into several regions. The segment length in each region

is different and precomputed based on some empirical analysis. Within each region, the segments are arranged in a staggered fashion. An appropriate routing algorithm is presented for the proposed model. Experimental results indicate substantial improvement on the longest and average net delay in the routing results obtained on this model compared to previous models. The idea was to increase the utilization of tracks and to reduce the average number of programmed switches (fuses) on routed portions of the nets. [273, 404] studied the problem of designing a segmented channel for improved routability.

When $k = 1$, the routing problem would be a one-segment routing problem and can be solved using the greedy approach. Let C, T represent the number of connections and tracks, respectively. Assign the connections in order of increasing left ends as follows. For each connection being processed, find the set of tracks in which the connection would occupy one (unoccupied) segment. Assign the connection to the track whose right end is farthest to the left. This simple scheme is guaranteed to find a solution for any set of connections in any segmented channel, if one exists. Since each track is checked for each connection, the running time of this algorithm is $O(CT)$.

For $k > 1$, it was shown in [118] that the k-segment routing problem belongs to the NP-complete class of problems. In the same paper the notions of frontiers and assignment tree are introduced, and an algorithm is presented that is guaranteed to find a solution for a given instance of k-segment routing problem if such a solution exists. The running time of the algorithm is exponential in the worst case; however it ends in time linear in C, when T (the number of tracks) is fixed. The algorithm works by constructing a data structure called an *assignment tree*. This tree is used to represent the effect of optionally assigning each connection to each track. The solution is directly constructed from this tree, when the tree is constructed to completion. A *frontier* is a function that shows how a valid routing of a set of connections $c_1, \ldots, c_i$ can be extended to include the next connection c_{i+1}. Note that a connection c_i is defined by its endpoints left(c_i) and right(c_i). The frontier of a track marks the leftmost column at which this track is still unoccupied. Apparently, connection c_{i+1} can be assigned to a track if the frontier of this track has not passed left(c_{i+1}).

Given a valid routing of $c_1, \ldots, c_i$, the frontier vector, X, is defined as a T-tuple $(x[1], x[2], \ldots, x[t])$, where $x[t]$ is the leftmost column in track t in which the segment present in that column is not occupied by any of the connections $c_1, \ldots, c_i$. Initially, the frontier vector is set to $X_0 = (0, 0, \ldots, 0)$. A frontier vector after step i is denoted by X_i.

The assignment tree is used to keep track of partial routing solutions. The root of the tree (at level 0) represents the frontier vector X_0. A node at level i corresponds to a frontier resulting from some valid routing of $c_1, \ldots, c_i$. This tree has a maximum of $C + 1$ levels, one for initial frontier (level 0), and one for each connection c_i (level i). If a valid partial routing for connections $c_1, \ldots, c_i$ exists, then there exists a corresponding node at level i of the assignment tree exists; otherwise level i (and also subsequent levels $i + 1, \ldots, C$) will be empty.

Once the assignment graph has been constructed to the end, a valid routing can be obtained by tracing back a leaf-root path in the assignment tree. If for some $i, 1 \leq i \leq n$, level i of the assignment tree is empty, then no valid routing of the connections exists. The algorithm to extend the assignment tree one level further from level i to level $i + 1$ is given below [118]:

> *for* each node X_i in level i;
>   *for* each track $t, 1 \leq t \leq T$;
>   *begin-1*
>   *if* $X_i[t] \leq left(c_{i+1})$ {;
>   *if* c_{i+1} would occupy $\leq K$ segments in track t;
>   *begin-2*
>   (* c_{i+1} can be assigned to track t *);
>   *let* X_{i+1} be the new frontier when c_{i+1} is assigned to track t;
>   *if* X_{i+1} is not yet in level $i + 1$ {;
>   *add* node X_{i+1};
>   *add* an edge from node X_i to node X_{i+1},label it with t;
>   *end-2*;
>   *else*
>   (* $X_i > left(c_{i+1})$ so c_{i+1} cannot be assigned to track t*);
>   *end-1*.

The presented approach can be slightly modified to minimize the summation of the lengths of the segments used to route the connections. To do this, an additional attribute w, for each edge, is computed as the algorithm proceeds. This attribute is the weight of the edge and corresponds to the total length of all segments assigned to $c_1, \ldots, c_{i+1}$ dictated by the path from the root to this node. In the construction of the assignment tree when considering track t for connection c_{i+1}, if a search at level $i + 1$ finds an already existing X_{i+1}, the weight of its incoming edge is compared against the weight w computed for the current assignment. If the latter is smaller, the edge entering X_{i+1} is replaced by track t coming from X_i. Thus, the solution traced back from unique node in level C (last level) will correspond to a minimal weight routing, which in turn corresponds to K-segment routing with minimized total length of the segments participating in the routing.

3.4.3 Discussion

The physical design aspect of FPGAs offers new challenges to researchers. The routing problem in the FPGA environment is significantly different from that of full-custom or semicustom chips. The wires, being prefabricated, lack the freedom to produce arbitrarily complex routing patterns. Furthermore, the quality of a router is particularly important because routing resources are scarce and costly.

The routing problem for FPGAs has been studied in [31, 118, 138, 212, 271, 392, 393]. Various problems in the computer-aided design of FPGAs were discussed in [138]. In [31], a branch and bound heuristic with pruning was used for routing. The global routing was based on a graph search technique guided by the channel density. The router tries to avoid congested area by incurring penalties to wires passing through congested areas. In [271] the detail routing for FPGAs was studied. However, the report did not address the global routing problem. In both routers, nets are routed sequentially. If some nets cannot be routed due to congestion, the routers rip up all nets on high density channels and reroute the nets with more aggressive congestion control parameters. It is difficult to predict the effects of earlier net routings on the routability of later nets. Often, the nets are randomly ordered due to the lack of a good ordering scheme. An algorithm proposed in [399] alleviates the net-ordering dependency of the router. The router considers the routability problem during global routing. Each net is assigned a horizontal or vertical backbone called an *axis*. An axis contains a major portion of the net wires. It also serves as an initial route during the detailed routing phase. The detailed routing can be done using any area router (see earlier sections), for example, a multiterminal maze router. In such a router, the routing originates from a terminal and tries to reach previously routed portions of the net. The routing phase is then repeated for other terminals. Because the axis constitutes a large portion of the net wires, it is necessary to arbitrate the nets competing for the same axis in order to optimize the global benefit. The net-axis contention process is solved using a weighted bipartite matching formulation [238], where a globally optimal solution can be obtained efficiently. Experimental results show that taking the routability into account during the global routing phase increases the routability and reduces route lengths as the number of nets participating in the matching increases.

As mentioned earlier, the routing problems for both array-based and row-based FPGA architectures have been shown to be NP-complete under certain routing models. Additional research is needed to explore improved routing architectures and models that allow fast and polynomial routing algorithms.

The foremost objective in the routing problem is to achieve 100% routing completion, because the routing resources are prefabricated and scarce. If a router cannot achieve 100% routing, then the previous stages in the automation process, namely the cell placement and technology mapping steps, will have to be redone, which is time-consuming. To avoid such recurrence, research is being directed toward devising more efficient and more successful FPGA routers, and the following two issues have caught the attention of many researchers:

- Analysis of the routability of existing routing structures and better design of routing resources in order to increase the routability and probability of successful routing. Furthermore, since routing switches consume significant chip area and introduce propagation delays, the design of the routing architecture can greatly influence both the area utilization and speed performance of an FPGA. FPGA routing architectures were studied experimentally in

[118, 295, 296]. In [33] a stochastic model to measure the routability on a wide range of FPGA architectures was introduced. [273] developed analytical models to facilitate the design of FPGA channel architectures and analysis of routability in row-based FPGA routers. [393] studied the array-based FPGA architectures from a theoretical standpoint. It formulated the routing problem as a two-dimensional interval packing problem and studied the complexity of that problem. Some directions to modify the routing architecture on array-based FPGAs to improve routability were also provided.

- Design of routability-driven technology mapping. The technology mapping step for FPGAs is the problem of mapping a digital circuit onto a given FPGA architecture. Several objectives of interest are: maximizing the size of the circuit that can fit in a given FPGA architecture (this is done by minimizing the number of logic cells that can implement a given network) [44, 97, 102, 254], maximizing the operation speed of the mapped circuit [67, 103, 255, 309], minimizing the power consumption of the mapped network [98], and improving the routability of the mapped network [20, 41, 311]. [311] introduces a routability-driven technology mapping algorithm for FPGAs. The algorithm tries to pack portions of design inside a cell such that the routing requirements of the packed cells are minimized. As a secondary objective the number of cells required to cover the whole design is minimized. A parametrical cast function is used to direct the packing step.

EXERCISES

3.1. Consider an n-vertex graph and a subset of m vertices to be connected. Give nontrivial lower and upper bounds on the number of distinct routings (i.e., the number of global routes).

3.2. What is the running time of Lee's maze router when there is only one two-terminal net in an $n \times n$ grid and the rectilinear distance between the two terminals is d? For what configuration of obstacles is the running time independent of n and depends only on d?

3.3. What is the running time of the line-searching example? Given an example that takes a long time for the line-searching algorithm to complete.

3.4. Apply the line-searching algorithm to the example shown in Figure 3.1. Apply the concept of track graphs to the same example.

3.5. Discuss the advantages and disadvantages of maze-running, line-searching, and search-based techniques on track graphs. Emphasize the quality (i.e., the length) and running time measures.

3.6. Prove that the total weight (i.e., cost) of a minimum-spanning tree in an edge-weighted graph is at most two times the length of an optimal Steiner tree in the same graph.

3.7. A rectilinear Steiner tree consisting of at most k vertical lines is called a k-*comb* Steiner tree. Design an efficient algorithm for finding a k-comb Steiner tree of a given set of n terminals in the plane. How bad could such a tree be; that is, what is the maximum ratio of the length of an optimal k-comb Steiner tree to the length of an optimal Steiner tree? Express your result in terms of k and n.

3.8. Consider a set of points where a point is distinguished as source. Design an algorithm for finding a Steiner tree interconnecting all points (including the source) such that the distance between the source and every other point in the tree is small. Elaborate on the quality of your solution.

3.9. Route the following channel consisting of 11 columns using the left-edge algorithm, where 0 indicates an empty position:

$$\text{TOP} = 3\ 4\ 0\ 1\ 2\ 4\ 3\ 5\ 2\ 1\ 0$$
$$\text{BOT} = 1\ 0\ 3\ 0\ 4\ 0\ 5\ 2\ 1\ 5\ 4$$

3.10. Consider two rows of points, where the vertical separation between the two rows is h grid units. Design an algorithm for finding a Steiner tree interconnecting the points. Analyze the running time of your algorithm and prove its correctness. (Hint: Employ a dynamic programming approach.)

3.11. Design a greedy algorithm to order the channel in a given placement so as to minimize the number of switchboxes.

3.12. What do you think is a good pattern of connection in a connection and in a switch block in an FPGA? Propose a specific pattern for each of them.

COMPUTER EXERCISES

3.1. Implement Lee's maze running algorithm.

Input format. Consider Figure 3.1. Input specifies the grid size, position of the two terminals, and the position of the obstacles (the northwest corner is grid point (1, 1)).

```
7 6;                    (* size of the grid *)
5 5, 2 2;               (* positions of source and target *)
3 1, 1 3;               (* positions of the obstacles *)
4 3, 5 3, 6 3, 3 4, 4 5.
```

Output format. The output is as shown in Figure 3.1. Highlight one possible path.

3.2. Consider a set of points in the plane. Find a minimum spanning tree interconnecting the points. Then design an efficient algorithm for finding a Steiner tree connecting the same set of points. Give a table comparing the length of a minimum spanning tree with the length of the resulting Steiner tree, for various values of n, where n is the number of points.

Input format. The input consists of the location of the given points in the plane.

```
3 0, 1 1, 2 3          (* there are 3 points *)
```

Output format. The output format is shown in Figure CE3.2. Edges of the spanning tree are shown as straight lines. However their length is a rectilinear length. Edges of the Steiner tree are shown as rectilinear lines (and the distances are also rectilinear).

3.3. Design a simulated annealing algorithm for solving the previous problem. Define your moves. Use the same input and output formats as the previous problem. Do you think simulated annealing is suitable for this problem? Explain.

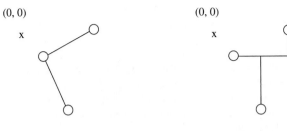

(0, 0)

x

Length = 6

(0, 0)

x

Length = 5

FIGURE CE3.2
Output format, a spanning and
Steiner tree.

3.4. Design and implement an algorithm for solving a given instance of the multicommodity flow problem. Given a graph where there is an integer assigned to each edge; the integer indicates the capacity of the edge. The goal is to minimize the total number of edges used (if there are k units of flow in an edge, that edge is counted k times). Note that the sum of flows in each edge should not exceed its capacity.

Input format. The input consists of the edges and their capacities. Then the set of commodities follows (the demand of each commodity is assumed to be one). For example (with reference to Figure 3.15), input consists of:

1 2 5, 1 14 2, . . . , 7 8 2;
2 8, 1 12, 3 7, 4 9;

meaning there is an edge of capacity 5 between vertices 1 and 2, there is an edge of capacity 2 between vertices 1 and 14, and so on. And there is a source at vertex 2 to be connected to a sink at vertex 8 (and so on).

Output format. The output should show the graph and highlight each flow with a different color. Also output the total number of edges used.

3.5. Implement the left edge algorithm. Input is the set TOP and BOT (terminals on the top row and the bottom row, respectively).

Input format.

1 2 0 3   (* Top *)
2 3 1 0   (* Bottom *)

Output format. The output is a drawing of the routed channel (see Figure 3.19a).

3.6. Solve an instance of the channel routing problem employing the greedy algorithm. Use the same input and output formats as in the previous exercise.

3.7. Consider a set of modules in an FPGA environment. Find a placement and routing. The main objective is to (find a routing and to) minimize the total length of the nets.

Input format. See Figure 3.28.

16   (* number of modules *)
4   (* number of rows and columns of cells *)
N1 2 3 ,   (* net 1 interconnects modules 2 and 3 *)
N2 1 3 5, . . .

You can place the terminals around a module in any form you want (note that due to the connection of channels your choices are limited). As shown in Figure 3.28, number of tracks in each channel is always 5 and the width of each cell is always 8.

Output format. The output format is a gate array placement and a complete routing as shown in Figure 3.32. Show all nets and write the total length of your placement.

3.8. Given a set of standard cells and a netlist, implement an efficient algorithm for placement and (global and detailed) routing of all the nets. The number of rows is given and the width of each row is a given number. The goal is to route all nets minimizing the height of the chip. Note that in each cell the same terminal appears on both the top side and the bottom side of the cell, and there are no terminals on the left and the right sides of the cell.

Input format.

```
3 7 200
1 3 0 4, ... ,
0 2 3 1 0 0
```

The above example contains a number of cells to be placed in three rows, each cell row has height 7, and the total number of columns is 200.

Two of the modules are shown, the first and the last one. The first module occupies four columns: net 1 has a terminal at the first column, net 3 has a terminal at the second column, the third column is empty, and net 4 has a terminal at the fourth column. Note that two modules should be placed at least one unit apart.

Output format. Show a placement (containing the name of the modules) and a complete routing of the nets. Indicate the height of your result (including the height of the cells and the total number of tracks used for routing). For example, your output can be similar to the one shown in Figure CE3.8.

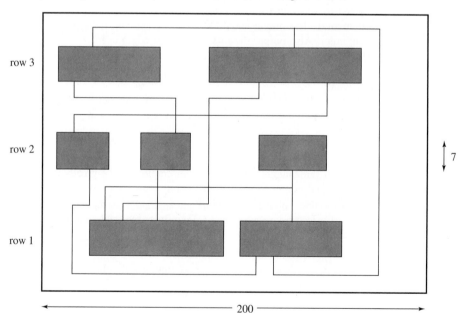

Number of tracks = 21 (cells) + 6 (routing) = 27

FIGURE CE3.8
Output format, placement and routing of standard cells.

CHAPTER
4

PERFORMANCE ISSUES IN CIRCUIT LAYOUT

Recent advances in VLSI technology allow the fabrication of millions of transistors in a single chip. As a result, interconnection delay has become increasingly significant in determining the circuit speed. Interconnection delay contributes up to 70% of the clock cycle in the design of high-performance circuits.

Therefore, performance-driven layout design has received increasing attention in the past several years. Earlier results were on the timing-driven placement problem, where a number of techniques were developed for placing blocks or cells in timing-critical paths closer together. More recently, timing-driven routing has also been studied.

This chapter discusses delay models and the timing-driven placement problem. Then the timing-driven routing and the clock skew problems are covered. Finally, the via minimization problem and the power minimization problem, along with other performance issues in circuit layout, are considered.

4.1 DELAY MODELS

Delay modeling of ICs can be done in several ways, resulting in trade-offs between computation times and analysis accuracy. Consider the following timing analysis

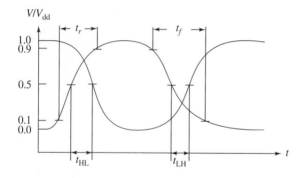

FIGURE 4.1
Definition of rise, fall, and delay times. Rise and fall times are t_r and t_f, respectively. The propagation delays $t_{50\%}$ are the output transition from high to low and the output transition from low to high, t_{HL} and t_{LH}, respectively.

methods:

- *Analog simulation* uses a simulator such as SPICE [257]. These simulators require models of the I-V characteristics of the devices. The analysis is done using numerical integration methods. Worse case delay analysis requires finding worse case input patterns. These methods are very time-consuming, and thus timing analysis using these methods is all but impractical, except for special problems.

- *Timing simulation* uses models of circuit and interconnect delays. As an example, consider the Crystal timing simulator [268]. In this case simple models of gates and interconnections are used, and thus timing analysis can be done efficiently on large circuits. However, these methods pay a price in terms of accuracy.

For the purposes of this chapter, only the timing simulation method of timing analysis will be considered. In the following sections, simple gate and interconnection delay models will be described. These models are intended to be used with other VLSI design tasks such as timing-driven placement and routing.

Before continuing, some assumptions about the signal waveforms and about the delay definitions must be stated. In static logic circuits, delay is measured between the 50% point of the waveforms, as shown in Figure 4.1. This choice assumes that the switching points of all gates are approximately at $V_{DD}/2$ [13]. The time from the 10% point of a waveform to the 90% point is called the *rise time* t_r. The time from the 90% to the 10% point is referred to as the *fall time* t_f. Here the term rise time will be discussed. The fall time is handled symmetrically.

4.1.1 Gate Delay Models

The simplest gate delay model sets a fixed propagation time, or gate delay T_D, from a gate input to a gate output. This model is useful particularly with combinatorial optimization algorithms such as technology mapping, and in simple logic simulators that are used early in the design process. However, this model is very inaccurate as it neglects many important influences on gate delay such as:

- **Output loading**. As capacitive loads caused by the input capacitance of the fanout gates increase, gate delay increases.

- **Wire loading.** As capacitive loads caused by the wiring network driven by the gate increase, gate delay increases.

- **Input waveform shape**. Slow input signals (signals with large rise times) cause long gate delays.

A more detailed model of gate delay is the on-resistance model shown in Figure 4.2. This model consists of the following components: the input gate loading or gate capacitance C_g, the output gate loading capacitance C_d, the intrinsic gate delay T_D, and the gate output on-resistance R_{on}. The function of a gate can be described as follows. Suppose the input patterns of the gate change, causing a change in the output of the signal. Suppose the latest change occurs at some time t. Then the gate output begins changing at time $t + T_D$. Suppose the gate is driving a load capacitance C_L, then the rise time constant of the output waveform is $\tau = R_{on}(C_d + C_L)$. Given some assumptions about the shape of the waveforms (exponential or ramp), these figures can then be used to compute gate delays.

Although inaccurate (the accuracy of this model is around 20%), the on-resistance model can describe many properties of physical designs. For example, the effects of increased capacitive loading are clear. Also, as shown in following sections, this model is suitable for efficient delay modeling of lossy

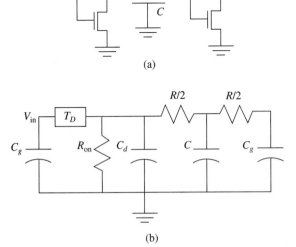

(a)

(b)

FIGURE 4.2
(a) Schematic for a simple CMOS inverter driving a distributed RC wiring load and a single gate load. (b) On-resistance buffer and interconnection delay model, the distributed RC wiring is modeled with lumped components.

interconnections. Additionally, this model embodies the effects of transistor sizing, since the on-resistance is a function of the transistor width and length,

$$R_{\text{on}} \approx \frac{1}{\mu C_{\text{gox}}(V_{\text{DD}} - V_T)} \frac{L}{W},$$

where μ is the conductivity constant and C_{gox} is the thickness of the gate's dielectric oxide.

The delay equation resulting from the on-resistance model can be shown to be a polynomial that is formally defined as $x = \{x_1, x_2, \ldots, x_n\}$, $f(x) = \sum_{j=1}^{k} C_j \prod_{i=1}^{n} x^i$ such that $C_j > 0$ [46, 89], which can be transformed into convex equations. Thus the on-resistance model is also suitable for transistor sizing. Notice that the on-resistance model assumes that the gate being timed is an inverter! It turns out that any gate can be transformed into an equivalent inverter for the purposes of simulation by assuming that the delay is dominated by the switching of a single input. Thus a worst case delay analysis of a gate requires two steps:

1. Find the longest delay path in the circuit. Currently this problem has not been solved exactly. However, as an approximation, the longest resistive path is computed. This is the path from the output of the gate to the power supply (or ground) with most resistance. Although the longest resistive path problem is NP-complete [96], polynomial solutions are known for practical circuits [368].

2. Given the latest arriving signal, convert the longest resistive path into an equivalent single on-resistance model.

Unfortunately, the on-resistance model neglects the effects of waveform shapes and output load on the gate-delay model, which can result in considerable loss of accuracy. Waveform shape can be parameterized, to first order, by the rise time, and the output load is the load capacitance. A more accurate model is as follows:

- The gate delay is a function of the input rise time of the driving signal and the output load capacitance: $T_D(t_{\text{in}}, C_L)$.

- The gate output rise time is also a function of the input rise time and the output load capacitance: $t_{\text{out}}(t_{\text{in}}, C_L)$.

- The gate on-resistance can be computed from the output rise time function: $R_{\text{on}} = t_{\text{out}}(t_{\text{in}}, C_L)/C_L$.

The equations for the gate delay and output rise time are empirical and are obtained by simulations and curve fitting. Timing analysis with these equations can be considerably more accurate. A comparison of the accuracy of several timing techniques can be found in [263]. Unfortunately, these equations are considerably more difficult to handle for tasks such as transistor sizing and timing optimizations.

4.1.2 Models for Interconnect Delay

Figure 4.3 shows a three-dimensional view of a wire and an electrical model including several parasitic effects. Interconnect delay is a result of these parasitic effects:

- Interconnect capacitance C_w, which arises from three components: the area component, also referred to as parallel plate capacitance; the fringing field component that accounts for the finite dimensions of wires; and the wire-to-wire capacitance between neighboring wires. In modern IC technologies the latter two are becoming comparable to parallel plate capacitance. Given a wire of width w, length ℓ, dielectric thickness t_{ox}, a dielectric constant ε_{ox}, and the fringing field factor, the wire capacitance is

$$C_w = \frac{\varepsilon_{\text{ox}} w \ell}{t_{\text{ox}}} f_r \ .$$

- Interconnect resistance R_w, which arises since conductors in modern ICs are not ideal. Given the resistivity of the wiring material ρ, and the thickness of the wire h, the wire resistance is

$$R_w = \frac{\rho \ell}{wh}.$$

- Other effects such as interconnect inductance L_w and shunt conductance G_w. These effects have been negligible up to date in IC technologies; however, they may become more important with decreasing dimensions.

This section considers various interconnect models that handle parasitic capacitance and resistance. The most accurate model of interconnect at high speed

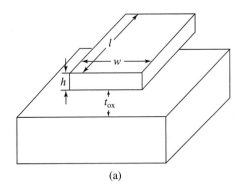

(a)

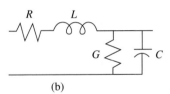

(b)

FIGURE 4.3
(a) Three-dimensional view of an IC wiring structure. The dimensions of the wire are its length ℓ, width w, thickness h, and its distance from a ground plane t_{ox}. (b) Electrical model of the parasitic effects of the wire.

is the transmission line model, where wires have capacitance, resistance, and inductance per unit length. These models are difficult to use and are unnecessarily complicated for present-day IC technologies. Transmission line analysis is only called for when the time of flight is longer than the signal rise times. This effect only occurs in the fastest of modern IC technologies such as gallium arsenide ICs. For this reason, this section focuses on two practical interconnect models, the distributed RC network model and the lumped RC network models.

The distributed line model considers the distributed nature of parasitic capacitance and resistance. The response of a distributed capacitance C_w and resistance R_w in the frequency domain is given by

$$V(s) = \frac{1}{s \cosh\sqrt{s R_w C_w}}.$$

There is no closed-form time-domain transfer for this function. A practical alternative is to approximate the distributed RC line with a lumped RC model (see Figure 4.4) such as the π and T models. The error in this approximation can be decreased by increasing the number of lumped sections for a given wire [300].

In addition to wires, other interconnect structures such as vias and terminals may have to be modeled. These structures are in reality three-dimensional interconnect structures that contribute parasitic capacitance and resistance to the networks. Depending on fabrication techniques, the contribution to the resistance of the network by these structures can be considerable. Also notice that the above formula for resistance and capacitance assumes that the wiring consists of long sections of wiring. When this is not the case, for example where thick wiring is used with many contacts or where wire width tapering occurs, the formula for parasitic resistance and capacitance may be very inaccurate. In these cases several approximations exist for special wire shapes [381], and in extreme cases finite element analysis may be required.

As a result of the previous discussion, the time-domain behavior of digital ICs using RC networks can be modeled. Since most network topologies are trees, an important and well-studied problem in timing analysis is the modeling of RC

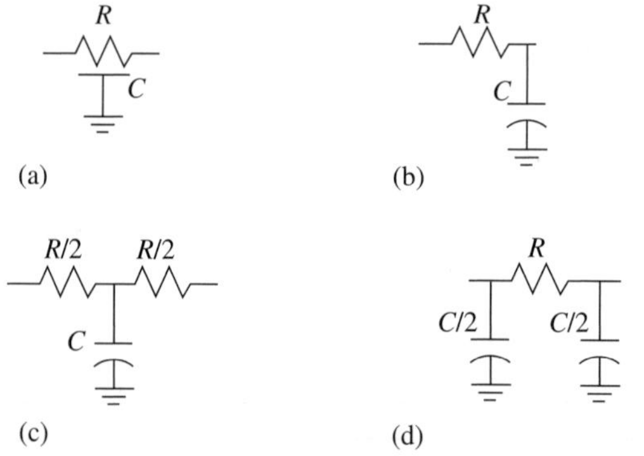

FIGURE 4.4
Lumped circuit approximations for distributed RC lines: (a) distributed RC model, (b) L model, (c) π model, (d) T model.

trees. The next sections show how this modeling can be done efficiently and how it can yield interesting results useful in other areas of VLSI design.

4.1.3 Delay in RC Trees

The general time-domain solution of any RC network is a well-known problem in numerical analysis. An RC network produces a linear system of differential equations that can be solved for an exact response [198]. In general, given a linear network with n storage elements the response of the Laplace transform (or frequency domain) of the network, $H(s) = \mathcal{L}(h(t))$ can be expressed in the form

$$H(s) = \frac{1 + a_1 s + \ldots + a_m s^m}{s(1 + b_1 s + \ldots + b_n s^n)}$$

$$H(s) = \frac{k_1}{s - p1} + \ldots + \frac{k_i}{s - p_i} + \ldots + \frac{k_n}{s - p_n},$$

which has the time-domain solution

$$h(t) = k_1 e^{p_1 t} + \ldots + k_i e^{p_i t} + \ldots + k_n e^{p_n t},$$

where p_i and k_i are the poles and residues, respectively, of the frequency domain expression. Unfortunately, the problem of computing the exact poles and residues from the frequency domain solution is a computationally expensive problem. Furthermore, it has been observed in practice that the response of practical RC networks is dominated by a few poles and residues.

As a result, several practical approaches have been proposed to analyze RC networks and trees:

- In [299] a method is proposed to bound the waveforms in an RC tree and compute a one-pole or dominant time-constant approximation of the network, an *Elmore delay*.
- In [145] this method is extended to include a second-order approximation, and the effects of nonlinearity on MOS resistors. Further extensions in [226] produce Elmore delays for any RC network including the effects of charge storage.
- In [287] the method of asymptotic waveform evaluation (AWE), a generalized kth order approximation to the dominant poles of the linear system, is proposed.

This section focuses on the computation of Elmore delays. Elmore delays are useful because of their relative simplicity, which allows them to be used in many timing-driven routing problems such as zero-skew clock tree routing and in timing-driven critical sink routing. Furthermore, the Elmore delay has been shown experimentally to be faithful to real delays [245]. Finally, as will be seen next, the Elmore delay formula gives further nontrivial insight into the effects of interconnect design on timing issues.

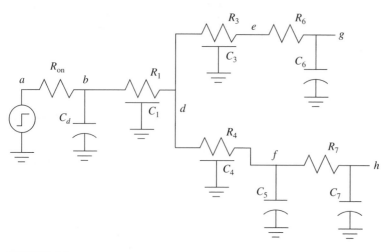

FIGURE 4.5
Example of an RC tree for a CMOS VLSI circuit with a fanout of 2, using distributed RC loads for wiring, lumped RC loads for contacts and transistor models.

First some RC tree terminology is defined. Let a tree $T(V, E)$ be a rooted RC tree, where the root is located at the driver of the tree (see Figure 4.5). A vertex $v_i \in V$ in the tree represents a voltage node in the RC tree with capacitance c_i, and an edge $e_{ij} \in E$ represents a resistance r_{ij} and connects a parent vertex v_i with a child vertex v_j. If a node v_i lies in the path from node v_j to the root (branch), then the node v_i is said to lie above (below) node v_j. The total capacitance seen under a node v_i is $C(i)$, and the total resistance seen above a node v_i is $R(i)$. Let P_i denote the set of vertices in the simple path from the root to vertex v_i. Given two vertices v_i and v_j let $a(i, j)$ denote the first vertex above both vertices in $P_i \cap P_j$. Then the Elmore delay $T_{ED}(i)$ at a node v_i is defined as

$$T_{ED}(i) = \sum_{v_k \in V} R(a(i, k))c_k = \sum_{e_{jk}, v_k \in P(i)} r_{jk} C(k).$$

Note that in using the Elmore delay model the delay contribution of an RC circuit driven by a gate is

$$T_{GATE}(i) = T_D + R_{on}(C_d + C_L) + \sum_{e_{jk}, v_k \in P(i)} r_{jk} C(k).$$

Several observations can be derived from this formula. First, if the wiring resistance is negligible, then to minimize delay one must minimize total wire length. However, if the contribution of the wire resistance is significant, then it might be more important to make short paths between the terminals of the tree and the sink [71]. This idea gives rise to important problems in timing-driven routing.

Second, consider the effect of widening a wire. A simple approximation of the time constant for a wire gives

$$t_r = R_w C_w = \frac{\varepsilon_{ox} w \ell}{t_{ox}} \frac{\rho \ell}{wh} f_r = \frac{\varepsilon_{ox} \rho \ell^2}{t_{ox} h} f_r.$$

According to this equation, wire widening has no effect on the delay! Now consider the Elmore delay of the same wire using the π model and a capacitance load C_L,

$$T_{\text{ELM}} = \frac{1}{2} R_w C_w + R_w C_L = \frac{1}{2} \frac{\varepsilon_{\text{ox}} \rho \ell^2}{t_{\text{ox}} h} f_r + \frac{\rho \ell}{wh} C_L.$$

Notice that in the first equation the width of the wire did not enter into the computation of the delay. However, in looking at the Elmore delay, it is clear that by increasing the width it may be possible to decrease the delay of the wire. These insights have made the Elmore delay a valuable tool in VLSI design.

Next timing-driven placement and routing problems will be considered using the models discussed in this section.

4.2 TIMING-DRIVEN PLACEMENT

Timing-driven placement incorporates timing objective functions into the placement problem. Depending on how important the timing is, designers can first try to fit the timing requirements and then minimize the area, or vice versa. Nets that must satisfy timing requirements are called *critical nets*. The criticality of a net is dictated by input and output timing requirements. In timing-driven placement, one tries to make critical nets timing-efficient and other nets length- and area-efficient. Timing can be represented by clock period T,

$$T = S + d_{\max} + T_0, \tag{4.1}$$

where S, $d_{\max}$, and T_0 represent clock skew, maximum delay, and the adjustment factor, respectively.

Placement is a very important step in designing timing-efficient circuits. If a placement is bad in terms of timing, even a very powerful router cannot help. There are basically three approaches to the timing-driven placement problem: the zero-slack algorithm, the weight-based placement, and the linear programming approach.

4.2.1 Zero-Slack Algorithm

Consider a circuit without any loops, represented by a directed acyclic graph (DAG). If a circuit has loops, it is transformed into one without loops by removing a small number of edges. These edges will later be placed back in the circuit and the timing is verified. Inputs to the circuit (i.e., vertices with in-degree zero) are called *primary inputs* (PIs) and outputs of the circuit (i.e., vertices with out-degree zero) are called *primary outputs* (POs). Assume the signal-arriving time at each PI and the time a signal is required at a PO are known. Gate delays are also given. By tracing a signal path starting from the PIs, the *arrival time* t_A of the signal at each pin (i.e., at inputs and the output of each gate) can be obtained. Here, it is assumed that each gate has single output; however, the fanout of a gate could be more than one. This is equivalent to saying that all outputs of a gate carry the same signal. This model can be extended to include one in which each gate has k distinct output signals. This is accomplished by duplicating the gate k times. The input to each of the duplicates is the same as the input to the original gate. However, each duplicate will have only one of the outputs of the original gate.

Similarly, by starting from the POs and tracing back the path of each signal, the *required time* t_R at each pin is obtained. Define

$$\text{slack} = t_R - t_A. \tag{4.2}$$

Note that the arrival time at the output of a gate is known only when the arrival time at every input of that gate has been computed. Thus, a topological sorting is performed on the gates (vertices) to find the arrival times. (A topological sort is the process of visiting the gates one by one. A gate can be visited if all its inputs are PIs or when all the gates providing signal to it have already been visited.)

The relationship between the arrival time of an output pin, t_A^j, and that of the output pins of the fanin cells, $t_A^1, t_A^2, t_A^3, \ldots, t_A^m$, is expressed in Equation 4.3. Terms $d_{n_1}, d_{n_2}, \ldots, d_{n_m}$ are net delays, and D is the delay of the cell containing pin j. The relationship is also shown in Figure 4.6.

$$t_A^j = \max_{i=1}^{m}(t_A^i + d_{n_i}) + D \tag{4.3}$$

Deciding the required times is done in a similar manner, except the signal direction (i.e., the edges of the DAG) is reversed when topological sorting is applied. The relationship between the required time of the output pin, t_R^j, and that of the related output pins of the fanout cells, $t_R^1, t_R^2, t_R^3, \ldots, t_R^m$, is shown in Figure 4.7 and expressed in Equation 4.4. The terms $D_1, D_2, D_3, \ldots, D_m$ are cell delays, and d_n is the delay of the net that connects the cell to its fanout cells.

$$t_R^j = \min_{i=1}^{m}(t_R^i - D_i) - d_n \tag{4.4}$$

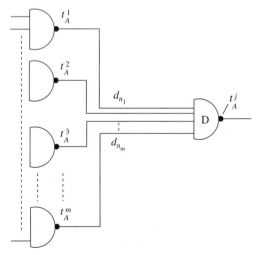

FIGURE 4.6
Computation of arrival time at an output pin.

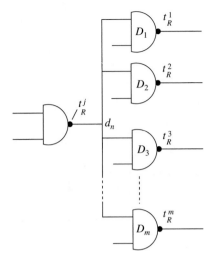

FIGURE 4.7
Computation of required time at an output pin.

The zero-slack algorithm associates a maximum allowable delay to each net in an iterative manner. In the first iteration, the delay of each net is assumed to be zero. The purpose of the zero-slack algorithm [259] is to distribute slacks to each net. It transforms timing constraints to net length constraints. Consequently, primary inputs, primary outputs, and the outputs of each cell will have zero slacks.

An example is demonstrated in Figures 4.8–4.14. A triplet t_A/slack/t_R is shown at each (input and output) pin. Figure 4.8 shows the input circuit with the delay of each net initially set to zero. The arrival times of primary inputs and the required times of primary outputs are given. Slacks are calculated according to Equations 4.2–4.4. The minimum positive slack is four and the corresponding

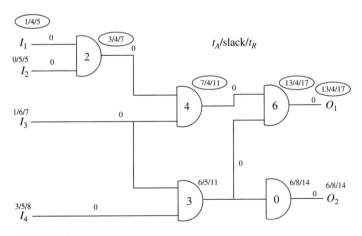

FIGURE 4.8
The zero-slack algorithm example, step 1.

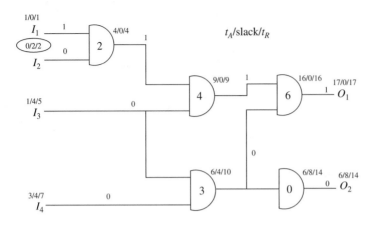

FIGURE 4.9
The zero-slack algorithm example, step 2.

path segment is indicated by the circled triplets—if there is more than one such path, then one is arbitrarily selected. In Figure 4.9, the four units of slack have been distributed along the path segment of Figure 4.8. It indicates that each of the segments on the path can have length proportional to the assigned length (in this case, 1) in the final layout. A unit-delay model is assumed; that is, the propagation delay along a wire of length ℓ is $O(\ell)$. After slack distribution, slacks on the path segment become zero. Then the new slacks are calculated. The new path segment is primary input I_2 and the net connected to it. Figures 4.10–4.14 demonstrate subsequent steps. Figure 4.14 shows the final result, where the output pin slack of each gate is zero. The just described algorithm is summarized as follows:

> *procedure* zero-slack
>   *begin-1*
>   *repeat until* no positive slack;
>   compute all slacks;
>   find the minimum positive slack;
>   find a path with all the slacks equal to the minimum slack;
>   distribute the slacks along the path segment;
>   *end-1.*

Theorem 4.1. The zero-slack algorithm assigns a delay to each net so that the resulting circuit has zero slack on every net.

Proof. Consider an intermediate step of the algorithm. Before this step, assume each pin has a non-negative slack. This is certainly the case before the algorithm starts; otherwise, a solution does not exist. Find a path P with the smallest positive slack. Let v_0 be the starting point of this path and v_f be the ending point of this path. v_0 is

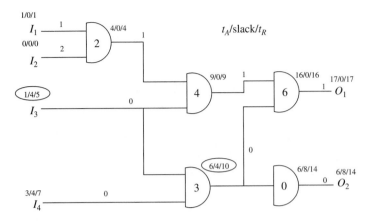

FIGURE 4.10
The zero-slack algorithm example, step 3.

an output pin (a primary input can be considered as output of another circuit) and v_f is an input pin (again, primary output can be considered as input of another circuit). The distribution of s units of slack along P may decrease the slack of many pins between v_0 and v_f; these pins may or may not be along P. First, the slack of pins with zero slack will not change. Also, since this decrease is by at most s units, and s was the smallest positive slack, all pins will have non-negative slacks. Furthermore, since at each step the slack of at least one pin is decreased by at least one unit, the process will terminate.

An advantage of the zero-slack-based algorithms is that it is easy to calculate slacks (i.e., the algorithm is simple). Furthermore, after slacks are computed, any traditional placement algorithm can be used to produce a timing-driven

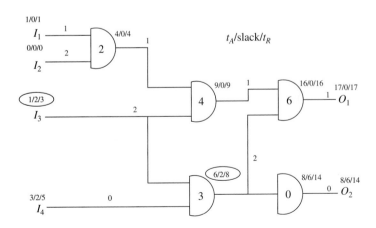

FIGURE 4.11
The zero-slack algorithm example, step 4.

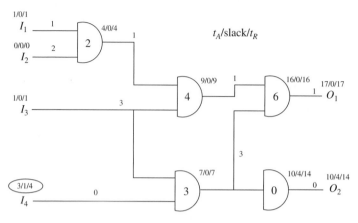

FIGURE 4.12
The zero-slack algorithm example, step 5.

placement. Indeed, weights on the nets are set inversely proportional to slacks: the smaller the slack of a net, the more critical it is to separate the terminals of it. An example of a placement corresponding to a given slack assignment is shown in Figure 4.15. Note that modules interconnected with nets having small slack (slack 1) are placed close to each other. Such paths with small slacks are called *timing critical paths* or, simply, critical paths. A main disadvantage of this approach is that it does not provide hints on how to use the slack distribution for placement in an iterative manner. Zero-slack algorithms have been widely adopted to generate timing constraints for nets as, for example, in [358].

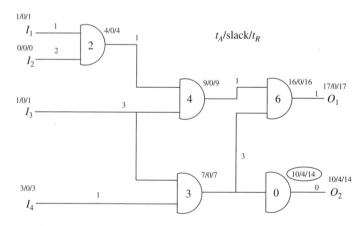

FIGURE 4.13
The zero-slack algorithm example, step 6.

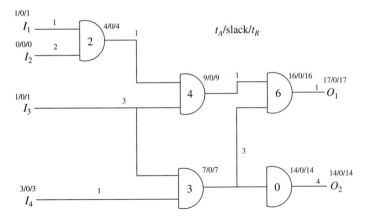

FIGURE 4.14
The zero-slack algorithm example, step 7.

4.2.2 Weight-based Placement

Most of the existing placement algorithms are designed to minimize certain cost functions (see Chapter 2). Therefore, if weights derived from timing considerations are incorporated into the cost function, a timing-driven placement algorithm is obtained (as described in the previous subsection). Two approaches for determining weights will be described. Motivated by that, a timing-driven placement technique can be summarized as follows [87].

Step 1. Find a layout using a placement algorithm that incorporates weights.

Step 2. Perform a timing analysis.

Step 3. Assign weights to critical paths or nets.

Step 4. Go to step 1, using the new weights.

The approach is widely used in hierarchical placement algorithms. The reason is that partitioning is a central step there and it is very easy to consider weights in the partitioning step. A disadvantage of the technique is that it is hard to predict the result of changing weights as new critical paths or nets may be created in subsequent steps. Moreover, the iteration behavior becomes uncontrollable, especially in aggressive designs, where most paths are critical.

One approach to deciding appropriate net weights is to use a measure of node criticality (where only the capacitive delay of the nets is considered) or measures of edge criticality (where the RC delay of the nets is considered) as weights. In [225], criticality was used as a guide to select cells from a cell library. (In a cell library, there are cells with different driving power and area that implement the

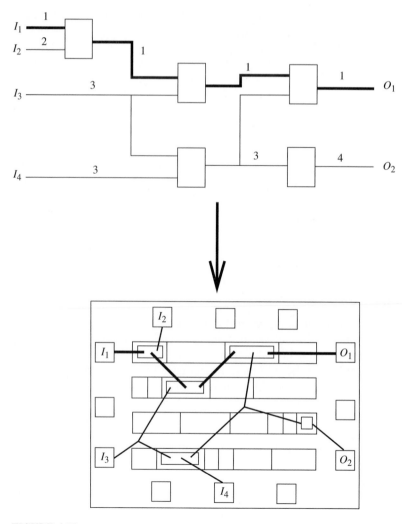

FIGURE 4.15
A placement corresponding to a given slack distribution.

same function.) This criticality can be used as a weighting function. For a module i, the criticality is estimated as

$$W_i^c = \frac{R_{out}^i * C_L}{S(i)/l(i)},$$
(4.5)

where R_{out}^i is the output resistance of module i, $S(i)$ is the slack of module i and its output net (assuming one output for each module), C_L is the loading capacitance, and $l(i)$ is the length of the longest path from the source modules to module i. Equation 4.5 implies that more weight is put on nets that have less slack, more delay, and are far from source modules. Note that the modules far from source modules have a better chance to affect slacks of others.

However, in the above formulation, the weight and the final satisfaction of timing constraints are not clearly correlated. Tsay and Koehl [358] described this relation as follows. They used the total squared wire length as the cost function P_c. The cost minimization problem is subject to timing constraints on nets. It can be formulated as

$$P_c = \min\left\{ L = \frac{1}{2} \sum_{i,j} c_{ij} l_{ij}^2 \mid l_{ij}^2 \leq u_{ij}^2, \forall n_{ij} \in \mathcal{N} \right\}, \tag{4.6}$$

where l_{ij} is the wire length between module i and j, c_{ij} is the connectivity between the two modules, n_{ij} is the set of nets between them, u_{ij} is a transformed length constraint from timing constraint, and $\mathcal{N}$ is the collection of all nets. It was shown that the optimum net weighting increase for the placement optimization is exactly the optimum Lagrange multiplier that satisfies the sufficient and necessary Kuhn-Tucker conditions [232]. Their condition is for optimality of the performance-constrained placement optimization problem.

Let $\mathcal{A}$ denote the index set of active constraints, that is, the set of nets n_{ij} that have the value of u_{ij} at the optimum solution. Then the Kuhn-Tucker conditions for optimizing the problem P_c require the following properties:

1. The Hessian matrix

$$\nabla^2 \left[\frac{1}{2} \sum_{ij} (c_{ij} + \lambda_{ij}) l_{ij}^2 \right] \tag{4.7}$$

   to be positive semidefinitive.

2. The gradient

$$\nabla \left[\frac{1}{2} \sum_{ij} (c_{ij} + \lambda_{ij}) l_{ij}^2 \right] \tag{4.8}$$

   to be equal to zero.

3. The Lagrange multiplier λ_{ij} to be non-negative for any active constraint, and to be zero otherwise. That is:

$$l_{ij}^2 = u_{ij}^2, \quad \lambda_{ij} \geq 0, \quad n_{ij} \in \mathcal{A} \tag{4.9}$$

$$l_{ij}^2 < u_{ij}^2, \quad \lambda_{ij} = 0, \quad n_{ij} \notin \mathcal{A} \tag{4.10}$$

From these conditions, it can be seen that Lagrange multiplier λ_{ij} appears in the same place in Equations 4.7 and 4.8 as connectivity c_{ij}, where weight is added. If $c'_{ij} = c_{ij} + \lambda_{ij}$ is used to substitute c_{ij}, no n_{ij} will be in $\mathcal{A}$. It means no constraint is needed since l_{ij} will be within a constrained range using a nonconstrained optimization algorithm. The calculation of Lagrange multiplier λ_{ij} is not an easy task. However, there is an approximation algorithm providing a simpler form of λ_{ij} that can provide more insight. Its result is

$$\lambda_{ij} = \frac{(l_{ij}^* - u_{ij})/u_{ij}}{1/a_{ii} + 1/a_{jj} - 2c_{ij}/(a_{ii}a_{jj})}, \tag{4.11}$$

where l_{ij}^* is the optimal solution with constraint, and a_{ii} is the total connectivity of all nets connecting to module i. From the denominator, we can see that it requires more weighting increase of the same percentage of length as a net whose module at either end has a heavier connection (large a_{ij}/a_{ii} or large a_{ij}/a_{jj}). This approximation algorithm produces an iterative method.

Step 1. Set $k = 0$, and initialize weight $c_{ij}^{(k)} = c_{ij}$.

Step 2. Solve the unconstrained optimization problem using $c_{ij}^{(k)}$ as weight.

Step 3. Compute $\lambda_{ij}^{(k)}$. If each $\lambda_{ij}^{(k)}$ is nonpositive, then stop and output the current placement. (Each net is within the time constraint range.)

Step 4. Compute a new weight $c_{ij}^{(k+1)} = c_{ij}^{(k)} + \lambda_{ij}^{(k)}$. If $c_{ij}^{(k+1)} \leq c_{ij}^{(0)}$, then let $c_{ij}^{(k+1)} = c_{ij}^{(0)}$. Go to step 2.

Again, the weight-based approaches work better in nonaggressive designs, where the percentage of critical paths (or nets) is small.

4.2.3 Linear Programming Approach

A linear programming- (LP)-based performance-driven placement algorithm was proposed in [165]. It is a hierarchical technique in which linear programming is an important step (at each level of hierarchy an LP is solved). The basic idea is to express spatial and timing constraints using a set of linear inequalities and to try to minimize a cost function that combines timing and area. Also, timing constraint information is considered by path delays instead of net delays. This expands the solution space being searched.

It is assumed that each cell has only one output pin, the delay of a net is proportional to the length of its bounding box (it treats wires as capacitors), and there is a single output resistance for all cells.

Let t_i, b_i, l_i, and r_i denote the top, bottom, left, and right coordinates of the bounding box of net n_i, respectively. For each pin p_j in the net n_i, with x-coordinate x_{ij} and y-coordinate y_{ij}, the spatial constraints are:

$$t_i \geq y_{ij} \tag{4.12}$$

$$b_i \leq y_{ij} \tag{4.13}$$

$$r_i \geq x_{ij} \tag{4.14}$$

$$l_i \leq x_{ij}. \tag{4.15}$$

The delay d_{n_i} induced by the net n_i is approximately

$$d_{n_i} = K_v(r_i - l_i) + K_h(t_i - b_i), \tag{4.16}$$

where both K_v and K_h are constants dictated by fabrication technology. To center the cells within the region in which they will be placed, constraints are introduced into the LP formulation. The first-order linear slot constraint [50] is used to force the center of mass of all cells in the region to be equal to the center of the region. In the x direction the center is

$$\bar{x} = \frac{\sum_{i=1}^{c} m_i X_i}{\sum_{i=1}^{c} m_i}, \tag{4.17}$$

where c denotes the number of cells, $\bar{x}$ is the center of the region in the x direction, X_i is the position of cell i, and m_i is the length of cell i in the x direction. The y direction is done similarly.

The relationship of input pins of adjacent cells is described by

$$\forall_{j\in \text{ fanin of } i} \quad t_A^i \geq (t_A^j + d_{n_j}) + D_i, \tag{4.18}$$

where t_A^i is the arrival time at the output pin of cell i and the index j is an output pin of the fanin cells of the cell to which i belongs. D_i is the cell delay of cell i. See Equation 4.18 and Figure 4.6. For primary outputs, the relationship differs by the addition of a variable M, representing the minimum slack of all the cells, and the required time constraints.

$$\forall_{j\in \text{ fanin of } i} \quad t_R^i \geq (t_A^j + d_{n_j}) + D_i + M, \tag{4.19}$$

where t_R^i denotes the required time of the primary output i. Note that both primary inputs and primary outputs can be treated as cells. Each I/O pad acts as a cell with a fixed position. The cost function is

$$\max(M - \alpha W), \tag{4.20}$$

where

$$\alpha = \frac{\kappa \bar{R}}{|N|}. \tag{4.21}$$

Here κ is a constant specified by the designer, $\bar{R}$ is the average output cell resistance in the circuit, and $|N|$ is the number of nets. W is the total circuit capacitance and therefore is a measure of the weighted sum of interconnections. This cost function is minimized when the longest path delay is small and the total wire length is shortened.

However, after linear programming is applied to all cells, the cells may overlap each other or cluster in some places. Linear programming can be performed in a hierarchical manner to minimize the cost function. Levels of hierarchy are defined by recursive partition until each cell is in a distinct slot. Each partition will divide the target region into two and move suitable numbers of cells, which are in a crowded region, to another region. The likelihood of this move depends on the distance between the two regions. At the next level, linear programming is applied to the cells that are located in the half-regions assigned at the previous level of hierarchy. The procedure is summarized formally as follows.

> *procedure* performance-driven placement (circuit specification)
> > *begin-1*
> > > apply linear programming to place the cells;
> > > > (* the placement may have overlaps *);
> > > partition the current region into two subregions:
> > > > left and right, or bottom and top, alternatively;
> > > evenly redistribute the cells;
> > > *call* performance-driven placement (left or top);
> > > *call* performance-driven placement (right or bottom);
> > *end-1.*

An advantage of this technique is that almost any constraint can be incorporated into the formulation. The main disadvantage of the approach is that it is time-consuming.

4.2.4 Discussion

Three basic timing-driven placement algorithms have been described in this section. Several other approaches, some being very similar to the three described here, have been proposed. For a history of the timing-driven placement problem, see [241].

Constructive approaches transform the timing information into the geometric domain [220]. The primary inputs and primary outputs are placed first; then the most critical path is placed and its length is minimized by laying it as straight as possible. At each stage, one focuses on one part of a circuit, called a window. A *window* represents a region in which all the modules along a given path can be placed without degrading the circuit performance. Another critical path (typically, the most critical one) is considered next. This approach places paths one by one.

The adaptive control approach [346] is also a constructive placement algorithm. The procedure is as follows. The algorithm chooses a critical module according to an adaptive look-ahead procedure, which decides what critical is. A module with large connectivity and delay will be chosen. Then it is placed according to an analysis of delay and routability by an incremental global router. This is done until all modules are placed.

Constructive algorithms are flexible and allow preferred positions for blocks, but they need precise timing-analysis tools. The main disadvantage of constructive techniques is the lack of a global view.

Gao, Vaidya, and Liu [111] used a convex program to derive a set of upper bounds on the nets' wire lengths. They modified the min-cut algorithm to reduce the number of nets that are over their upper bounds at each hierarchical level. The main feature is that if the final result cannot fit the current set of upper bounds, an iterative method is employed to modify the set of upper bounds obtained from previous placements.

Srinivasan [339] proposed an algorithm with a quadratic cost function subject to path time constraints. It incorporates Kuhn-Tucker conditions and, for example, it places a circuit with 1418 cells within 7 minutes. In [223], fuzzy logic is applied to optimize the placement problem. Multiple objectives such as area efficiency, routability, and timing were considered simultaneously and balanced by fuzzy logic algorithms.

4.3 TIMING-DRIVEN ROUTING

Timing-driven placement has been given much attention because it is of fundamental importance. That is, if attention is not given to timing issues at the placement level, not much can be done in the routing phase. However, timing-driven routing can also improve circuit performance. Earlier work in timing-driven routing is as follows. In [87], net priorities were determined based on static timing analysis, that is, nets with high priorities are processed earlier using fewer feedthroughs. In [166], a hierarchical approach to timing-driven routing was outlined. However, these results do not provide a general formulation of the timing-driven global routing problem. Moreover, their solutions are not flexible enough to provide a trade-off between the interconnection delay and the routing cost (e.g., length of the routing trees).

Linear delay models assume that delay along a wire of length ℓ is $O(\ell)$. While less accurate than the distributed RC tree delay model, the linear delay model has been effectively used in clock tree synthesis. A more accurate delay model is the Elmore delay model, in which the delay calculation is recursive and somewhat involved.

An instance of timing-driven routing consists of a set of points in a two-dimensional plane. One of the points is the source and the rest are the sinks. The distances are measured in L_1 metric or rectilinear metric, the metric commonly used in VLSI environments. Most algorithms can be adapted in other geometries, such as 45° geometry (allowing horizontal, vertical, and plus and minus 45° lines) and Euclidean geometry (allowing arbitrary angles).

The next two subsections focus on timing-driven interconnections, in particular, delay minimization, clock routing, and the clock skew problem, that is, the problem of asymmetric clock distribution.

4.3.1 Delay Minimization

Delay minimization can be used to obtain an effective timing-driven global router. The procedure can also be used as a detailed router.

Given a (routing) region Ψ and a point set (a set of terminals) P interconnected as a net N, a shortest path in Ψ between two terminals $x, y \in N$, denoted by minpath$_\Psi(x, y)$, is the path connecting x and y with a minimum total length. The length of minpath$_\Psi(x, y)$ is denoted by dist$_\Psi(x, y)$. The length of a set $\mathcal{P}$ of paths is denoted by dist$_\Psi(\mathcal{P})$.

Let the radius R of a signal net be the length of a shortest path from the source s to the furthest sink, that is, $R = \max_{x \in N} \text{dist}(s, x)$. The radius of a routing tree T, denoted by $r(T)$, is the length of a shortest path in T from the source s to the furthest sink. Clearly, $r(T) \geq R$, for any routing tree T. According to the linear RC delay model, the interconnection delay of a net is minimized by minimizing the radius of the routing tree, that is, the maximum interconnection delay between the source and the sinks. Furthermore, the goal is also a routing tree with small total cost. Here, the total length of the tree is used as the cost. Other cost measures might be a measure of the crowdedness of the channels (see Chapter 3) or the number of bends, proportional to the number of vias, in the net.

In order to consider both the radius and the total length in the routing tree construction, *bounded radius minimum routing trees* (BRMRT) are used. The reason for minimizing the length is that it has been observed that global routers aiming to minimize wire length produce small chips. Given a parameter $\epsilon \geq 0$ and a signal net with radius R, a BRMRT is a minimum-length routing tree T with radius $r(T) \leq (1 + \epsilon) \times R$.

The parameter ϵ controls the trade-off between the radius and the length of the tree. When $\epsilon = 0$, the radius of the routing tree is minimized and thus an SPT (shortest path tree) for the signal net is obtained. On the other hand, when $\epsilon = \infty$, the total length of the tree is minimized and an MST (minimum spanning tree) is obtained. In general, as ϵ grows, there is less restriction on the radius, allowing further reduction in tree cost. Figure 4.16 shows an example where three distinct spanning trees are obtained using different values of ϵ.

An algorithm for finding a bounded radius spanning tree (BRST) of a net N was proposed in [70]. The algorithm shows that for a given value of a parameter ϵ, the radius is at most $(1 + \epsilon) \times R$ and the total length is at most $(1 + \frac{2}{\epsilon})$ times the minimum spanning tree length.

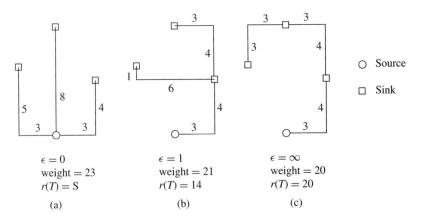

FIGURE 4.16
Three different spanning trees with distinct ϵ values.

To find a BRST of a net N, find a graph Q that has both a small total length and small radius. Thus, an SPT of Q will also have a small length and radius, and corresponds to a good routing solution. The algorithm is described below (note that the algorithm works in general graphs).

Step 1. Compute the minimum spanning tree MST_P of the given point set P. Also, initialize the graph Q to MST_P.

Step 2. Let L be the sequence of vertices corresponding to a depth-first tour of MST_P, where each edge of MST_P is traversed exactly twice (see Figure 4.17). The total (edge) length of this tour is twice that of MST_P.

Step 3. Traverse L while keeping track of the total S of the traversed edge weights. As this traversal reaches each node x_i, check whether S is greater than $\epsilon \times \text{dist}(s, x_i)$ (this is the rectilinear distance). If so, then reset S to 0 and merge $\text{minpath}(s, x_i)$ into Q. Continue traversing L until every node in the tree is visited.

Step 4. The final routing tree is SPT_Q, the shortest path tree of Q.

A formal description of the algorithm is given in procedure LAYOUT-BRST.

procedure LAYOUT-BRST(N, ϵ)
begin-1 compute MST_P;
    $E' = $ edges of MST_P;
    $Q = (V, E')$ $(* V = N *)$;
    $L = $ depth-first tour of MST_P;
    $S = 0$;

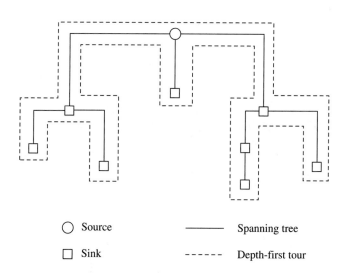

  ◯ Source   ——— Spanning tree

  □ Sink   - - - - - Depth-first tour

FIGURE 4.17
A spanning tree and a depth-first tour.

> *for* $i = 1$ *to* $|L| - 1$ *do*
>> *begin-2*
>>> $S = S + dist(x_i, x_{i+1})$;
>>> *if* $S \geq \epsilon \times dist(s, x_{i+1})$ *then*
>>>> *begin-3*
>>>>> $E' = E' \cup minpath(s, x_{i+1})$;
>>>>> $S = 0$;
>>>> *end-3*;
>>> *end-2*;
>> $T = $ shortest path tree of Q;
> *end-1*.

The following theorem shows that for any fixed ϵ this algorithm produces a routing tree with a radius and total length simultaneously bounded by small constants times optimal.

Theorem 4.2. For any point set P in a routing region and parameter ϵ, the routing tree T constructed by procedure LAYOUT-BRST has radius $r(T) \leq (1+\epsilon) \times R$ and length $(T) \leq (1 + \frac{2}{\epsilon})$ length (MST_P).

Proof. Let $v_1, v_2, \ldots, v_m$ be the set of nodes to which the procedure LAYOUT-BRST added shortest paths from the source node (the source is denoted by v_0). For any $v \in V$, let v_{i-1} be the last node before v on MST_{G_P} traversal L for which minpath (s, v_{i-1}) was added to E' in the algorithm. By the construction of the algorithm, it is known that dist $(v_{i-1}, v) \leq \epsilon \times R$. Then:

$$dist_T(s, v) \leq dist_T(s, v_{i-1}) + dist_L(v_{i-1}, v)$$

$$\leq dist(s, v_{i-1}) + \epsilon \times R$$

$$\leq R + \epsilon R = (1 + \epsilon) \times R.$$

The next focus is the total length of the constructed tree. Let $v_0 = s$, $\mathcal{P} = \{P_1, P_2, \ldots, P_m\}$, where P_i is the path from v_{i-1} to v_i. Since length$(\mathcal{P}) = \sum_{i=1}^{m} dist(s, v_i)$,

$$\text{length}(T) \leq \text{length}(MST_P) + \text{length}(\mathcal{P})$$

$$= \text{length}(MST_P) + \sum_{i=1}^{m} dist(s, v_i).$$

T is a subtree of the union of the MST and the added shortest paths. By the algorithm construction, $dist(v_{i-1}, v_i) \geq \epsilon \times dist(s, v_i)$, and so

$$\text{length}(T) \leq \text{length}(MST_P) + \sum_{i=1}^{m} \frac{1}{\epsilon} \times dist(v_{i-1}, v_i)$$

$$\leq \text{length}(MST_P) + \frac{1}{\epsilon} \times dist(L).$$

Since $\text{dist}(L) \leq 2 \times \text{dist}(\text{MST}_P)$,

$$\text{length}(T) \leq \text{length}(\text{MST}_P) + \frac{2}{\epsilon} \times \text{length}(\text{MST}_P)$$

$$= \left(1 + \frac{2}{\epsilon}\right) \times \text{length}(\text{MST}_P).$$

An example of LAYOUT-BRST is given in Figure 4.18. Observe that if node x is connected to v_1, then the radius of the tree becomes too large. Thus, v_1 is connected to the source directly, certainly, this decision depends on the value ϵ. The same situation occurs at v_2.

One approach for obtaining bounded radius Steiner trees is to employ traditional MST-based Steiner tree constructions; that is, to modify the MST obtained in procedure LAYOUT-BRST. There is a more effective way to accomplish this task. Introduce Steiner points on the tour L whenever $s = 2\epsilon \times \text{dist}(s, L_{i+1})$, instead of when $s = \epsilon \times \text{dist}(s, L_{i+1})$. From each of these Steiner points construct shortest paths to the source and add them to Q, as in the original algorithm. The advantage of using 2ϵ rather than ϵ is that every such new Steiner point can service terminals on either side of it in the traversal of L. Thus the same bound is maintained on path-lengths from the source, while adding only half as many shortest paths. This removes the factor of 2 from the $\frac{2}{\epsilon}$ ratio in the routing length bound of Theorem 4.2, yielding an improved routing length bound of $(1 + \frac{1}{\epsilon}) \times \text{length}(\text{MST}_P)$. Moreover, as is well-known, the total length of MST is at most two times of the total weight of a rectilinear Steiner tree (RST).

Theorem 4.3. Given a set of terminals N and a real number ϵ, the proposed algorithm will produce a rectilinear Steiner tree T with $r(T)$ bounded by $(1 + \epsilon)$ times the optimal radius, and with length bounded by $2 \times (1 + \frac{1}{\epsilon})$ times the optimal length.

It might be the case that different sinks have distinct delay requirements, depending on the type of terminals they are connected to. It is possible to have

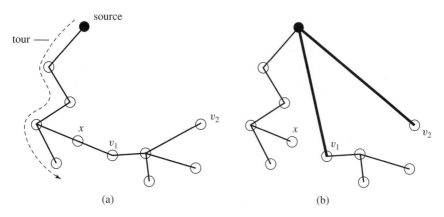

(a) (b)

FIGURE 4.18
An example of LAYOUT-BRST. (a) Input points and MST; (b) a BRMST.

a different value ϵ_i at sink x_i. Basically, when the radius of the tree gets larger than $\epsilon_i R$ at point P_i, then a direct connection is made.

Most global routers that employ a Steiner tree construction (see Chapter 3) can make use of the bounded radius Steiner trees described here. For example, consider a maze-router-based global router that routes the nets one by one. The routing of each net can be performed by the algorithm described in this section, and experiments have verified that a reduction in chip delay can be obtained in this manner (see [70]).

4.3.2   Clock Skew Problem

In present and future circuit design, speed is of fundamental importance. As discussed in the previous subsection, delay on the paths is a major factor in determining the speed. Another important factor is the clock skew that is caused by asymmetric clock distribution. Clock skew increases the total clock cycle time. Consider a clock tree consisting of a source and a set of sinks. The delay of the path with longest delay between the source and a sink is denoted by d_{max}. Similarly, the delay of the path with smallest delay between the source and a sink is denoted by d_{min}. The value $d_{max} - d_{min}$ is the *clock skew*. In most cases, clock trees with skew equal to zero (and small total length) are most desirable. See Figure 4.19, where an example of a tree with a large skew (left) and one with a small skew (right) are shown. The operation shown in the figure is called *H-flipping*.

The minimization of clock skew has been studied in the past few years. First, researchers proposed to use the H-tree construction in systolic arrays. The H-tree construction can be used when all components have equal sizes and are regularly placed. When all blocks are organized hierarchically, a clock distribution technique was proposed in [285]. It was assumed that the number of blocks at each level of hierarchy was small since an exhaustive search algorithm was employed to enumerate all the possible routes.

More recently, two recursive algorithms were proposed. They are both based on the linear delay model; recall that the delay along a path of length ℓ in a given routing tree is $O(\ell)$. First, Jackson, Srinivasan, and Kuh [167] proposed a top-down approach to the problem. Their algorithm (the JSK-algorithm) recursively partitions the problem into two equal-sized (i.e., with equal number of points) subproblems. At each step, the center of mass of the whole problem is directly

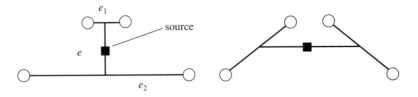

FIGURE 4.19
Transforming a clock tree with large skew to one with small skew.

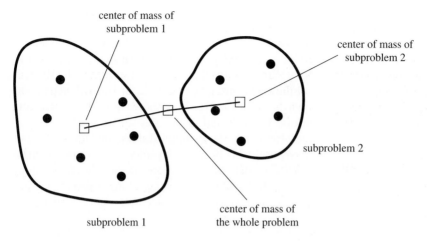

FIGURE 4.20
Example of the top-down JSK-algorithm.

connected to the center of mass of the two subproblems, as shown in Figure 4.20. The *center of mass* of a point set $\{P_1, \ldots, P_n\}$, where P_i has coordinates x_i and y_i, is defined as a point with x-coordinate $x = (x_1 + \ldots + x_n)/n$ and y-coordinate $y = (y_1 + \ldots + y_n)/n$. It is shown that clock skew $(d_{\max} - d_{\min})$ is bounded by $O\left(\frac{1}{\sqrt{n}}\right)$ for a point set uniformly distributed in the unit square, where n is the total number of points.

The second recursive algorithm was proposed by Cong, Kahng, and Robins in [69]. Their algorithm is a bottom-up technique based on recursive matching. Let $\{P_1, \ldots, P_n\}$ be the set of sinks, each being a point in a two-dimensional Euclidean plane. The set of points are matched—a minimal (length) matching has been employed as shown in Figure 4.21a. Then the centers of the just matched points, denoted by $\{P_1^1, \ldots, P_{n/2}^1\}$, are found (Figure 4.21b). Here, the *center* is

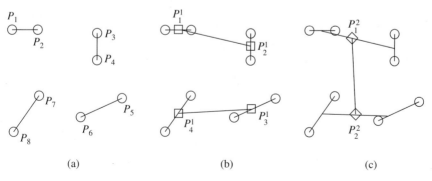

FIGURE 4.21
Recursive steps in the bottom-up matching algorithm.

defined as the point that minimizes the length from it to the furthest node minus the length from it to the closest point. That is, let c be the selected center, l_{max} the length of the longest path from c to a sink, and l_{min} the length of the shortest path from c to a sink. The center c is selected to minimize $l_{max} - l_{min}$. At this stage the generated center points are also matched as shown in Figure 4.21b. This procedure is repeated until all the points are matched (i.e., for $[\log_2 n]$ steps). The final tree is obtained as shown in Figure 4.21c. It is shown that the total length of the final tree for a set uniformly distributed in the unit square is $O(\sqrt{n})$.

Many known techniques for clock skew minimization, including the two techniques just described, have used the linear delay model. A more accurate delay model is the Elmore delay model. In the Elmore model, the delay along an edge is proportional to its length. However, the delay along a path is defined recursively. Although the Elmore model is a more accurate delay model, it is more difficult to obtain analytical results based on this model.

An example illustrating the Elmore delay model is in Figure 4.22. Let T be an RC tree consisting of points $\mathcal{P} = \{P_1, \ldots, P_n\}$. The root of the tree being a source is denoted by P_1. A subtree T_i consists of point P_i and all its successors. Branch i, denoted by B_i, is the edge between P_i and its immediate predecessor. Let c_i denote the capacitance of P_i, and r_i denote the resistance of B_i. The total subtree capacitance $\mathcal{C}_i$ of T_i is defined recursively as

$$\mathcal{C}_i = c_i + \sum_{j \in S_i} \mathcal{C}_j,$$

where S_i is the set of all the immediate successors of P_i.

Let $\mathcal{P}(i, j)$ be the set of the nodes of T, identifying the path between P_i and P_j, excluding P_i and including P_j. The delay to a branch node P_i, by the Elmore delay model, is

$$t_{1i} = \sum_{j \in \mathcal{P}(1,i)} r_j \mathcal{C}_j.$$

With reference to Figure 4.22, the total subtree capacitance is

$$\mathcal{C}_8 = c_1 + c_3 + c_6 + c_8$$

and the delay by the Elmore delay model is

$$\begin{aligned}
t_{18} &= r_3 \mathcal{C}_3 + r_6 \mathcal{C}_6 + r_8 \mathcal{C}_8 \\
&= r_3(c_1 + c_3) + r_6(c_1 + c_3 + c_6) \\
&\quad + r_8(c_1 + c_3 + c_6 + c_8)
\end{aligned}$$

Based on the Elmore model, a zero clock skew algorithm was proposed by Tsay [357]. The point set is recursively partitioned into two subsets and trees are constructed in a bottom-up manner. Assume, inductively, that every subtree has achieved zero skew. Given two zero-skew subtrees, merge them by an edge to achieve zero skew on the new tree. To do so, it is necessary to decide the position of the connecting points, called the *tapping points*. A general picture is

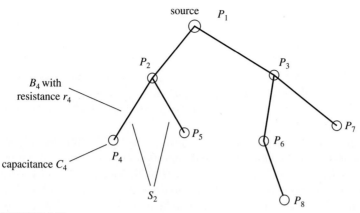

FIGURE 4.22
Calculation of parameters in the Elmore model.

shown in Figure 4.23. The tapping point separates the wire interconnecting the two subtrees. To ensure that the delays from the tapping point to all other points are equal, it must be the case that

$$r_1\left(\frac{c_1}{2} + C_1\right) + t_1 = r_2\left(\frac{c_2}{2} + C_2\right) + t_2.$$

Let l denote the total length of the interconnecting wire. Let $\xi \times l$ be the length of the wire from the tapping point to the root of subtree 1. Then $r = \alpha l$, $r_1 = \alpha \xi l$, $r_2 = \alpha(1 - \xi)l$, and $C = \beta l$, $c_1 = \beta \xi l$, and $C_2 = \beta(1 - \xi)l$, where

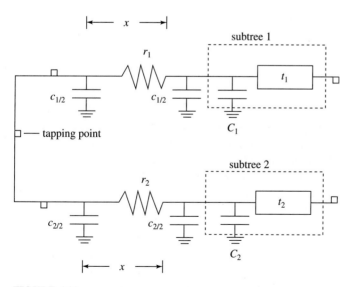

FIGURE 4.23
Parameters in the Elmore model.

α and β are the resistance and capacitance per unit length of wire, respectively. Then,

$$\xi = \frac{(t_2 - t_1) + \alpha l(C_2 + \frac{\beta l}{2})}{\alpha l(\beta l + C_1 + C_2)}.$$

If $0 \leq \xi \leq l$, then the tapping point is somewhere in the interconnecting segment. Otherwise, the length of the interconnecting segment has to be altered from l to l', where $l' > l$.

Later Chao et al. in [42] proposed another zero-skew clock routing technique. Besides minimizing the clock skew, they made an attempt to minimize the total wire length. Essentially, the same type of merge step as the one proposed by Tsay was proposed there. However, a balanced partitioning scheme was employed to heuristically minimize the wire length. A comprehensive study of clock design and routing appears in [173].

4.3.3   Buffered Clock Trees

Modern high-speed digital systems are designed with a target clock period (or clock frequency), which determines the rate of data processing. A clock network distributes the clock signal from the clock generator, or source, to the clock inputs of the synchronizing components, or sinks. This must be done while maintaining the integrity of the signal, and minimizing (or at least upper bounding) the following clock parameters:

- the *clock skew*, defined as the maximum difference of the delays from the clock source to the clock pins;
- the *clock phase delay* (or *latency*), defined as the maximum delay from the clock source to any clock pin;
- the *clock rise time* (or *skew rate*) of the signals at the clock pins, defined as the time it takes the waveform to change from a V_{LO} to a V_{HI} value;
- the sensitivity to the parametric variations of the clock skew, clock phase delay, and rise time. The most important sensitivity is to the clock skew.

Additionally, since in modern microprocessors the clock networks can consume a large portion of the total microprocessor's power (up to 50%), the clock design objectives must be attained while minimizing the use of system resources such as power and area. In high performance systems, these constraints can be quite severe. Consequently, the layout of a good clock distribution network is difficult and time-consuming. This section discusses some of these problems in buffered clock trees.

A very basic concern in the construction of clock trees is the integrity of the clock signal. Since the chip interconnect is lossy, the clock signal degrades as it is transmitted through the network. The longer the path from the source to the sink, the longer the rise time of the clock signal. If the rise time of the clock signal exceeds the clock period, then the clock signal is unable to reach suitable logic high and low values. This problem can be remedied by several different clock network design strategies.

Other work on buffered clock trees has focused on minimizing the phase delay of the clock tree [327, 371] on the assumption that reducing clock tree delays also reduces skew and skew sensitivity to process variation. In these works, the delay model usually consists of a fixed on-resistance and delay for each buffer. More recently, work in [282] considers the minimization of skew and delay in the presence of process variations in buffered clock trees. For a survey on clock network construction issues, see [13, 106].

With the objectives of clock tree design in mind, several effective clock tree construction techniques exist. Section 4.3.2 showed the zero-skew merge algorithms. These algorithms only consider the zero-skew objective and sometimes the phase delay minimization objective. Unfortunately, as explained previously, these algorithms do not produce trees that will satisfy all the clock construction objectives. In particular, if the clock frequency is high enough, a wiring solution will not produce a functional clock tree.

Various wiring-oriented approaches exist to resolve these problems. First, it is possible to taper the wire widths of the clock tree such that the rise times at the terminals are decreased and perhaps satisfied (Figure 4.24a). Another alternative is to increase the capacitance and decrease the resistance on the source-to-sink paths by using wide bus techniques (Figure 4.24b). This approach can be taken to extremes by constructing a fat clock bus from which other clock pin connections are obtained (as shown in Figure 4.24c). This last type of a clock tree has to be driven by a very large buffer, as in Figure 4.25a; thus large amounts of power are consumed by the clock driver and the clock interconnect. However, since the delays of the clock tree are dominated by the capacitance of the wide bus, the bus-to-clock pin interconnection can be done simply and efficiently.

There are also buffer-oriented approaches to clock network constructions. Buffer-oriented approaches partition the clock tree into sections using buffers. Consider a *buffered clock tree* or *clock power-up tree*, which contains buffers in its source-to-sink paths. Several examples of buffered clock trees are shown in

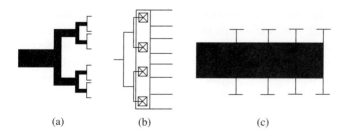

(a)   (b)   (c)

FIGURE 4.24
(a) Clock tree with tapered wire widths. (b) Clock tree using a wide bus approach. (c) Fat bus clock tree construction.

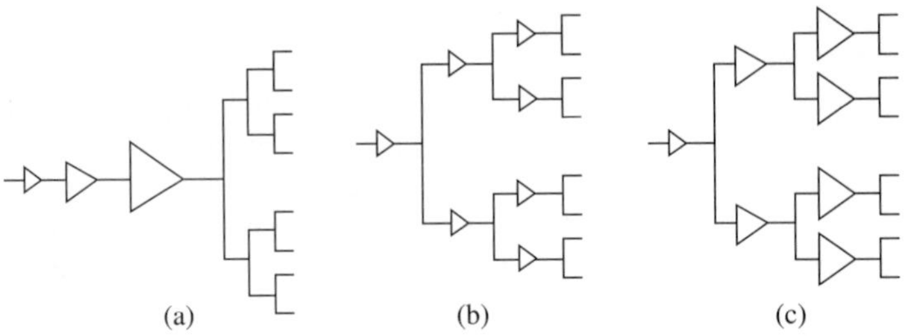

FIGURE 4.25

(a) A buffer chain with tapered sizes driving a clock tree. The buffer size in the schematic corresponds to a buffer size in the physical design. (b) Clock power-up tree, with buffers of a single size. (c) Clock power-up tree, with buffers of tapered sizes. Note that all buffers on the same level have the same size.

Figure 4.25. In this case, to maintain the integrity of the clock signal, it is sufficient to restrict all the rise times in the network to values smaller than the clock period [1]. The addition of buffers in source-to-sink paths reduces the resistance in these paths, and therefore reduces the signals' rise times. As a result, the number of buffers needed by the clock tree is bounded below by the clock's target clock period. In [353], algorithms are proposed to compute the minimum number of buffers required to satisfy a given clock period.

Buffer-oriented and wiring-oriented clock designs exhibit several trade-offs.

- Wiring-oriented solutions are not very sensitive to fabrication variations, while buffer-oriented solutions may be more so.
- The construction of wiring-oriented networks requires careful analysis and modeling of the interconnect networks, while buffer-oriented trees must also take into account detailed buffer delay modeling.
- Buffer-oriented solutions potentially consume less power than wiring-oriented solutions.
- Embedding the buffers into the placement may not be as easy as constructing a wiring-oriented solution.

Consider buffered clock tree constructions along with process variations and realistic buffer delay models. To obtain zero skews while reducing sensitivity to process variations, buffered clock trees usually have equal numbers of buffers in all source-to-sink paths. Furthermore, at each buffered level the clock tree buffers are required to have equal dimensions (Figure 4.25c) to avoid buffer parameter variation and buffer delay model accuracy problems. Using the buffer delay model described earlier, two additional steps are required to obtain zero-skew clock trees: the capacitive loads of the buffers in one level must be the same (this is called load matching), and the rise times at the inputs of the buffers at one level must also match. Note that wiring-oriented techniques may be applied at each buffer stage to produce load and rise time matching.

The clock network construction problem presents trade-offs between wire length and skew, and between variation of parameters under manufacturing conditions (effectively affecting chip yield) and power consumption. These trade-offs present a challenge to the designer and to those seeking to automate the clock design process.

4.4 VIA MINIMIZATION

Vias (or contact holes) between different layers of interconnection on dense integrated circuits can reduce production yield, degrade circuit performance (by increasing propagation delay), and occupy a large amount of chip area (as the physical size of a via is larger than the wire's width). Also, vias on multilayer printed circuit boards raise the manufacturing cost and reduce product reliability. Thus, much research effort has been focused on minimizing the number of vias.

There are basically two categories of via minimization problems, constrained via minimization (CVM), and unconstrained via minimization (UVM). In constrained via minimization, given a routing region, a set of modules and terminals, and the layout of the nets (the detailed routing), the question is how to assign the wire segments to the layers such that no two wires of different nets intersect in the same layer and the number of vias is minimum. In an unconstrained via minimization problem, the layout of the nets is not given, and the question is how to find a topological (or global) routing such that the minimum number of vias is needed.

4.4.1 Constrained Via Minimization

This section focuses on constrained via minimization in two-layer environments on circuits involving two-terminal nets. A circuit involving multiterminal nets can be partitioned into one with two-terminal nets, as discussed in Chapter 2. Figure 4.26a shows a layout before via minimization. The objective is to find the

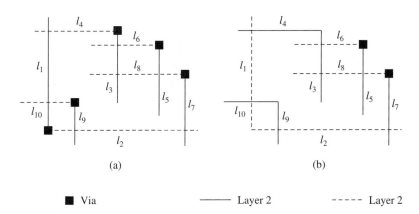

(a)   (b)

■ Via   ——— Layer 2   ----- Layer 2

FIGURE 4.26
CVM. (a) Before via minimization; (b) after via minimization.

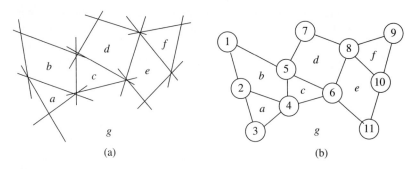

(a)   (b)

FIGURE 4.27
An example of CVM and its crossing graph.

position for vias such that no two wires of different nets intersect in the same
layer and the number of vias is minimum. Figure 4.26b shows an optimal solution.
An algorithm is presented below for finding a minimum via solution, along with
several examples.

Consider an instance of CVM shown in Figure 4.27a. A *crossing graph* is
one where the vertices represent the intersection of nets and an edge is present
between two adjacent intersections. The crossing graph of the layout shown in
Figure 4.27a is shown in Figure 4.27b. The faces of the graph are labeled as a,
$b, \ldots, g$, where face g is the infinite (or boundary) face. From the result of CVM,
represented by a plane graph G, the dual graph G' is obtained. A face with an
odd number of edges on its boundary in G is an *odd-cycle face*. An *even-cycle
face* is similarly defined.

Theorem 4.4. In an instance of CVM, the number of odd-cycle faces is even [23].

It is clear that a set of wires forming an odd-cycle face cannot all be assigned
into two layers. Thus, for each odd-cycle face, at least one via is needed. However,
a via placed on the boundary of an odd-cycle might force another via in an
even-cycle face, as shown in Figure 4.28. To circumvent this, it is necessary to
determine a path (in dual graph G') between two odd-cycle vertices (an odd-cycle
vertex corresponds to an odd-cycle face). The number of edges of G' in the path,
being equal to the number of edges cut in G, should be small to reflect a small
number of vias. Thus, the goal is to find a set of paths, each interconnecting a pair
of odd-cycle vertices (indicated by the thick lines in Figure 4.28). The total number
of edges on the paths is to be minimized. First, a complete weighted graph G_w
is constructed. Each vertex of G_w corresponds to an odd-cycle vertex in G'. Each
pair of odd-cycle vertices in G' represents an edge between the corresponding
vertices in G_w. Each edge has a weight, where the weight is the number of edges
of the shortest path between the corresponding vertices in G'. In other words, each
edge in G_w corresponds to a shortest path between the corresponding vertices in
G'. Thus finding a minimum weighted matching of G_w corresponds to finding
a set of odd-cycle vertex pairing of G' with minimum cost path. Therefore, a

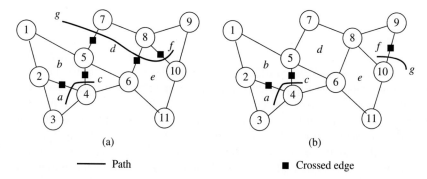

————— Path   ■ Crossed edge

FIGURE 4.28
Examples of via assignment. (a) A bad via assignment; (b) an optimal via assignment.

minimum weight matching algorithm (e.g., running in $O(n^3)$ time [90]) is applied to find a matching pair of odd-cycle faces, and vias are placed along the paths that connect each pair of odd-cycle faces. An optimal solution of the graph given in Figure 4.27b is shown in Figure 4.28b.

Consider the layout given in Figure 4.26a. Since the vertical line segment l_1 is assigned to layer I (e.g., layer 1), then line segments l_4 and l_{10} crossing it must be assigned to layer II. l_1, l_4, and l_{10} are said to form a *cluster*, meaning assigning one of them to a layer dictates the assignment of the rest. A new graph G^*, called a *cluster graph*, is constructed from the original layout. Each vertex of G^* represents a cluster, where a cluster is a maximal set of mutually crossing wire segments. An edge in the graph represents a connection between the clusters and is weighted according to the number of wires. The construction of cluster graph G^* of Figure 4.26a is shown in Figure 4.29. A set of vias are essential and cannot be eliminated from the routing. The via minimization problem can then

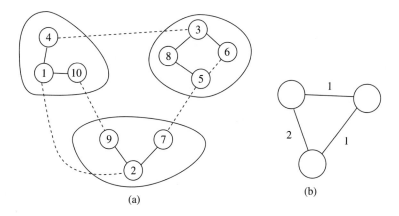

FIGURE 4.29
(a) A clustering. (b) The corresponding cluster graph.

be represented as a graph partitioning problem. The nodes of graph G^* can be partitioned into two groups, A and B. If a vertex is in group A, all the horizontal segments in the corresponding cluster are on layer I, and the vertical segments are on layer II. Conversely, if a vertex is in group B, then all of its horizontal wires are on layer II, and the vertical wires are on layer I. In construction, the horizontal segments in one group are on the same layer as the vertical segments in the other group. Therefore, an edge that separates the two groups represents an unnecessary via and can be eliminated. Finding a partition with the maximum number of such edges will result in maximum reduction of vias in the final layout. This is referred to as the *max-cut* problem. Note that the max-cut problem in a general graph is NP-complete [113]. The max-cut problem in planar graphs can be transformed into a maximum matching problem (see [121]).

Recently, Kuo, Chern, and Shin [194] proposed an efficient algorithm for the constrained via minimization problem. The time complexity of their algorithm is $O(n^{1.5} \log n)$. The algorithm first constructs a cluster graph $G^* = (V, E)$ of the given layout. A weight $w(e)$ associated with each edge e of the cluster graph is defined as $w(e) = \sigma - (\kappa - \sigma)$, where κ is the number of via candidates. A via candidate is a maximal wire segment that does not cross or overlap any other wires and can accommodate at least one via. σ is the number of vias introduced by the known layer assignment connecting two clusters. Note that a wire segment is defined as a piece of a wire connecting two via candidates. For example, in Figure 4.30, the wire segments are labeled by numbers 1 to 18. Wire segments 4, 5, 6, 12, and 14 form one cluster; wire segments 7, 8, 9, and 18 form another cluster. The cluster graph for the layout in Figure 4.30 is shown in Figure 4.31a.

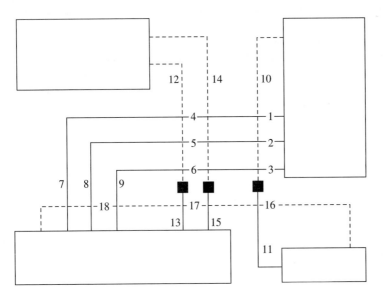

FIGURE 4.30
Layout before via minimization.

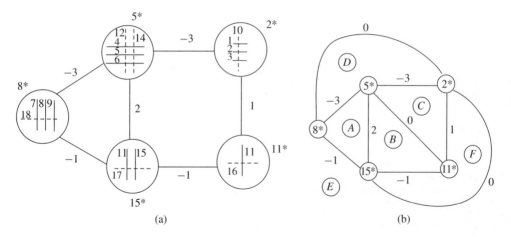

FIGURE 4.31
(a) Cluster graph. (b) The corresponding triangulated graph.

Note, for instance, that the weight of the edge connecting clusters 5* and 15* is 2 because, for this edge, $\kappa = \sigma = 2$.

Without loss of generality, assume G^* is connected. Then a triangulated graph $G_t = (V, E_t)$ of G^* is obtained by adding new edges to G^*. Zero weight is assigned to each new edge in $E_t - E$. An example of a triangulated graph of Figure 4.31a is shown in Figure 4.31b. After that, a dual graph G_d of G_t is constructed in the same fashion as before, as shown in Figure 4.32. Then a new graph G' is obtained by replacing each vertex of G_d with a star. The configuration is shown in Figure 4.33.

Finally, a minimum weight complete matching M of G' is obtained. A minimum weight complete matching of $G' = (V', E')$ can be found in the same way as a maximum weight matching of the same graph is found, except that the weight $w(e)$ of each edge $e \in E'$ must be replaced by a new weight $Ww(e)$, where W is a big constant. The edge set $E_d - M$ of G_d corresponds to a maximum cut of G_t and thus a maximum cut of G^*. The result is shown in Figure 4.34.

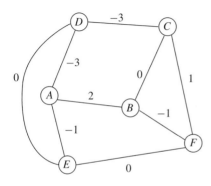

FIGURE 4.32
Geometric dual graph.

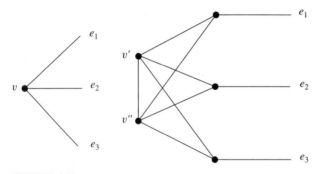

FIGURE 4.33
The star substituting for a vertex.

A formal description of the algorithm follows.

Algorithm *MaxCut(L)*
  begin-1
    construct the cluster graph G^* of L;
    construct the triangulation graph G_t of G^* by adding
      new edges to G^* ;
    construct the geometric dual graph G_d of G_t;
    construct the graph G' from G_d; (each vertex of G_d
      is replaced by a "star")
    find a minimum complete matching M of G';
  end-1.

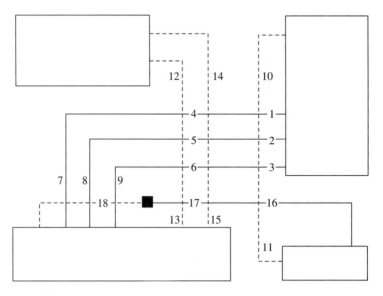

FIGURE 4.34
An optimal solution.

4.4.2   Unconstrained Via Minimization

In contrast to the constrained via minimization discussed in the previous section, an unconstrained, or topological, via minimization (UVM) considers topological (or relative) routing of the wires and layer assignment simultaneously. Techniques for solving UVM [150, 239] first decide the topological information among the signal nets such that the number of required vias is minimum. Then a geometrical mapping of the topology to the layers is performed. An example is shown in Figure 4.35. The topology for the nets interconnecting the given terminals is shown in Figure 4.35a, and the layout after geometrical mapping is shown in Figure 4.35b.

Consider a routing region and a set of terminals. The routing region is characterized as a circle and the nets are drawn as chords. The intersection graph of the chords is called a *circle graph*, as shown in Figure 4.36a. The vertices of the net intersection graph (see Figure 4.36b) represent the nets, and the edges represent intersections between the corresponding nets. That is, an edge (i, j) exists if nets i and j intersect in the circle graph. The goal is to delete a minimum number of vertices from the intersection graph, such that the resulting graph is bipartite; see Figure 4.36c. This creates two sets of vertices that can be placed into two layers without any vias. In the figure, L_1 represents a set of nets assigned to layer 1, and L_2 represents a set of nets assigned to layer 2. The sets L_1 and L_2 define the maximum sets of wires that can be routed on two layers without vias. The deleted vertices are referred to as the *residual nets*. The problem now reduces to one of routing the residual wires.

Lemma 4.1. Any residual net can be routed with only one via [239].

Proof. The basis of this proof is that wires of one of the sets, L_1 or L_2, can be moved so that they no longer interfere with the residual wire being routed. See Figure 4.37. Consider the case where wires from set L_1 are chosen to be moved. Then, starting from one of its end terminals, the residual wire is routed on the same layer as the set L_1, with all wires in set L_1 being pushed ahead until all

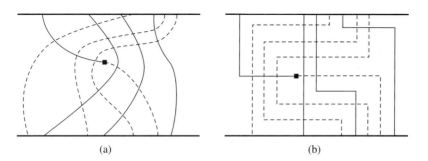

| (a) | (b) |

FIGURE 4.35
UVM. (a) Topology of the net with one via; (b) geometrical mapping.

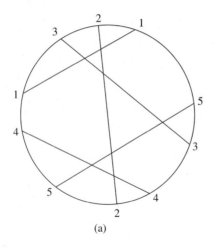

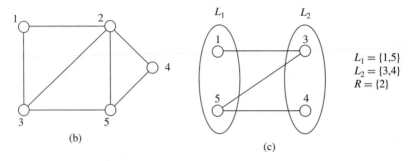

FIGURE 4.36
Topological routing graphs. (a) A circle graph; (b) net intersection graph; (c) a max-cut.

L_2 wires have been crossed. The residual wire is then assigned to the other layer and routed across the L_1 wires. A via is placed at the point where the layers change.

Once the residual nets and the sets L_1, L_2 are defined in the circle graph, this method uses only one via for each residual wire. Thus maximizing the cardinality of the set $L_1 \cup L_2$ minimizes the number of residual nets, and thus, the number of vias is minimized.

This UVM problem was shown to be NP-complete in [305]. However, if the routing region is constrained to be a two-shore channel (see Section 3.3.1 for a definition), then a polynomial time algorithm was proposed to find an optimal solution. Consider a set of two-terminal nets $S = \{N_1, N_2, \ldots, Nn\}$ with the given two-layer topological (or homotopic) routing of these nets. A net that does not use a via in a given homotopy is called a *solid net*. Denote the set of solid nets by U, where $U \subseteq S$. Clearly, the number of vias used in the routing region is $\kappa = |S| - |U|$.

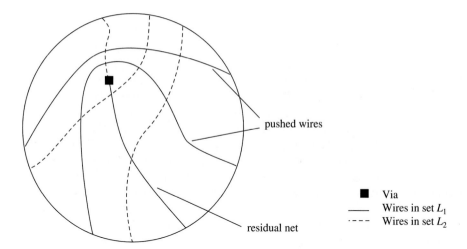

pushed wires

■ Via
—— Wires in set L_1
- - - Wires in set L_2

residual net

FIGURE 4.37
Using one via in routing residual wire.

> **Corollary 4.1.** A solution to the two-shore channel UVM is an optimal solution if it contains a maximum number $|U^*|$ of solid nets over all possible solutions, where U^* is a set of solid nets of maximum cardinality over all homotopies.

Thus to obtain an optimal solution to the two-shore channel UVM problem, it is necessary to find a maximum set of nets U^* such that each net $N \in U^*$ can be placed entirely either on layer 1 or on layer 2. Then a homotopy is obtained by using one via for each net $N' \in (S - U^*)$ or a total of $\kappa^* = |S| - |U^*|$ vias. The problem of finding U^* can be transformed into the problem of finding a maximum-two-chain [305], which can be solved in $O(n^2 \log n)$ time. Recently, the time complexity was improved to $O(n \log n)$ [230].

Recently, the k-layer UVM problem, which is to determine the topology of each net using k routing layers such that a minimum number of vias is used, was discussed in [73, 307]. They showed that both the general k-layer planar subset (k-PSP) problem and the k-layer topological via minimization (k-UVM) problem are NP-complete. Moreover, they showed that both problems can be solved in polynominal time when the routing regions are channels. They transformed the k-PSP problem and k-UVM problem for crossing channels to the problem of finding minimum cost flows in a network. Thus the k-UVM and k-PSP problems can be solved in $O(n^2 \log n + nm)$ time, where n is the number of nets and m is the number of intersections. Note that m is bounded from above by n^2.

4.4.3   Other Issues in Via Minimization

As VLSI technology advances, particularly as more than two layers are being used in industry, via minimization will become increasingly important. Researchers in [343] provided a more general formulation of UVM, and considered variations that arise due to differences in multilayer technologies such as VLSI and high-performance packaging. Their formulation allows pins to be assigned to a layer, more than two pins per net, and regions of unordered pins that occur in channel routing problems. Various restricted classes are discussed; problems in most classes are NP-hard. The case resulting from the traditional switchbox or channel routing with all pins assigned to layers is solvable in $O(kn^2)$, where k is the maximal number of pins of a net on a layer and n is the number of pins.

Another class of via minimization problems, allowing modification of the given layout, was proposed in [354]. Two procedures are used to modify the layout. The first one is to search for an alternate route to eliminate a via. The second one is to improve the result by changing the layer of wire segments by shifting vias along wire segments. The time complexity of the algorithm is $O(vn)$, where v is the number of vias and n is the number of grid points in the layout. In general, the algorithm obtains better results compared with conventional CVM algorithms. The technique can be extended to handle multilayer layouts.

4.5   POWER MINIMIZATION

Power consumption in CMOS circuits is due to the dynamic power consumption of charging and discharging capacitive loads during output transitions at gates, the short circuit current that flows during output transitions at gates, and the leakage current. The last two factors can be made sufficiently small with proper device and circuit design techniques; thus, research in design automation for low power use has focused on the minimization of the first factor, the dynamic power consumption.

The average dynamic power consumed by a CMOS gate is given below, where, C_l is the load capacity at the output of the node, V_{dd} is the supply voltage, T_{cycle} is the global clock period, N is the number of transitions of the gate output per clock cycle, C_g is the load capacity due to input capacitance of fanout gates, and C_w is the load capacity due to the interconnection tree formed between the driver and its fanout gates.

$$P_{av} = 0.5 \frac{V_{dd}^2}{T_{cycle}} C_l N = 0.5 \frac{V_{dd}^2}{T_{cycle}} (C_g + C_w) N \tag{4.22}$$

Logic synthesis for low power attempts to minimize $\sum_i C_{g_i} N_i$, whereas physical design for low power tries to minimize $\sum_i C_{w_i} N_i$. Najm [260] presented an efficient technique to propagate transition densities at PIs into the network to compute the transition densities at every node inside the network (ignoring the correlation between nodes due to existence of reconvergent fanouts). The shortcoming of his model is that it assumes that inputs do not make concurrent transitions. Ghosh et al. [116] address the problem of estimating the average power consumption in VLSI combinational and sequential circuits using a symbolic simulation technique. Tsui et al. [361] present a power analysis technique for CMOS circuits that accounts for the correlation due to reconvergent paths. An algorithm called PCUB, a performance-driven placement algorithm for low power [369], solves a given problem in two phases, global optimization and slot assignment. The objective in both phases is the total weighted net length, where net weights are calculated as the expected switching activities of gates driving the nets.

The two familiar objective functions for the cell-based placement problem are the total wire length and the wire density. The objective function for power minimization can be formulated in a similar manner as follows:

$$\mathcal{I} = \{i_1, \ldots, i_I\} \quad \text{set of primary inputs}$$

$$\mathcal{M} = \{m_1, \ldots, m_M\} \quad \text{set of internal nodes}$$

$$\mathcal{O} = \{o_1, \ldots, o_O\} \quad \text{set of primary outputs}$$

$$\mathcal{N} = \{n_1, \ldots, n_N\} \quad \text{set of nets} \tag{4.23}$$

The total number of nets, N, is given by $N = I + M$. A six-tuple $(N_i, C_{\text{in}_i}, a_i, r_i, C_{w_i}, C_{g_i})$ is associated with each node i, where N_i is the switching rate at output of gate i, C_{in_i} is the input capacitance of gate i, a_i is the arrival time at the output of gate i, r_i is the required time at the output of gate i, C_{w_i} is the wire load capacity due to net n_i, and finally, C_{g_i} is the gate load capacity due to nodes attached to net n_i. The total power consumption for the circuit is given by the summation of (4.22) over all nets. By dropping the placement independent terms and constant factor $0.5V_{\text{dd}}^2/T_{\text{cycle}}$, the following objective function for low power is obtained:

$$L_1 = \sum_{i \in (\mathcal{I} \cup \mathcal{M})} (C_{w_i} N_i).$$

In order to avoid the excessive amount of computation time required for linear functions and also to allow efficient quadratic optimization techniques, the quadratic approximation of this summation is used instead as the objective function, which is

$$L_2 = \sum_{i \in (\mathcal{I} \cup \mathcal{M})} (C_w^2 N_i^2).$$

Using matrix notation, L_2 is rewritten to obtain an objective function that is similar to the one derived in PROUD [359], GORDIAN [187], and RITUAL [340]. It is shown that as long as the whole circuit is connected and there are some fixed modules, this is a convex objective function and the quadratic optimization techniques can be applied to obtain a global optimal solution to this problem.

The quadratic optimization techniques are combined with iterative partitioning of the circuit. After each global optimization step, the circuit is further partitioned, and the partition information is introduced in the subsequent global quadratic optimization step as center-of-mass constraints. Unlike the divide-and-conquer approach, the above mechanism maintains a global view of the circuit beyond the partition boundaries and allows migration of gates across boundaries. The hierarchical approach is applied until five to ten gates remain in each partition. Then, assuming that each partition has at least as many slots as gates, the problem of assigning a gate to a slot reduces to a linear assignment problem. For each partition a cost matrix is created, with as many rows as gates and as many columns as slots. Element C_{ij} in the cost matrix reflects the power cost if gate i is assigned to slot j, and is given by

$$C_{ij} = \sum_{i \in v_k} \left(N_k \frac{\text{Steiner-approx}(v_k, i, j)}{|v_k|} \right)$$

where Steiner-approx(v_k, i, j) calculates a rectilinear single trunk Steiner approximation for the net v_k when gate i is assigned to slot j. To allow gate migration across partitions even in the slot assignment phase, slot assignment iterations are interleaved with the shifting of partition boundaries. To capture the performance requirements, arrival/required time constraints are introduced in the objective function L_2, which can then be formulated as a Lagrangian function optimization and can be solved using Lagrangian relaxation techniques [324, 340].

4.6 DISCUSSION AND OTHER PERFORMANCE ISSUES

Timing-driven placement, timing-driven routing, the via minimization problems, and the power minimization problem are the problems given the most attention in the literature. Other problems have been briefly studied from a timing point of view. A summary of these results follows.

Timing-driven partitioning was considered in [329]. They considered both timing and capacity constraints in the partitioning phase. The main application is to assign functional blocks into slots on multichip modules during high level design. It provides fast feedback on impact of high level design decisions. In the first phase, clustering is done to ensure the timing constraints are satisfied. Then a packing, followed by a traditional partitioning (e.g., a Kernighan-Lin technique), is done to satisfy capacity constraints. Compared to a classical partitioning scheme, experiments show that the new technique has no timing violation if the capacity is increased by a small percentage (about 5 to 6% on the average).

Some factors that determine the suitability of a decomposition are the geometry of the modules, the connectivity among modules, and timing constraints. A hierarchical-clustering-based algorithm that leads to a small number of superior candidate hierarchies was introduced in [108]. An approach to perform simultaneous placement and routing of hierarchical IC layouts is presented in [350]. The method is based on the concept of slicing, which introduces a specific regularity to the problem with certain constraints on the chip floorplan. The placement part consists of a number of interrelated linear ordering problems, corresponding to the ordering of slices. The routing part outlines a hierarchical pattern router, applicable to slicing structures. The IC layout construction progresses top down in small design increments, corresponding to the processing of the individual slices. The placement and routing data are acquired at intervening time slots and influence each other at each level of hierarchy.

Timing-driven floorplanning was considered in [276]. Their technique is a mathematical programming (i.e., a constrained nonlinear programming) approach that incorporates both timing and geometric information in the formulation. The three steps employed are timing minimization with module overlap, module separation, and timing minimization without module overlap. To obtain a fast algorithm, techniques for removing redundant paths are used; thus, the number of constraints is reduced. The information obtained on the initial placement is fed into a floorplan package to find the final shapes, sizes, and pin locations.

A new three-layer over-the-cell channel routing algorithm with a timing consideration was proposed in [262]. This router minimizes the channel height by using over the cell areas and achieves the net's timing requirements; each net has a maximum allowable upper bound on its length. In this work, 45° wire segments are used to route the nets over the cell to further reduce the net length. Experimental results show that the same area as used by previous over the cell routers can be obtained while satisfying length constraints.

Other performance issues in physical design include regularity, layout synthesis systems, and automated generation. Regularity is an important issue in VLSI design. Ishikawa and Yoshimura [164] presented a new module generator for generating functional modules with structural regularity. Unlike most module generators that are only able to place cells, the proposed generator can generate functional modules with structural routers for regular structure layouts such as multipliers and RAMs. A synthesis system for the automatic layout of NMOS gate cells is described in [233]. The cells are based on multigrid cell models and are intended for use as part of a chip synthesis system. An outline of the basic concepts of a CAD procedure for the layout synthesis of such cells is given. The main objective is to generate correct and compact cells with a controlled growth in area when they are subjected to modified speed requirements. Both the layout synthesis procedure itself and algorithms are discussed. The paper [14] investigated technical issues concerning the automated generation of highly regular VLSI circuit layouts (e.g., RAMs, PLAs, systolic arrays) that are crucial to the designability and realizability of large VLSI systems. The key is to determine the most profitable level of abstraction for the designer, which is accomplished by the introduction of macro abstraction, interface inheritance, delay binding, and the complete decoupling of procedural and graphical design information. These abstraction mechanisms are implemented in the regular structure generator, an operational layout generator with significant advantages over first-generation layout tools. Its advantages are demonstrated using a pipelined array multiplier layout example.

Recently, a new class of placement strategies, incorporating the knowledge of underlying circuit structure targeted for high-performance circuits, has been examined [401]. The net regularity is measured using the standard deviation of the locations of its terminals. Experimental results showed that exploiting circuit structural information led to substantially more regular nets, compared to placement without structural information. Benchmark circuits (from MCNC) showed 1.9 to 26 times the number of nets with zero standard deviation. Furthermore, the length of the longest net in a chip was reduced from 5 to 22%. The placement strategy is crucial for high-performance circuits since regular nets have relatively fewer vias and reducing the longest net length improves signal delay.

The problem of minimizing the number of power pads in order to guarantee the existence of a planar routing of multiple nets was studied in [141, 243]. A general lower bound was established and a heuristic for minimizing the number of pads was given. The general pad minimization problem is NP-complete. Given tree topologies for routing power/ground nets in integrated circuits, [59] formulated and solved the problem of determining the widths of the branches of the trees. Constraints are developed in order to maintain proper logic levels and switching speed, to prevent electromigration, and to satisfy certain design rules and regularity requirements.

A new approach for optimizing clock trees, especially for high-speed circuits, was proposed in [58]. The approach provides a useful guideline to a designer; by user-specified parameters, various design objectives can be satisfied. These objectives are:

1. To provide a good trade-off between skew and wirelength, a new clock tree routing scheme was proposed. The technique is based on a combination of hierarchical bottom-up geometric matching and minimum rectilinear Steiner tree.

2. When a clock tree construction scheme is used for high-speed clock distribution in the transmission line mode (e.g., multichip modules), there are several physical constraints to ensure correct operation. One of the crucial problems is that given a clock-net topology with hierarchical buffering, the best way to redistribute the buffers evenly over the routing plane avoiding congestion is at the expense of wirelength increase. An effective technique for buffer distribution was proposed. Experiments show that on the average, congestion is reduced by 20% at the cost of a 10% wirelength increase.

3. A postprocessing step offering a trade-off between skew and phase-delay was also proposed. The technique is based on a combination of hierarchical bottom-up geometric matching and a bounded radius minimum spanning tree.

Electrical performance and area improvements are important parts of the overall VLSI design tasks. Given designer-specified constraints on area, delay, and power, EPOXY [263] will size a circuit's transistors and will attempt small circuit changes to help meet the constraints. Since the sum of the transistor area is a better measure of dynamic power than cell area, a more accurate model is presented. Optimization of a CMOS eight-stage inverter chain illustrates this difference; a typical minimum power implementation is 32% larger than the one for minimum area. The combination of a TILOS-style heuristic and an augmented Lagrangian optimization algorithm yields quality results rapidly. EPOXY's circuit analysis is from 5 to 56 times faster than Crystal.

Chen and Wong [48] proposed a channel-based thin-film wiring methodology (using two layers) considering cross-talk minimization between adjacent transmission lines. [56] proposed a new layer assignment algorithm for high performance multilayer packages, such as multichip modules (MCMs). They focus on assigning nets to layers to minimize the cross-talk between nets, while simultaneously minimizing the number of vias and layers. A novel net interference measure based on potential cross-talk and planarity is used to construct a net interference graph (NIG), and a new graph coloring and permutation algorithm is used to find an interference-minimized subset in each layer and a minimum cross-talk between layers.

A comprehensive theory for multilayer critical area determination was developed in [81]. While it is important to exhaustively verify IC designs for their functional performance, it is equally important to verify their robustness against spot defects. *Spot defects* are local disturbances of the silicon layer structure caused by process variabilities, dust particles, and other contaminations of the fabrication equipment. IC sensitivity to spot defects can be studied by extracting critical areas from the layouts. A *critical area* in the layout is an area such that if spot defects are centered there, a malfunction in the respective critical circuit arises. The design's defect sensitivity is the ratio of the critical area for a defect size to the total layout area. Initial attempts to perform this verification were based on a critical area extraction of one layer at a time. However, this extraction neglects the electrical significance of the interrelationships between layers.

The paths in adjacent layers can be overlapped in the same region. Thus, it is most likely that a circuit short can occur only with layers that are adjacent. To determine which pairs of paths cross or are too close, the paths are considered to be neighbors and the distance between two paths to be the edge-weight on edges of a graph [52, 114].

Bend minimization was studied in [179, 183, 284, 398]. With signals in the gigahertz range, the electrical characteristics of the packages require that the signal lines be treated as transmission lines. This involves careful consideration of minimizing transmission line effects, such as cross-talk, reflections, and the effects of crossings, bends, and vias. In practice, transmission lines are not perfectly uniform. That is, in the package level, significant reflections can be generated from capacitive and inductive discontinuities along the transmission lines. Moreover, in a multilayer ceramic substrate of MCM, wires at different levels do not have exactly the same impedance. Such mismatches of line impedance can cause reflections from the junction points such as vias and bends. These discontinuities must be controlled in order to keep the resulting reflections to a minimum. Therefore, propagation delays associated with discontinuities (e.g., see [76, 157, 158, 159]) should be minimized through careful design. Since high-performance circuits are usually designed aggressively (i.e., most of the nets are considered to be critical nets), it is preferable to minimize the number of bends and vias.

EXERCISES

4.1. What is the running time of the zero-slack procedure? What if the input circuit is a tree?

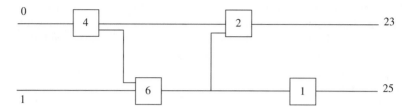

FIGURE E4.2
A circuit.

4.2. Use the zero-slack algorithm to
   (a) find arrival times, required times, and slacks in the circuit of Figure E4.2.
   (b) distribute the slack along the nets.
   (c) explain how you use the zero-slack algorithm to obtain a timing-driven placement.

4.3. Consider Figure 4.8. Change the arrival and required times as follows: $t_A^1 = 6$, $t_A^2 = 2$, $t_A^3 = 1$, $t_A^4 = 0$, and $t_R^1 = 22$, $t_R^2 = 24$. Now find all slacks (gate delays remain unchanged).

4.4. Describe an algorithm for removing the minimum number of edges to make a directed cyclic graph into a DAG. Is your algorithm optimal? Why? Analyze the time complexity of your algorithm.

4.5. Consider a directed acyclic graph with one source consisting of n vertices. Present nontrivial upper- and lower-bounds on the number of directed spanning trees of it.

4.6. Show a minimum delay tree that does not have minimum skew (use points in the plane and employ the rectilinear metric).

4.7. Design an algorithm with bounded delay and small skew, using the linear delay model.

4.8. (a) Show an instance where a minimum spanning tree (MST) of a point set has skew that is almost equal to the MST length.
   (b) Show an instance (i.e., a point set) where the length of a hierarchical matching tree is two times the length of MST.
   (c) Show an instance where the length of a hierarchical matching tree is $O(\log n)$ times the length of MST, where n is the number of points.

4.9. Prove that any geometric matching of minimum length is planar, that is, no two edges intersect.

4.10. Prove Theorem 4.3.

4.11. In the weighted constrained via minimization (WCVM), each edge is assigned a positive weight (1 to ∞). A via on an edge with weigh w costs w units. The goal is to find a minimum weight via assignment. Design an efficient algorithm for solving WCVM.

4.12. Design a heuristic algorithm for solving the three-layer version of CVM (the problem is known to be NP-hard).

4.13. Give an example showing that a solution to unconstrained via minimization (UVM) may produce very bad solutions in terms of area. Design an efficient heuristic for obtaining an area-efficient routing with a small number of vias in a channel.

4.14. When would you use UVM, and when is CVM more desirable?

COMPUTER EXERCISES

4.1. Consider a circuit, represented by a directed acyclic graph. Given a signal arrival time at each PI, a signal required time at each PO, and gate delays, implement the zero-slack algorithm.

> **Input format.** The circuit corresponding to Figure 4.8 is represented as:
>
> I1 1, I2 0, I3 1, I4 3,
> O1 17, O2 14
> G1 2, G2 4, G3 3, G4 6, G5 0;
> I1 G2, I3 G4, I3 G3, I4 G3, G2 G4, ...

In the above example, first all primary inputs and their arrival times are specified. For example, I1 1 means primary input 1 arrives at time 1. Next, all primary outputs and their required times are specified. For example, O1 17 means primary output 2 is required at time 17. Then all gates and their delays are specified (G1 2 means gate 1 has 2 units of delay). Finally, the set of connections is specified: I1 G2 means primary input 1 is connected to input of gate G2.

> **Output format.** The output is a drawing of the circuit and the corresponding information, the same as the one shown in Figure 4.14.

4.2. Consider a circuit consisting of a set of primary inputs, outputs, and a set of gates (see Figure 4.14). Given a parameter k, design a clustering and slack-assignment algorithm such that there are fewer than k gates per cluster. Each net is assigned a weight, MAX-slack, where MAX is a given constant and slack is the slack assigned to it by your algorithm. Your algorithm should minimize the sum of the weight of nets that have terminals in more than one cluster.

> Use the same input and output formats as CE 4.1. In addition, identify the clusters in a separate figure or in the same figure.

4.3. Consider a module adjacency graph G (it is a graph and not a hypergraph). Each vertex (modules) v_i is assigned a weight $A(i)$. Indicate the minimum area it requires. Assume that each edge is assigned a budget obtained from a timing analyzer and reflecting the maximum length of the nets (actually, the maximum length of the most critical nets) between the two modules. Design and implement an algorithm that finds a floorplan F corresponding to G. The distance (center to center) between two modules must be less than the corresponding budget. Note that any adjacency that is not maintained is charged twice its cost.

> **Input format.** Input is specified by a set of vertices and their areas, followed by a set of edges and their budgets. With reference to Figure CE 4.3a, the following input is given:
>
> a 12, b 24, c 3, d 8,
> ac 6, bd 4, ab 9, bc 2, cd 100, ad 7

This example indicates that module a should have an area at least 12, module b should have area at least 24, and so on. The budget on the distance between the center of module a and c is 8, and so on. The objective is to satisfy the maximum number of budget contraints.

> **Output format.** The output is shown in Figure CE4.3b. Output how many constraints are not satisfied.

4.4. Consider a set of modules. Each module has a set of terminals on the upper side of its horizontal edge. One terminal per net is specified as the source. Each net is

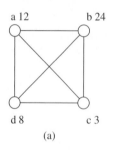

(a)

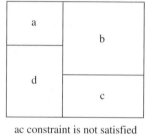

ac constraint is not satisfied

(b)

FIGURE CE4.3
Input and output formats, timing-driven floorplanning.

assigned a timing constraint; that is, the length between the source and the sink of the net should be less than the given constraint. Find a linear placement satisfying all timing constraints. Among all such placements, find one with small density.

Input format.

3
M1 6, 3 1, 5 4;
M2 4, 2 1; M3 7, 3 1, 2 4;
N1 1 7, N2 3 12

The first line indicates that there are 3 modules. The second line indicates that module 1 occupies 6 grid points, at the 3rd grid point there is a terminal of net 1 and at the 5th grid point there is a terminal of net 4. The other two modules are similarly specified. Then it is specified that net 1 (N1) has its source on module 2 and its budget length is 7 units, and so on.

Output format. The output format is shown in Figure CE4.4. (In Figure CE4.4, grid units corresponding to M1 are shown. You don't need to show it in your output.). Show all nets and their routing. Highlight portions of nets whose timing is not satisfied.

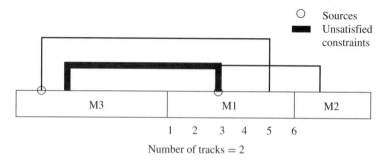

FIGURE CE4.4
Output format, a linear placement satisfying timing constraints.

4.5. Implement the algorithm LAYOUT-BRST for obtaining a timing-driven routing.

Input format. The input specifies x- and y-coordinates of all points. The first point is the source. For example,

   2 4, 3 7, ...

indicates that the source has x-coordinate 2 and y-coordinate 4. The first sink has x-coordinate 3 and y-coordinate 7, and so on.

Output format. The output can be similar to the one shown in Figure 4.18. Also output the length of the final tree and its radius. Experiment with different values of ϵ and different point sets. State your conclusion.

4.6. Consider a routing problem given by a source, a set of sinks $v_1 \ldots v_n$, and a value ϵ_i for each sink v_i. Design and implement an algorithm that finds a routing tree with a distance of source to v_i (in the tree) bounded by $(1 + \epsilon_i)R$, where R (as in the text) is the maximum geometric distance between a sink and the source. Your algorithm should minimize the total length.

Input format. The input specifies the x- and y-coordinates of all points followed by an ϵ for each sink. The first point is the source. For example,

   2 4, 3 7 0.2, 4 1 1.3 ...

indicates that the source has x-coordinate 2 and y-coordinate 4. The first sink has x-coordinate 3 and y-coordinate 7 and ϵ equal to 0.2, and so on.

Output format. The output should show the constructed tree and the length of the tree.

4.7. Implement the JSK algorithm for finding a zero-skew tree. Also implement the matching-based algorithm. Compare the two techniques in terms of skew and total length.

Input format. The input specifies the x- and y-coordinates of all points. The first point is the source. For example,

   2 4, 3 7 0.2, 4 1 1.3 ...

indicates that the source has x-coordinate 2 and y-coordinate 4. The first sink has x-coordinate 3 and y-coordinate 7, and so on.

Output format. The output should show the constructed trees and their skew and length.

4.8. Design and implement an algorithm for the two-layer constrained via minimization problem.

Input format. The input is specified by a set of horizontal or vertical segments (see Figure 4.26a). For example, the following is a typical input:

   1 2 h4, 4 2 v3, ...

indicating that there is a segment with one of its endpoints having x-coordinate 1 and y-coordinate 2, the segment is horizontal and has length 4, and so on. Two segments that share an endpoint belong to the same net.

Output format. Your output should show a two-layer layer assignment and position of vias (as in Figure 4.26b). Also, output the number of vias.

4.9. Given a set of standard cells and a net list, implement an efficient algorithm for placing and routing (global and detailed) all the nets and satisfying all timing constraints. The number of rows is given and the width of each row is given. The goal is to route all nets minimizing the height of the chip. Note that in each cell the same terminal appears on both the top side and the bottom side of the cell, and there are no terminals on the left and the right sides of the cell.

Input format.

3s25 7 200
1s18 3 0 4, ...,
0 2 3 1 0 0 .

The above example contains a number of cells to be placed in 3 rows, each cell row has height 7, and the total number of columns is 200.

Two of the modules are shown, the first and the last one. The first module occupies four columns: net 1 has a terminal at the first column, net 3 has a terminal at the second column, the third column is empty, and net 4 has a terminal at the fourth column. Note that two modules should be placed at least one unit apart. An s next to a terminal indicates that this is a source and specifies the allowed budget. For example, terminal 1 on the second module is a source and has a budget of 18 grid units (i.e., from the source to each sink should be 18 units or less).

Output format. Show a placement (containing the name of the modules) and a complete routing of the nets. Indicate the height of your result (including the height of the cells and the total number of tracks used for routing). Also indicate the number of nets that do not satisfy the given timing constraints. For example, your output can be similar to the one shown in Figure CE4.9.

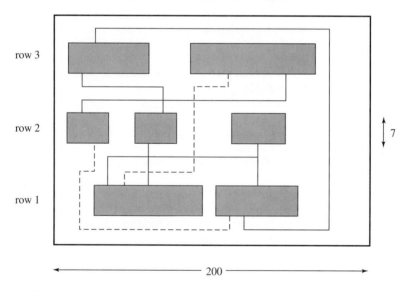

FIGURE CE4.9
Output format, timing-driven placement and routing of standard cells.

CHAPTER
5

SINGLE-LAYER ROUTING AND APPLICATIONS

In VLSI layout design, critical nets such as timing-critical nets and power-ground nets are to be placed on a preferred layer of lower resistance. A layer with lower resistance allows fast propagation of signal delays. In a multilayer environment, reserving a layer for critical nets is affordable. Other multilayer environments, for example, multichip modules, systematically employ single-layer routing. These facts motivate the study of single-layer, or planar, routing.

This chapter discusses the problem of finding a maximum-weighted planar subset, the problem of single-layer global routing, and the detailed routing problem in single-layer regions. The focus is on two subproblems, when the routing region is a bounded region (e.g., a switchbox) and when the routing is in more general regions based on a given global routing. An important post-processing algorithm, wire-length and bend minimization techniques, and applications of the problem of over-the-cell routing and routing in multichip modules, are also discussed.

5.1 PLANAR SUBSET PROBLEM (PSP)

Given any circuit, there is the problem of finding a planar subset of nets. Such a subset can be routed in one layer, and thus will have no vias established on them. The problem to be considered here is finding a maximum-weighted planar subset (PSP) of n multiterminal nets in a global routing with w modules. As the general problem is NP-hard, the focus will be on cases with no modules or with a fixed small number of modules.

5.1.1   General Regions without Holes

Consider a restricted class of PSP called circle PSP (or, CPSP). An instance of the problem consists of a set $\eta = \{N_1, \ldots, N_n\}$ of multiterminal nets in a routing region without any modules. The terminals are located on the external boundary, for example, on the boundary of a switchbox. A point $(x, y) \in N_i$ denotes that net N_i has a terminal at point (x, y), $1 \leq i \leq n$. Each net N_i has weight $W(i)$, indicating its degree of criticality. Nets with larger weights are more likely to be selected in the planar subset. An optimal solution to CPSP is a maximum-weighted planar subset of η in the routing region (i.e., the interior of the boundary).

Consider a continuous deformation of the external boundary into the boundary of a circle C. This transformation preserves the ordering of the terminals in a clockwise scan of the boundary. Let P_i be a convex polygon inside C with vertices corresponding to terminals of $N_i \in \eta$. In Figure 5.1, the shaded region P_1 corresponds to net N_1. The set $\wp = \{P_1, \ldots, P_n\}$ of convex polygons corresponds to the set η of nets as demonstrated in Figure 5.1. Each polygon has the same weight as the corresponding multiterminal net.

The following algorithm was first proposed in [34]. Here a modification of it is used that handles multiterminal and weighted nets. The technique is reported in [215]. The algorithm is based on dynamic programming that obtains a maximum-weighted pairwise nonoverlapping polygon (MNP). Let the vertices of the polygons, on the boundary of the circle C, be numbered 1 to t in a clockwise scan of the boundary starting from the origin O, where O is arbitrarily selected and t is the total number of terminals of η. That is, $t = \sum_{1 \leq i \leq n} |N_i|$. Call the arc of C, containing vertices $i, i+1, \ldots, j$, interval ij and denote it by $[i, j]$. Let $\wp_{ij} \subseteq \wp$ denote the subset of polygons with all vertices in $[i, j]$. A set of maximum-planar nonoverlapping polygons in $\wp_{ij}$ is denoted by MNP(i, j). The plan is to find MNP(i, j) for all i and j $(i < j)$. Consider processing a vertex k $(1 \leq k \leq t)$, where P_k is the polygon containing k.

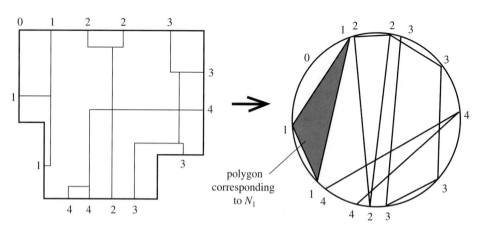

FIGURE 5.1
A continuous deformation from a boundary $\mathcal{J}$ to its corresponding circle C.

Case 1. If P_k has a vertex in $[k + 1, t]$, then P_k cannot contribute to MNP$(1, k)$. Therefore, MNP$(1, k) =$ MNP$(1, k - 1)$.

Case 2. All vertices of P_k are in $[1, k]$. There are two possibilities. If P_k is in the solution, then MNP$(1, k) = P_k \cup (\cup_i$ MNP$([v_i, v_{i+1}]))$, where v_i's are vertices of P_k. Otherwise, MNP$(1, k) =$ MNP$(1, k - 1)$. The maximum of the two values is MNP$(1, k)$.

An optimal solution to the problem is stored in MNP$(1, t)$. It is necessary to evaluate all intervals ij in $[1, t]$ $(1 \le i < j \le t)$ to obtain MNP$(1, t)$. As there are a total of t^2 $(t \le mn)$ intervals (assuming each polygon has at most m vertices), the processing of each point takes $O(m)$ time. Thus, the algorithm runs in $mt^2 \le m(m^2 n^2)$ time, or, $O(m^3 n^2)$ time.

However, a careful implementation of this technique (considering only the endpoints of the polygons) results in an $O(nt)$-time algorithm. To calculate MNP(v_1, v_{m_j}) of polygon P_j with vertices $[v_1, v_2, \ldots, v_{m_j}]$, proceed as follows. Here, v_1 and v_{m_j} are the endpoints; that is, v_1 is the first vertex of P_j in a clockwise scan starting from O, and v_{m_j} is the last vertex. Consider interval $[v_i, v_{i+1}]$ and test all polygons in $\wp$ to see if they are inside the interval. Store the weight of the maximum-weight polygons dominated by (i.e., are entirely inside) (v_i, v_{i+1}). There are a total of $t = \sum_1^n m_i$ intervals. A formal description of the algorithm follows.

> *Procedure* MNP-ALG$(\wp)$;
> begin-1
>   (* $\wp = \{P_1, \ldots, P_n\}$ *);
>   (* $\mathcal{I} = \{I_1, \ldots, I_n\}$,$I_i = (l_i, r_i)$ where l_i, r_i correspond to the
>   first and last vertex of P_i; without loss of generality
>   assume $r_i > r_{i-1}$ *);
>   (* VERTX(P_i): return vertex array $V(= \{v_1, \ldots, v_{m_i}\})$
>   of polygon P_i *);
>   (* MWIS is the weight of maximum-weighted independent set
>   of intervals $\mathcal{I}$ *);
>   (* PRED(i): the largest r_i to the left of i, i.e., $max_{r_i}(r_i < i)$ *);
>   for $i = 1$ **to** n do
>   begin-2
> $V(= \{v_1, \ldots, v_{m_i}\}) :=$ VERTX(P_i) ;
>   (* note : for each polygon, $l_i k = v_1, r_i = v_{m_i}$ *);
> $TEMP1_l :=$ MNP$(1, $PRED$(l_i))$;
> $TEMP1_r :=$ MNP$(1, $PRED$(r_i))$;
>   for $k = 1$ **to** $m_i - 1$ do
>   begin-3
>   for $j = 1$ **to** n do
>   begin-4
>   if $(v_k < l_j < v_{k+1})$ *and* $(v_k < r_j < v_{k+1})$
>   then $W(I_j) :=$ MNP(l_j, r_j);
>   else $W(I_j) := \emptyset$;

$$TEMP2_k := \text{MWIS}(\mathcal{I});$$
   end-4;

   end-3;
   $$TEMP2 := \sum_k TEMP2_k + \text{W}(P_i) ;$$
   $$\text{MNP}(1, r_i) := \text{MAX}(TEMP1_r, TEMP2 + TEMP1_l);$$
   $$\text{MNP}(l_i, r_i) := TEMP2;$$
 end-2;

end-1.

Theorem 5.1. MNP-ALG finds a maximum-weighted nonoverlapping polygon set in $O(nt)$ time, where n is the total number of polygons and t is the total number of vertices.

Proof. The algorithm takes a dynamic programming approach. MNP(i, j) is computed for each pair i, j $(i < j)$; MNP(i, j_1) is computed before MNP(i, j_2) if $j_1 < j_2$ (where i is the first vertex of polygons and $j, j_1,$ and j_2 are the last vertices of polygons). Finally, MNP$(1, t)$ is readily obtained. Since $\wp = \wp_{1,t}$, then MNP$(1,t)$ is a maximum-weighted nonoverlapping polygon set of $\wp$.

Examine either 1 (P_i is not in the solution) or m_i (P_i is in the solution) intervals (i.e., the interval between two adjacent vertices of a polygon) per polygon. Here, m_i is the number of vertices of polygon P_i and $\sum_i^n m_i = t$. Therefore, the total number of intervals visited is at most t. For each interval, check all n polygons, and obtain MWIS$(\mathcal{I})$ in $O(n)$ time [11]. Therefore, the entire algorithm runs in $O(nt)$.

5.1.2   PSP with a Fixed Number of Modules

An instance of a general PSP (GPSP) is a global routing consisting of n weighted multiterminal nets $\eta = \{N_1, \ldots, N_n\}$ and w modules $M_1, \ldots, M_w$. As was discussed earlier, an instance of PSP can be transformed into a corresponding problem where each module and the external boundary $\mathcal{J}$ is a circle. The technique involves merging the modules and therefore reducing the number of modules to zero, that is, reducing the problem to a CPSP. Then the algorithm for CPSP can be used to solve the corresponding CPSP. For example, consider a net N_i connecting modules $M_1, M_2,$ and M_3 in Figure 5.2a. If it is assumed that N_i is in the solution, then the other nets (e.g., N_x and N_y) that cross N_i should be removed, and the corresponding modules $M_1, M_2,$ and M_3 can be joined to form a new module M', as shown in Figure 5.2b.

Note that when there is a path connecting the external boundary and a module without intersecting any nets or modules, the module can be thought of as part of the external boundary. A formal description of the algorithm follows.

Procedure GPSP-ALG(G: global route, Sol_list)
   (* G is the given global routing along with nets and modules *);
   (* MAXSET is a global variable with initial value empty *);
   (* Sol_list is a local variable that records the nets already
      selected in the solution *);

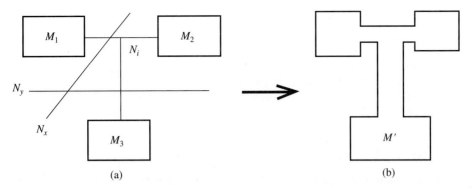

FIGURE 5.2
An example of the merging operation. (a) Before merging; (b) after merging.

> *begin-1*
>   initialize Sol_list;
>   NODE_OP(G, Sol_list);
>   return(MAXSET);
> *end-1.*

> *Procedure* NODE_OP(G, Sol_list)
>   *begin-1*
>   *if* (G is CPSP)
>   *then return* (MNP-ALG(G) ∪ Sol_list);
>   *else*
>   *begin-2*
>   (* do in breadth-first manner *);
>   *for each item (nets and modules) in G*
>   *begin-3*
>   *if* (item is net) *then*
>   Sol_list := Sol_list ∪ item;
>   G' := Merge(G,item) ;
>   MAXSET :=
>   MAX(MAXSET,NODE_OP(G',Sol_list));
>   *end-3*;
>   *end-2*;
>   *end-1.*

Theorem 5.2. An arbitrary instance of PSP, with w modules, n weighted multi-terminal nets, and a total number of t terminals, can be solved in $O(n^{w+1}t)$ time.

Proof. Only an outline of the proof is given. At the first stage, try all nets as possible solutions. Each net interconnects two or more modules, in which case the number of modules decreases as shown in Figure 5.2, or has all its terminals on the same module. As one branch of the search space, consider the class of solutions where no net has terminals in more than one module. Then obtain w independent instances of

CPSP that can be solved in $O(n^2)$ time. This branch of the search procedure does not continue any further.

Therefore, at each step, the number of modules decreases by one. After $w - 1$ steps, after spending $O(n \times (n - 1) \times \ldots (n - w + 2)) = O(n^{w-1})$ time, one module remains. Then an additional factor of $O(n^2)$ time is needed to find a planar subset within one module, that is, to solve an instance of CPSP. Since all nonredundant possibilities have been tried, an optimal solution is thus obtained.

5.2 SINGLE-LAYER GLOBAL ROUTING

This section considers the single-layer global routing problem (SLGRP). Given a set of nets, first the routing sequence is determined. Then a routing path, being a sequence of tiles, is found for each net. The size of tiles, $w \times w$, is an input parameter to the algorithm. For $w = 1$, the global router serves as an efficient detailed router. A post-processing algorithm, minimizing the wire length and number of bends, will also be presented.

The SLGRP has similarities to the two-layer global routing problem, except that now all nets must be routed on one layer without crossing. This distinction makes the two problems inherently different. The following notations and definitions, demonstrated in Figure 5.3, will be used throughout this section.

A w-tessellation is a square tessellation of the routing plane (with size $X_{\max} \times Y_{\max}$) with each tile's size being $w \times w$, $1 \leq w \leq \min(X_{\max}, Y_{\max})$. A tile located at the ith column and the jth row in the w-tessellation is denoted by $\tau(i, j)$. Consider a set $\eta = \{N_1, \ldots, N_n\}$ of multiterminal nets. The layout environment (plane grid) is a w-tessellation with $m_1 \times m_2$ tiles. For simplicity, $m_1 = m_2 = m$ is used. The

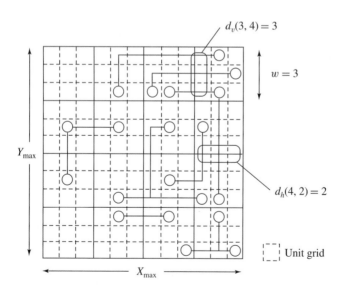

FIGURE 5.3
An instance of the SLGRP with three-tessellation.

algorithm can be easily extended to arbitrary m_1 and m_2. Each k-terminal net N is specified by a k-tuple $[(x_1, y_1), \ldots, (x_k, y_k)]$, where (x_i, y_i), $1 \leq i \leq k$, are the coordinates of tiles containing terminals of net N. Terminals in the same tile are mapped to one central point in the tile.

In an instance of the global routing problem (GRP) with w-tessellation, let $c_h(i, j)$ denote the capacity of the border of $\tau(i, j)$ and $\tau(i, j + 1)$, $1 \leq i \leq m$ and $1 \leq j \leq m - 1$. That is, $c_h(i, j)$ is the maximum number of nets that can cross this border. Similarly, let $c_v(i, j)$ denote the capacity of the border of $\tau(i, j)$ and $\tau(i + 1, j)$. In an output of the global routing, or a partial global routing, let $d_h(i, j)$ denote the number of nets crossing the border of $\tau(i, j)$ and $\tau(i, j + 1)$, $1 \leq i \leq m$ and $1 \leq j \leq m - 1$. Similarly, let $d_v(i, j)$ denote the number of nets crossing the border of $\tau(i, j)$ and $\tau(i + 1, j)$. An example is given in Figure 5.3.

In an output of GRP, each routed net is realized as a *wire tree*. Given a forbidden set D (which consists of obstacles, e.g., modules, terminals, and wires), two wires W_1 and W_2 in the routing plane are *homotopic* if they can be transformed continuously to each other without intersecting D. Also, W_1 is a *homotopy* of W_2 if W_1 and W_2 are homotopic, and vice versa. A solution S of GRP is called a *planar solution* if there exists a routing S' homotopic to S such that in S' no two wires cross each other. Note that planarity does not depend on the capacities. A tile τ in a solution of GRP is called a *planar tile* if wires that go through this tile form a *planar pattern* (i.e., no two wires cross each other). Note that the pattern in each tile is specified by the ordering of wires on the four boundaries of the tile and the position of terminals within that tile. Each wire is specified by the set of tiles it goes through and its relative position on the tile boundaries. The proof of the following theorem is straightforward.

Theorem 5.3. A solution of an arbitrary instance of the global routing problem is planar if and only if every tile is planar.

An instance of the SLGRP is specified by a set η of multiterminal nets, a w-tessellation, and the two capacity matrices C_v and C_h, corresponding to the vertical and horizontal capacities, respectively. Note that with different values of w, there are different C_v and C_h. A solution to an instance of the SLGRP(η, w, C_v, C_h) is a collection of wires ϖ, where each wire connects all the terminals of a net, such that

- $d_h(i, j) \leq c_h(i, j)$ and $d_v(i, j) \leq c_v(i, j)$, for all i and j; and
- every tile is a planar tile.

Next, an algorithm [217] is described that solves an arbitrary instance of the SLGRP involving only two-terminal nets. First, the estimated congestion is established, and stored in a data structure, *congestion-map* (CM). Second, the sequence of nets to be processed is determined. Then a global routing is found (one net at a time) such that it

- avoids passing through congested areas (according to the information in CM),
- satisfies the capacity constraint on each tile's boundary, and
- maintains the planarity in each tile.

Let the length of a wire be the number of tile boundaries it passes through. A trivial lower bound on the length of a two-terminal net is the Manhattan-distance of its two terminals, when length of a tile is considered to be one unit. A net with two terminals (x_1, y_1) and (x_2, y_2) can be routed with no bends (a bend being a 90° turn) if and only if either $x_1 = x_2$ or $y_1 = y_2$ (i.e., the two terminals are colinear). If $x_1 \neq x_2$ and $y_1 \neq y_2$, then there are exactly two ways to connect these two terminals using only one bend. These two ways are shown in Figure 5.4 and are called the upper-L and lower-L routing. The *net-box* of a net N_i (denoted by $box(N_i)$) is the union of its upper-L and lower-L routings. CM is the result of superimposing a box(N_i), for all i, onto the routing plane. Each tile $\tau(i, j)$ of CM, denoted by CM(i, j), records the number of net-boxes passing through it.

The sequence of nets is determined by three factors: $L(N_i)$, the Manhattan-distance of the net; $G(N_i)$, the congestion along box(N_i); and $P(N_i)$, the priority of the net. Define the congestion of a net N_i, denoted by $G(N_i)$, as the number of net-boxes that cross box(N_i). Here, a linear combination of $L(N_i)$, $G(N_i)$, and $P(N_i)$ (i.e., $f(N_i) = \alpha_1 \cdot L(N_i) + \alpha_2 \cdot G(N_i) + \alpha_3 \cdot P(N_i), 0 \leq \alpha_1, \alpha_2, \alpha_3 \leq 1, \alpha_1 + \alpha_2 + \alpha_3 = 1$) is used to order the nets.

The variable w in a given instance of the SLGRP(η, w, C_v, C_h) plays an important role, since w determines how much information is provided for detailed routing. When $w = \min(X_{max}, Y_{max})$, the solution of the SLGRP provides no information for detailed routing (since all nets are in the same tile). As w gets smaller, more precise information is provided. When $w = 1$, the solution of the SLGRP is indeed a detailed routing.

The two terminals of each net are arbitrarily called the source terminal and the target terminal. Consider the processing of a net N_a. To each tile $\tau(i, j)$ associate a cost $C(i, j)$, which represents the cost of the cheapest path from $\tau_s(N_a)$ to $\tau(i, j)$, where $\tau_s(N_a)$ is the tile that contains the source terminal of net N_a. Also, each tile $\tau(i, j)$ has a pointer $P(i, j)$ to its predecessor in the cheapest path. Since the proposed algorithm runs in a breadth-first manner, a data structure, *LIST*, is used to record all tiles that will be processed at each step. For each tile $\tau(i, j)$ in *LIST*, the *directional cost* (denoted by $DIR\_C(i, j)$, where $DIR \in \{N, E, S, W\}$) is the cost of tile $\tau(i, j)$ plus the cost of passing through the boundary between

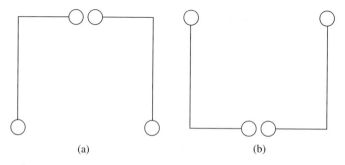

(a)   (b)

FIGURE 5.4
Two ways to route a net with one bend. (a) Upper = L; (b) lower = L.

$\tau(i, j)$ and its neighbors (denoting the passing through cost by $DIR\_PC(i, j)$; that is, $DIR\_C(i, j) = C(i, j) + DIR\_PC(i, j)$). The algorithm is based on Dijkstra's Shortest Path Algorithm [85] and aims to maximize the number of routed nets.

When evaluating the cost (of wire W) of passing through the boundary of tile $\tau(i', j')$ and tile $\tau(i'', j'')$, the following rules are used.

1. If there is not enough boundary capacity, set the passing cost to infinity.
2. If the passing path does not maintain the local planarity in $\tau(i', j')$, set the passing cost to infinity.
3. If neither rule 1 nor rule 2 applies, then $DIR\_PC(i', j') = \beta_1 \cdot l + \beta_2 \cdot CM(i'', j'')$, where l is the length of the cheapest path from $\tau_s(N_a)$ to $\tau(i', j')$, $0 \le \beta_1, \beta_2 \le 1, \beta_1 + \beta_2 = 1$.

When applying rule 3, we also record the relative position of W on the boundary that it passed through (so that the trace-back process can be done efficiently). Constants β_1 and β_2 in rule 3 reflect the importance of path length relative to area congestion. For example, when net length is more important, β_1 is larger than β_2.

Elements in CM are updated before each net N_i is processed, where the value of each element that was crossed by box(N_i) is subtracted by one to reflect the current congestion (since a net cannot block itself).

5.3 SINGLE-LAYER DETAILED ROUTING

A number of solutions to the single-layer detailed routing problem have been obtained; it is one of the best understood problems in VLSI physical design. There are three classes of solutions. The first class assumes a set of modules, terminals, and nets are given and tries to obtain a detailed routing. The second class assumes that, in addition to the previous information, the global routing, a homotopy, is given. The third class considers restricted regions such as channels and switchboxes, that is, a bounded routing region. Whereas the general problem in the first class is NP-hard, the problems in the latter two classes can be solved in polynomial time.

5.3.1 Detailed Routing in Bounded Regions

In the boundary single-layer routing (BSLR) problem, there is a planar graph (e.g., a grid), a collection of terminals on the boundary of the infinite face, and a set of multiterminal nets. Typical examples are channels and switchboxes. A solution of a BSLR consists of a set of vertex-disjoint Steiner trees interconnecting the terminals belonging to the same multiterminal net.

The earlier problems along these lines consider a channel environment and assume all nets are two-terminal nets. The corresponding problem is called the *river-routing problem*. Necessary and sufficient conditions, along with efficient algorithms, for this problem have been proposed [62, 210, 250, 274]. The main result shows how to calculate the optimal channel width based on the position of terminals.

The following describes a technique that handles multiterminal nets and routing in (or around) arbitrarily shaped regions. The technique also handles arbitrary routing environments, that is, arbitrary graphs, including square-grid and 45° grid.

The BSLR problem can be described as follows. Consider a planar undirected graph $G = (V, E)$ with a fixed embedding into the plane (e.g., a grid). The embedding represents a routing region. There is a set $N^* = \{N_1, \ldots, N_n\}$ of multiterminal nets, where each net $N_i \in N^*$ is a sequence of vertices $\{t_{i,1}, \ldots, t_{i,k_i}\}$ on the infinite cycle. k_i is the number of terminals of N_i, that is, $k_i = |N_i|$. Call this cycle the boundary, and $N_i \cap N_j = \emptyset$, for $i \neq j$. The boundary single-layer routing problem is solvable if and only if there is a set $T^* = \{T_1, \ldots, T_n\}$ of pairwise vertex-disjoint trees in G, such that T_i interconnects the terminals of net N_i.

In a BSLR problem, two concepts must be considered, the topological realization and the detailed realization. In the topological realization, the configurations of nets (i.e., the ordering of the nets' vertices along the boundary) are examined to see whether a single-layer routing exists, assuming unlimited capacity is available. The detailed realization is concerned with whether there is enough space (i.e., capacity) to route the nets in one layer such that no two nets cross each other. Figure 5.5 gives examples of unsolvable, topologically realizable, and solvable instances. It is assumed that boundary edges can also be used for routing; if they cannot, they are removed before the routing process.

The test for the topological realization is as follows. Arbitrarily choose a vertex O_p on the boundary as the starting point. The first terminal $t_{i,1}$ of a net N_i is the one closest to O_p in a clockwise scan of the boundary of G. The jth terminal $t_{i,j}$ of a k-terminal net N_i is the jth terminal of N_i visited in a clockwise scan of the boundary of G, starting from O_p. Scan the terminals of N^* starting from O_p in a clockwise manner. If the jth terminal of a k-terminal net N_i, $1 \leq j < k$, is visited, push it on STACK—a first-in-last-out data structure. If the kth terminal of N_i is visited and all the top $k - 1$ elements of STACK are terminals of N_i, then pop the top $k - 1$ elements; otherwise, take no action. FINAL-STACK is a

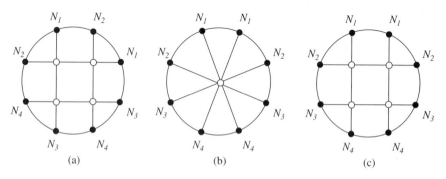

FIGURE 5.5

Possible conditions for an instance of BSLR. (a) Unsolvable; (b) topologically realizable; (c) solvable.

configuration of STACK corresponding to an instance of the BSLR after all the terminals are processed. The algorithm is called the STACK algorithm.

Lemma 5.1. If FINAL-STACK corresponding to an instance of the BSLR is not empty, then the instance is not solvable.

The correctness of the lemma can be verified from the following observation. Consider the situation in which FINAL-STACK is not empty. Since FINAL-STACK is not empty, there must be at least two nets left in FINAL-STACK after all the terminals are processed. No matter which net is routed first, the other net cannot be routed without crossing the previously routed net. Thus the instance of the BSLR is not solvable. Note that in the STACK algorithm every terminal is visited at most twice, once when it is pushed and once when it is popped. Thus the following corollary is established.

Corollary 5.1. It is possible to test in linear time whether an instance of the BSLR is topologically realizable.

If an instance of the BSLR problem is topologically realizable, proceed in the following manner. The nets are processed as dictated by the sequence in which they were popped. When a net is popped from STACK, route that net using the (current) outermost available edges (i.e., of the edges on the boundary) and delete those edges after they are assigned to a tree T for connecting the terminals of the popped net. Also, to ensure vertex-disjoint paths (i.e., single-layer routing), delete all edges directly connected to the tree T; those edges are denoted by T^{adj}. An edge is *directly connected* to T if one of its vertices is in T. If there is no available edge before the end of the scan process, then report that the given instance is unsolvable and stop the process. As an example, consider Figure 5.6. The net N_i is routed on edges a, b, c. Then edges $1, \ldots, 6$ are removed and net N_i is routed. This method is called greedy-routing. An informal description follows.

Greedy-routing:
begin-1

Step 1: Call STACK algorithm to check if the instance
of the BSLR is topologically realizable;
if not, then report "Unsolvable" and stop.

Step 2: Process nets according to the sequence they were
popped in STACK algorithm. For each net N_i,
do as follows:

Step 2.1: Route N_i using the (current) outermost available
edges and delete those edges (denoted by T_i).

Step 2.2: If there is no available edge for routing N_i, then
report "Unsolvable" and stop.

Step 2.3: Edges that directly connect to T_i are also deleted.

Step 3: Report "Solvable," output the result, and stop.

end-1.

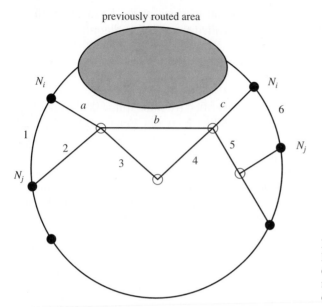

FIGURE 5.6
Illustration of greedy-routing:
edges $\{a, b, c\}$ are assigned to
net N_i; edges $\{1, \ldots, 6\}$ are
deleted.

Theorem 5.4. Greedy-routing finds a solution in $O(e)$ time to an instance of the BSLR if and only if it is solvable [5, 216].

Proof.

   only if. Trivially true (by definition).

   if. Let $S_g = \{T_1, \ldots, T_n\}$ be the result of greedy-routing, where T_m (being a tree) denotes the interconnection of the mth routed net N_m (choosing an arbitrary terminal as O_p). A solution $S_g(m) = \bigcup_{i=1}^{m} T_i$ is called a *boundary-dense solution* of the first m nets if none of the edges used in $S_g(m)$ can be used by nets in $\{N_{m+1}, \ldots, N_n\}$. If $S_g(n)$ is boundary-dense, then the entire solution is a boundary-dense solution. Note that the boundary-dense solution is not necessarily unique. Denote a boundary-dense solution by $S_{bd} = \{T'_1, \ldots, T'_n\}$, where T'_i connects the same set of terminals as T_i does, $1 \leq i \leq n$. Define interval $I_{i,j}$ as the boundary vertices between the jth and the $(j+1)$st terminals of net N_i (starting from O_p, in a clockwise manner).

   Clearly an instance of the BSLR is solvable if and only if at least one boundary-dense solution exists. Let $S_g(m) = \bigcup_{i=1}^{m} T_i$. It will be proved by induction on m that there exists a boundary-dense solution S_{bd} such that $S_g(m)$ is equivalent to $S_{bd}(m)$, where $S_{bd}(m) = \bigcup_{i=1}^{m} T'_i$. Therefore, the result of greedy-routing is a boundary-dense solution (if the given instance of the BSLR is solvable).

   When $m = 1$, no other net has terminals in the intervals $I_{1,1}, \ldots, I_{1,k_1-1}$ ($k_1 = |N_1|$) of net N_1; otherwise, net N_1 will not be popped from STACK. According to greedy-routing, net N_1 is connected by the outermost available edges on the boundary of G_0 (i.e., the input graph G). After N_1 is routed, delete those edges (i.e., the edges of T_1) and the edges directly connected to T_1 from G_0. The resulting graph is denoted by G_1. By definition, $T'_1 \in S_{bd}$ is also connected by the edges on the boundary; thus $S_{bd}(1)$ is equivalent to $S_g(1)$.

Assume $S_{bd}(m)$ is equivalent to $S_g(m)$, for all m, $m \leq a - 1$. Then consider the situation when $m = a$. All nets that have terminals in the intervals of N_a must be routed before N_a if the instance of the BSLR is solvable. T_a is the tree obtained by greedy-routing, that is, T_a is the interconnection of edges on the boundary of G_{a-1}. Assume $T'_a \in S_{bd}$ is not connected by the edges on the boundary of G_{a-1}. Since $T'_1, \ldots, T'_{a-1}$ are already routed, there exists a tree T''_a that connects N_a by edges on the boundary of G_{a-1}. This is a contradiction to the assumption; therefore, T'_a (which is equivalent to T''_a) is connected by the edges on the boundary of G_{a-1}, and thus T_a is equivalent to T'_a. By definition, $S_g(a) = S_g(a - 1) \cup T_a$ and $S_{bd}(a) = S_{bd}(a-1) \cup T'_a$. This shows that T_a is equivalent to T'_a and $S_{bd}(a - 1)$ is equivalent to $S_g(a - 1)$ (by induction); that is, $S_g(a)$ is equivalent to $S_{bd}(a)$.

Since greedy-routing finds a boundary-dense solution at each step, then no edge is wasted. Thus a solution will be found, if one exists.

Since the STACK algorithm takes $O(n)$ time and each edge is visited at most once during the detailed routing process, greedy-routing needs $O(e)$ time, where e is the number of edges of the input graph G.

5.3.2   Detailed Routing in General Regions

Two other types of detailed routing algorithms have been proposed. The first type obtains a detailed routing from a global routing, a homotopy. The second type considers detailed routing, given a placement.

Consider a set of modules and a collection of nets, where for each net a global routing (or homotopy) is given. See Figure 5.7a. The goal is to obtain a detailed routing as dictated by the given global routing, as shown in Figure 5.7b. As discussed in previous subsections, the given global routing must be planar; otherwise, a solution does not exist.

This problem was studied in [209]. A *straight cut C* is a line segment connecting two modules (i.e., the endpoints of the cut are on the boundary of two

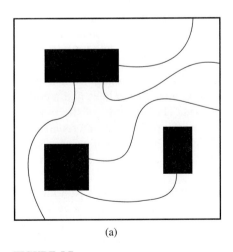
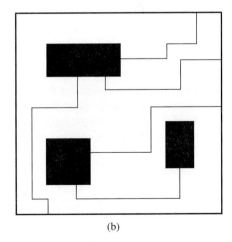

(a)   (b)

FIGURE 5.7
A given single-layer global routing. (a) Input; (b) output.

modules) and not crossing any other modules. A *degenerate* cut is one whose endpoints are on the same module or on the two terminals of a net; otherwise, a cut is *nondegenerate*. The capacity $K(c)$ of each cut is the maximum number of nets that can go through the cut. The density $d(c)$ of each cut is the number of nets, in the given global routing, that pass through the cut.

> **Theorem 5.5.** A layout is realizable (i.e., there exists a detailed routing) if and only if no nondegenerate straight cut has a density exceeding its capacity [209].

The theorem states that the global realization problem can be transformed into a set of local realization problems. It also alleviates the difficulty of considering all the nets that could possibly pass through a cut (as a detour). Furthermore, it was shown that there is a realization with the length of each net at its minimum. The algorithm proposed in [209] runs in $O(n^2 \log n)$ time, where n is the number of terminals plus the number of modules.

A technique for generating rubber-band sketches is described in [77]. A sketch represents one layer of a design. It consists of a set F of rigid objects, called *features*, and a finite set W of flexible interconnecting wires. A wire in W is a simple path between two points in F (these points are on the objects of F). No two wires may intersect each other and no wire may cross itself. A sketch represents a topological routing of one layer. In a rubber-band sketch, a rubber-band wire is a wire with minimum-length routing that is a polygonal path whose vertices are the points in F.

The first step for producing a legal routing of a rubber-band sketch is to satisfy the width and spacing design rules by pulling the wires from the points. The geometric routing may be rectilinear, octilinear, or even curvilinear. In order to perform refinements on a layout, some basic updating mechanism must be provided to move a point in F without changing the routing topology. In other words, as a point in F is moved, the mechanism should update the sketch as if rubber bands were being pushed/pulled about a sketch. Constrained Delaunay triangulation (see [280] for a definition) is used for the basic data representation for the rubber-band sketch.

Now consider the problem of obtaining a detailed routing from a given placement. Consider a routing region, specified by a set of modules and a collection of terminals on the boundary of the modules. The positions of the modules are fixed; that is, the placement is given. Consider the problem of interconnecting the terminals, as specified by a set of nets, in one layer. The solution consists of a set of pairwise noncrossing wires realizing the nets. The main algorithm was proposed by Marek-Sadowska and Tarng [243].

Consider a graph $G = (V, E)$, where each vertex corresponds to a module, an isolated terminal, or the external boundary—for simplicity, all three will be referred to as modules. There is an edge between two vertices if there is a net interconnecting the corresponding modules.

Finding if the given circuit (i.e., placement) is planar is equivalent to testing for planarity of the graph G, that is, to testing if a planar drawing of G can be obtained. Different planarity tests are needed depending on the assumptions made, for example, if the modules are flippable or not. Here, routing a net is equivalent to merging two modules; that is, after routing the two modules can be thought of as one module. Once all the modules are merged, the problem is transformed to routing in a bounded region.

An example is shown in Figure 5.8. Figures 5.8a, b show a placement and the corresponding graph. A planar embedding of the graph is shown in Figure 5.8c. The embedding dictates a (topological) planar routing, as shown in Figure 5.8d. The topological routing can be used to produce a detailed routing.

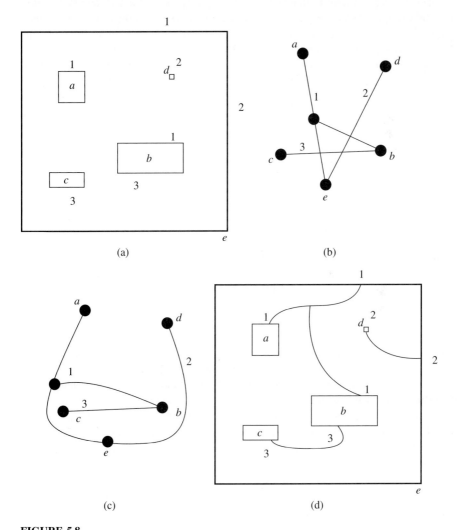

FIGURE 5.8
An instance of a circuit and the corresponding graph. (a) Input; (b) hypergraph; (c) an embedding; (d) a planar (topological layer) layout.

5.4 WIRE-LENGTH AND BEND MINIMIZATION TECHNIQUES

This section discusses two post-processing problems for single-layer routing, the problems of minimizing the length and the bend of a given single-layer layout.

5.4.1 Length Minimization

A drawback common to most single-layer routing algorithms is the introduction of long wires. The reason is that good single-layer routing techniques try to maximize the area usage or, equivalently, to minimize wasted areas. To do so, they start routing close to the boundaries. This minimizes wasted areas; indeed, this strategy produces area-optimal routings. However, it also produces long wires. Heuristics to minimize total wire lengths were proposed [149]. Another application of single-layer wire-length minimization is in multilayer routing environments (e.g., in multichip modules). Assume a routing strategy assigns a set of wires into layers such that the total length of wires in each layer is minimized. Later, the algorithm may move one wire from layer i to another layer. After that, the total length of wires in layer i may not be minimum. Thus a technique to minimize their length is needed. This section discusses the single-layer wire-length minimization problem involving two-terminal nets (SLWP); see Figure 5.9.

In a given single-layer layout $\mathcal{L}$ there is a collection of terminals and modules. Allow the terminals to be either on the boundary of the modules or isolated in the plane (or both). There is also a collection of pairwise non-intersecting wires, each of which interconnects a subset of terminals. Think of the wires as rubber bands and of the modules as obstacles; the transformations to minimize wire lengths that are allowed are those permitted by stretching and shrinking the rubber bands without crossing the modules. Such transformations are homotopic transformations. In other words, only the possible detailed routings that are consistent with the original global routing are considered. If more general transformations are allowed, then the problem becomes equivalent to the problem of finding a single-layer routing from scratch, and, thus, the problem becomes NP-hard.

Part of a wire that contains a sequence of three wire segments forming the shape of the letter U is called a *U* (see Figure 5.10a). The middle segment of a U is called the *ceiling*, and the others the *legs*. A U is said to be *empty* if nothing exists one unit under the ceiling. An empty U can be pushed. A push operation on an empty U deforms the wire by sliding down its ceiling along its legs and shortening the excessive wires on the two legs (see Figure 5.10b). A push operation thus decreases the length of a wire. A *greedy push operation* on an empty U refers to the deformation of the U by sliding its ceiling down the legs as much as possible. The greedy push operation pushes the ceiling until it hits the other end of the shorter leg or a feature (i.e., an obstacle or a fixed wire), as shown in Figure 5.10b. Here the greedy push operation will also be referred to simply as the push operation. Note that the push operation as defined is a homotopic transformation; that is, it continually deforms a wire without crossing any other features.

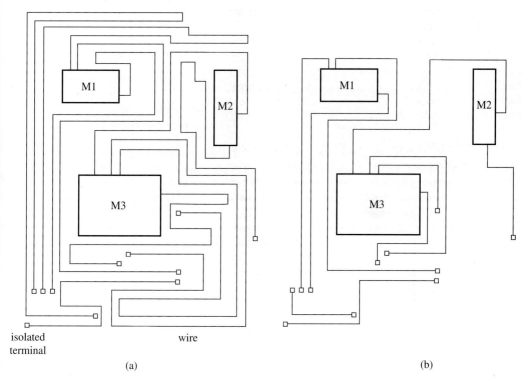

isolated
terminal

wire

(a)

(b)

FIGURE 5.9
An instance of the SLWP and a solution. (a) An input layout; (b) an optimal layout.

At any instant, a U that cannot be pushed is called a *fixed* U and is classified as a feature. For example, the U in Figure 5.10a is not fixed, and, thus, it is not a feature. However, the U in Figure 5.10b is a fixed U, and, thus, it is a feature. The terms pushing a U and fixing a U will be used here interchangeably. Certainly, a fixed U cannot be pushed. A single-layer layout with minimum total wire length, under homotopic transformations, is an optimal layout.

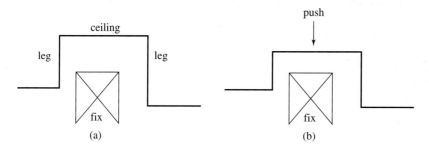

FIGURE 5.10
Definition of an empty U and the push operation. (a) An empty U; (b) pushing an empty U.

Lemma 5.2.   Minimization Lemma. A single-layer layout is optimal if and only if it does not contain any empty U's [139].

Proof.

if. Certainly, if a layout contains an empty U, then the length of at least one wire can be reduced (by being pushed).

only if. Assume there is no empty U, and the length of the given layout $\mathcal{L}$ can be reduced to obtain a new layout $\mathcal{L}'$. There must be at least one net N whose length has been reduced in going from $\mathcal{L}$ to $\mathcal{L}'$. Let q denote the layout of N in $\mathcal{L}$ and q' denote the layout of N in $\mathcal{L}'$. Layout q was transformed to q' by means of homotopic transformations. Think of homotopic transformations as stretching and shrinking the nets without crossing any features. To transform q to q' it is necessary to transform (push) at least one of the U's of q. Let that U be called u_1. There must be another U below u_1. Denote that U by u_2. The length of the ceiling of u_2 is smaller than the length of the ceiling of u_1. Inductively, it is concluded that a U, denoted by u_s, with a ceiling of length 1 or 2 must have been pushed in this process, that is, in the process of transforming q to q'. By inductive hypothesis, such a U was not pushable, and there is nothing under its ceiling. A contradiction is reached.

Procedure GREEDY ($\mathcal{L}$);
   begin-1
        while there exists an empty U *do*
        begin-2
            u := an empty U;
            push(u)
        end-2;
   end-1.

Note that GREEDY is a very simple algorithm. However, it is not clear how to implement it efficiently. If the sequence of push operations is not carefully chosen, then GREEDY requires $O(n^3)$ time. In general, it is necessary to identify those U's that should be pushed first in order to avoid redundant push operations.

Focus on U's. At any step of the algorithm, a U may exist that is not classified yet. Such a U becomes an empty U if the segments immediately below it are pushed away, and it becomes fixed if one of them is found to be fixed. An overview of the algorithm is given first and then the details of the implementation are discussed.

Consider a collection of vertical line segments S in the plane.[1] Two line segments s_i and s_j are said to be horizontally visible if a horizontal line segment can be drawn that intersects only s_i and s_j (and no other line segments of S). The horizontal visibility graph $G_h = (S, E_h)$ has a vertex for each vertical segment. An edge indicates that the corresponding segments are horizontally visible. If s_i is to the left of s_j, then we say s_i is *left-visible* from s_j, and s_j is *right-visible* from s_i. Vertical visibility and the corresponding graph $G_v = (S, E_v)$, along with *up-visibility* and *down-visibility*, are symmetrically defined. It is known that G_h and

---

[1] A point is a degenerated line segment.

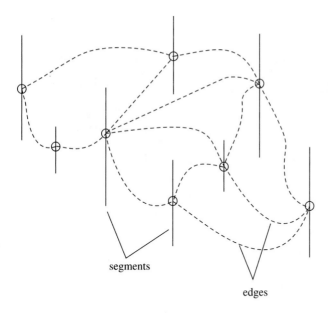

segments

edges

FIGURE 5.11
An example of vertical visibility.

G_v are planar graphs, and, thus, each has $O(n)$ edges [313], where $n = |S|$. G_h can be shown to be planar by starting with a drawing of vertical segments and visibility edges (see Figure 5.11) and then shrinking each segment into a point (i.e., one vertex). The resulting figure is a drawing of G_h without any edge crossings. Thus it is planar. Certainly, the same is true for G_v. In [313], the authors also gave an $O(n \log n)$-time algorithm to construct the (vertical or horizontal) visibility graph of S. Here, it is assumed that the segments are closed, i.e., that each segment contains both of its endpoints.

Consider the horizontal visibility graph of the wire segments in a single-layer layout $\mathcal{L}$. With reference to Figure 5.12a, consider a *nes*-type U (i.e., north-east-south are the three directions defining the U, as shown in Figure 5.10) consisting of ceiling t and legs l and r. The set of line segments that are right-visible from l are shown by dashed directed edges. The topmost of these is r. The next one is segment s, which dictates how far the U can be pushed. Such a segment s may not exist (see Figure 5.12c). The information about these segments and how far it is possible to push a U can be extracted from the corresponding visibility graphs, as will be shown. After a U is pushed, however, it is necessary to update the graphs (see Figures 5.12b–d). Note that in Figure 5.12, vertical visibility remains unchanged except, as in Figure 5.12c, where it is necessary to merge two segments. Here, the corresponding records (each segment is associated with a record) and, for that matter, the corresponding vertical visibility lists need to be merged.

A U with one leg completely visible (i.e., one leg in one direction can see only the other leg) to the other leg is called a *box*, for example, the U with ceiling

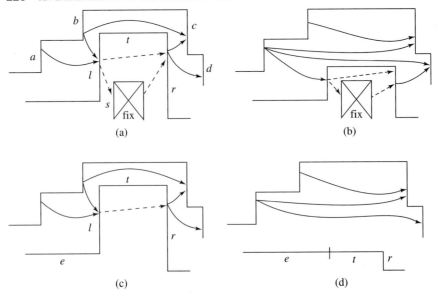

FIGURE 5.12
Visibility of empty U's when they are pushed. (a) Case 1; (b) pushing case 1; (c) case 2; (d) pushing case 2.

t in Figure 5.12c. Removing a box is one way of fixing a U. As indicated in GREEDY, our strategy is to push U's, one by one; it includes removing boxes. A linked list, called *fixable-list*, stores all fixable U's. As will be described, boxes have priority in the pushing operation; thus, boxes will be stored in front of the list and other U's at the end. U's to be processed will be selected from the front of the list.

Fixable-list is initialized in the following manner. Consider the segments down-visible from the ceiling of a nes-type U. If all these segments are fixed and at least two units below the ceiling, then the U is added to fixable-list. Also, if one leg of the U is left-visible from the other leg and not left-visible from any other segments, then that U is classified as a box and is added to fixable-list; symmetrically, if one leg is right-visible from the other leg and not right-visible from any other segments, the same process holds. Similar analysis holds for other types of U's. Thus, to initialize fixable-list, visit every edge of G_h and G_v once. That is a total of $O(n)$ time. As a fixable U is pushed, to become fixed, it is necessary to update fixable-list and both (vertical and horizontal) visibility graphs. As discussed before, priority is given to boxes when fixing fixable U's. Let l and r be the legs and t be the ceiling of the U that was fixed.

When a nes-type U is fixed:

- Vertical visibility will not change.
- Horizontal visibility must be re-examined for segments that were left-visible from l and segments that were right-visible from r.

- Fixable-list must be updated. The current U is removed from fixable-list. If there is only one segment that is up-visible from t, it has to be checked, as it may have become a fixable U. In addition, if the current U is a box, then a new U may be created with t as part of its ceiling (see Figure 5.12c). The new U must also be examined, as it may have to be added to fixable-list.

A more explicit description of GREEDY, based on the notion of visibility, follows.

> *Procedure* GREEDY2 $(\mathcal{L})$
>   *begin-1*
>   *initialize* visibility graphs;
>   *initialize* fixable-list;
>   *while* fixable-list is not empty *do*
>   *begin-2*
> $u :=$ an empty U;
>   *push*(u);
>   *update* fixable-list and visibility graphs;
>   *end-2*;
>   *end-1.*

The following theorem can be proven by observing that each push operation takes $O(n)$ time, where n is the total number of segments in the given layout. Note that after each push operation it may be necessary to update all visibility information, and there are $O(n)$ visibility pairs.

> **Theorem 5.6.** GREEDY2 finds an optimal layout of $\mathcal{L}$ and runs in $O(n^2)$ time, where n is the number of bends in $\mathcal{L}$.

5.4.2 Bend Minimization

This section shows how to minimize the number of bends in a single-layer layout, assuming a length-minimization algorithm has been applied. The homotopy will not be changed nor will the length be increased in the process.

A *section* is a subset of connected line segments in a wire denoted by $S(p_1, p_2)$, where p_1, p_2 are the endpoints. A *monotone section* is a section of a wire with the following properties: it is cut by any vertical line or horizontal line at most once, and no other monotone sections properly contain it. See Figure 5.13. The wire that contains the minimum number of bends among all its homotopies is called an *optimal-bend routing*. To reduce the total number of bends in a single-layer layout, an optimal-bend routing is found for each wire, one at a time, using the greedy-slide algorithm.

The greedy-slide algorithm first divides the given wire into a set of monotone sections. For example, in Figure 5.13, an artificial point p_{mid} is introduced in the middle of segment (d, e). Then sections $S(a, p_{\text{mid}})$ and $S(p_{\text{mid}}, h)$, are processed as follows.

In the following discussion, it will be assumed (without loss of generality) that the monotone section S goes from the northeast toward the southwest. The

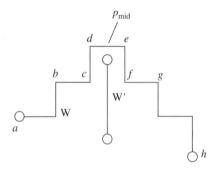

FIGURE 5.13

An example of monotone sections. $S(a, e)$ and $S(d, h)$ are monotone sections of wire W, while $S(b, g)$ is not a monotone section.

starting and ending points of S are denoted by p_s and p_e, respectively. (For example, in Figure 5.13, $p_s = p_{mid}$ and $p_e = p_a$.) The bends of S are numbered from 1 to k, where bends with smaller indices are closer to p_s. By slightly abusing the definition of bend, denote p_s by b_0 and p_e by b_{k+1}. Let $R(b_{i-1}, b_{i+1})$ $(0 \le i \le k)$ be the rectangle region having bends b_{i-1} and b_{i+1} as corners. $R(b_{i-1}, b_{i+1})$ is empty if it contains no obstacle. Denote the line segment with b_{i-1} and b_i as endpoints by $L(b_{i-1}, b_i)$.

For each monotone section, slide every bend, from b_1 to b_k, one bend at a time. When we slide bend b_i toward b_{i+1}, two cases are possible:

Case 1. If b_i meets b_{i+1} while $L(b_{i-1}, b_i)$ does not cross any obstacles, then the operation on b_i is a *full-slide*.

Case 2. If $L(b_{i-1}, b_i)$ meets an obstacle before b_i meets b_{i+1}, the operation is a *partial-slide*, and $L(b_{i-1}, b_i)$ is placed at the position (the *stopped-point*) one unit before it meets the obstacle.

It is clear that full-slide can be applied on b_i only when $R(b_{i-1}, b_{i+1})$ is empty. Figure 5.14 gives an example of the greedy-slide algorithm. The segment shown as a thick segment corresponds to the ceiling of a U (e.g., segment de in Figure 5.13). Examples of full-slide are shown in Figure 5.14a, b. Figure 5.14c illustrates partial-slide on b_4.

Theorem 5.7. The greedy-slide algorithm finds an optimal-bend routing for a wire, without changing its homotopy and length.

Now these techniques can be generalized to routing planes with w-tessellation. When $w = 1$, the atomic operation for deforming a wire (to reduce the wire length or the number of bends) is a *one-unit shift* (see Figure 5.15 for an example). Each push operation (on an empty-U) or slide operation is a combination of a finite number of atomic operations. In a one-tessellation environment (as is assumed for the techniques being discussed), no wire (or terminal) exists between two wires in adjacent tiles. Thus, a shift of one unit will maintain the homotopy if the tile that the new bend will occur in is not occupied by any other wires or terminals. However, when $w > 1$, more consideration is needed.

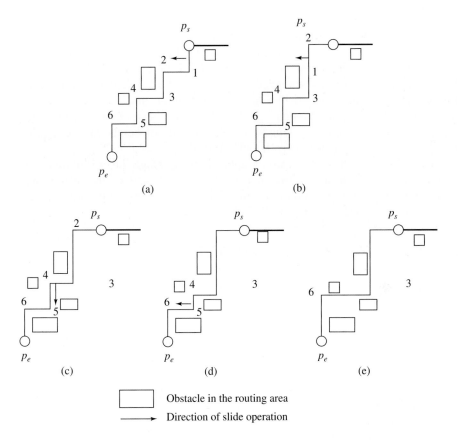

Obstacle in the routing area

Direction of slide operation

FIGURE 5.14
An example of processing a monotone section in the greedy-slide algorithm.

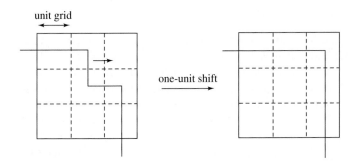

FIGURE 5.15
An example of a one-unit shift.

When $w > 1$, the atomic operation becomes a w-*unit shift*. Note that a w-unit shift can be specified by the tiles where the bend initially occurred and intended to occur. To ensure that the homotopy is maintained during the deformation of a wire in a w-tessellation environment, extra caution is necessary. Both tiles where the bend initially occurred or intended to occur must be checked so that no other wires or terminals are crossed during the deformation, and the tile where the bend intended to occur has enough capacity. With these restrictions, the proposed techniques to reduce the wire length and the number of bends can be applied directly to the environment with w-tessellation, where $w > 1$.

The next two sections review two applications of single-layer routing: over-the-cell routing, how to utilize the area over the cells to reduce the overall area of the chip; and the applications of single-layer routing in a high-performance packaging environment, namely, in MCMs.

5.5   OVER-THE-CELL (OTC) ROUTING

As discussed in earlier chapters, one important routing subproblem in standard cells is the channel routing problem. Many two-layer and three-layer channel routers have been developed, several of which can produce very good solutions (see Chapter 3). Despite this fact, a large part of the area in a typical standard cell layout is still consumed by routing. For this reason, several researchers have suggested use of area outside the channel to obtain a further reduction in channel height [72, 74, 119, 144, 191, 221, 333]. This routing style is referred to as over-the-cell (OTC) routing. Routing over the cell rows is possible due to the limited use of the second (M2) and third (M3) metal layers within the cells.

Over-the-cell channel routing is NP-hard, as it is a generalization of the channel routing problem. The class of two-layer OTC channel routing problem has been studied, and efficient algorithms for the three-layer OTC have been proposed [262]. Here the two-layer OTC problem and the subproblems that arise are discussed.

5.5.1   Physical Model of OTC Routing

In this section, a physical model for over-the-cell channel routing is presented using two metal layers for interconnection. The model assumes that power (VDD) and ground (GND) lines are routed on the second metal (M2) layer in the center of the area over the cell rows leaving two rectangular over-the-cell routing regions, one region bordering the channel below the cell and the other bordering the channel above the cell. The number of tracks available for routing in the rectangular regions is determined by the cell height as well as design rules, such as width of power and ground lines and the minimum allowable spacing between two adjacent wires. Let k denote the number of tracks available for routing in each of the two routing regions. Clearly, the net segments routed in the area over the cell rows must be planar since only the M2 layer can be used for over-the-cell routing. The M1 layer is typically allocated for completing intracell connections.

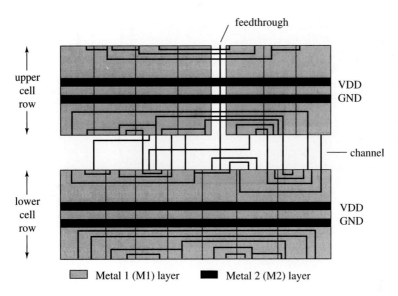

FIGURE 5.16
Physical model.

In the channels, the vertical wires are routed in the M2 layer, and the horizontal wires are routed in the M1 layer. The physical model is depicted in Figure 5.16.

5.5.2 Basic Steps in OTC Routing

This section reviews the basic subproblems that arise in OTC routing and presents an overview of OTC routing algorithms. There are two issues to be considered in OTC routing, the use of vacant terminals to increase the number of nets that can be routed over the cells and the selection of nets to be routed over the cells.

Consider the example in Figure 5.17a. If the area over the cells is not utilized, then a routing with four tracks can be obtained as shown in Figure 5.17b. It is also easy to verify that if the vacant terminals are not employed, the area over the cells cannot be effectively used; that is, the best solution still requires four tracks in the channel. However, the employment of vacant terminals and vacant abutments (i.e., two aligned vacant terminals) reduces the channel width, as shown in Figure 5.17b. In selecting the nets to be routed over the cells, the goal is to minimize the maximum density of the channel (i.e., the maximum clique in the vertical constraint graph) and the longest path in the vertical constraint graph (see Chapter 3 for definitions).

The basic steps in OTC routing follow.

1. **Net decomposition.** Each multiterminal net is partitioned into a set of two-terminal nets. A very simple decomposition scheme is used. The set of terminals of a net N on each side of the channel is considered. The (upper and lower)

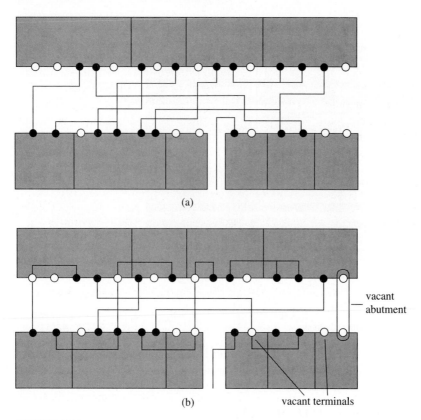

FIGURE 5.17
An example demonstrating over-the-cell routing. (a) Greedy channel router; (b) OTC routing.

subnet is replaced by a set of two-terminal nets. The first net connects the left-most terminal of N to the second terminal of N. The second terminal is connected to the third one by another two-terminal net, and so on (see Figure 5.18a). Note that before the channel routing phase starts, one of the terminals of each net on the upper side of the channel has to be connected to one of the terminals of the net on the lower side of the channel.

2. Net classification. Each net is classified as one of three types; these intuitively indicate the difficulty involved in routing this net over the cells.

Type 1. There is a vacant terminal directly opposite to one of the terminals of the net.

Type 2. There is a vacant terminal between the two terminals of a net.

Type 3. The net does not satisfy either of the above two conditions.

It is easier to route nets of type 1 over the cells than nets of type 2. Similarly, it is easier to route nets of type 2 over the cells than nets of type 3. This step is straightforward.

3. **Vacant terminal/abutment assignment.** Vacant terminals/abutments are assigned to each net depending on its type and weight. The weight of a net indicates the improvement in channel congestion possible if this net can be routed over the cells. Other weights, such as a weight dependent on the timing criticality of a net, can also be included. The basic algorithm is based on a bipartite graph matching. One set of vertices (called net vertices) corresponds to nets, and the other set of vertices (called vacant vertices) corresponds to vacant terminals and abutments. The weight of an edge between a net vertex and a vacant vertex indicates the importance of that assignment. A maximum-weight matching in the graph dictates the assignment.

4. **Over-the-cell routing.** The selected nets are assigned exact geometric routes in the area over the cells. Among all the nets that are assigned to the area over the cell, a planar subset is obtained (using the algorithm CPSP described earlier). Among such nets, a subset with density less than k (the available capacity over the cell) is selected and routed. Such a subset can be obtained employing a dynamic programming algorithm.

5. **Channel segment assignment.** For multiterminal nets, it is possible that some net segments are not routed over the cells and, therefore, must be routed in the channel. In this step, select the best segments for routing in the channel to complete the net connection. The best segment is one that does not increase the longest path in VCG (the vertical constraint graph) and the cliques in HCG (the horizontal constraint graph).

6. **Channel routing.** The segments selected in the previous step are routed in the channel using any of the channel routers discussed in Chapter 3.

5.6   MULTICHIP MODULES (MCMs)

An important aspect of chip design is packaging. The basic concept of packaging was introduced in Chapter 1. This section applies single-layer routing to packaging. A multilayer routing strategy for high-performance multichip modules (MCMs) is discussed whose objective is to route all nets optimizing routing performance and to satisfy various design constraints (e.g., minimizing coupling between vias as well as between signal lines and minimizing discontinuities such as vias and bends).

After an overview of the MCM technique, the pin prewiring and redistribution problem, which redistributes the pins or prewired subnets uniformly over the MCM substrate using pin redistribution layers, is introduced. Pin redistribution is very important in MCM design. It not only provides a global distribution for the pins congested in the chip site over the chip layer so as to ease the future routing difficulty, but also reduces the capacitive coupling between vias induced by many layers by separating the pins far apart. The goal of the problem is to minimize the number of layers required to redistribute the entire set.

5.6.1 An Overview of MCM Technology

Packaging is becoming a limiting factor in translating semiconductor speed into system performance. In high-end systems such as supercomputers, mainframes, and military electronics, 50% of the total system delay is usually due to packaging, and by the year 2000, the share of packaging delay is expected to rise to 80% [13]. Moreover, increasing circuit count and density in circuits have been continuing to place further demands on packaging. In order to minimize the delay, chips must be placed close together. Thus the multichip module (MCM) [22] technology has been introduced to improve system performance significantly by virtue of the elimination of the entire level of interconnection. An MCM is a packaging technique that places several semiconductor chips, interconnected in a high-density substrate, into a single package. This innovation led to major advances in interconnection density at the chip level of packaging. Compared with single chip packages or surface mount packages, MCMs reduce circuit board area by 5 to 10 times and improve system performance by 20% or more. Therefore, MCMs are used in a large percentage of today's mainframe computers as a replacement for the individual packages. The size of MCMs varies widely [105]: 10–150 ICs, 40–1000 I/Os per IC, 1,000–10,000 nets, where the low end is ceramic/wirebond and the high end is thin film/flip chip (the maximum linear dimension is now up to 4–6 inches for thin-film MCMs, up to around 8.5 inches for ceramic MCMs, and up to 18 inches for laminated MCMs).

An automatic layout of silicon-on-silicon hybrid packages developed at Xerox PARC was presented in [278]. Placement determines the relative positions of the ICs automatically or interactively and organizes the routing areas into channels. The hybrid routing uses the topological model that reduces the complexity of hybrid routing by abstracting away the geometrical information. The computation of geometry is deferred until needed. Global routing attempts to find a minimum Steiner tree based on symmetric expansion from all pins. An improvement phase follows to minimize the hybrid area by selecting and rerouting critical nets. Detailed routing is based on the enhanced dogleg router that guarantees routing completion even in the presence of constraint cycles. Once two-dimensional global paths for all nets are defined, the next problem is to assign wires to specific layers, such that some objective function is minimized and the specific constraints are satisfied. This problem is referred to as the *constrained layer assignment* problem. [140, 342] investigated the layer assignment problem that arises in MCM and presented approximation algorithms on the number of xy-plane-pairs. A multilayer router for generating rubber-band sketches is described in [77]. The router uses hierarchical top-down partitioning to perform global routing for all nets simultaneously. Layer assignment is performed during the partitioning process to generate a routing that has fewer vias and is not restricted to one layer and one direction. The local router generates a shortest-path routing.

Another version of layer assignment is to generate a graph based on interference analysis without actually performing the routing in the layer assignment stage. Some metric of pairwise interference between nets can be used to generate a weighted graph. Given a net interference graph (NIG), the problem is to assign vertices (nets) to move the minimum number of colors (layers) so that the total edge weight (interference between nets) is minimized. This problem is referred to as *topological multilayer assignment* [160, 239, 305, 343]. Chen and Wong [47] proposed a channel-based thin-film wiring methodology (using two layers) considering cross-talk minimization between adjacent transmission lines. Then [57] proposed a topological multilayer routing strategy for high-performance MCMs, focusing on cross-talk minimization between nets, while simultaneously minimizing the number of vias and layers.

Another type of performance-driven multilayer routing for MCMs is shown in [182]. The routing problem is formulated as the k-layer planar subset problem, which is to choose a maximal subset of nets such that each net in the subset can be routed in one of k preferred layers.

5.6.2   Requirements on MCM Routers

The primary goal of MCM routing is to meet the high-performance requirements and design objectives, rather than overconstraining the layout area minimization. In general, an efficient MCM multilayer routing algorithm for high-speed systems should have the following features.

- **A way to control propagation delays associated with discontinuities.** Unlike conventional approaches, the electrical characteristics of the packages require the signal lines to be treated as transmission lines. In practice, transmission lines are not perfectly uniform. That is, in the package level, significant reflections can be generated from capacitive and inductive discontinuities along the transmission lines. Moreover, in a multilayer ceramic substrate of MCM, wires at different levels do not have exactly the same impedance. Such mismatches of line impedance can cause reflections from the junction points such as vias and bends. These discontinuities must be controlled in order to keep the resulting reflections to a minimum. Therefore, producing a via/bend-minimum layout for MCM is very important. Furthermore, to reduce the voltage drops at the voltage (or ground) vias, the vias have to be placed far apart from each other. Note that the redistribution vias are neglected in the calculation of the voltage drop because they are five times shorter than the signal vias.

- **A way to reduce cross-talk between signal lines.** Cross-talk noise is a parasitic coupling (i.e., mutual capacitances and inductances) phenomenon between neighboring signal lines. This noise becomes serious in a high-density wiring substrate with narrow line pitch. The closer the lines, the higher they are from the ground plane, and the longer they are adjacent, the larger the amount of coupling that results. The coupling between the lines can be minimized by making sure that no two lines are laid out in parallel or next to each other for longer than a maximum length. Noise can be reduced by keeping the lines far apart. The coupling can also be reduced by placing ground lines between signals. For the signal lines running in parallel in adjacent planes, the impedance of the signal lines can be controlled by placing a ground plane between the planes. The problem can also be eliminated by forcing wires of one layer to be laid out orthogonally (0° and 90°) and wires of its adjacent layers to be laid out diagonally (45° and 135°).

- **A way to reduce the skin effect in thin-film interconnections for ULSI/ VLSI packages.** As the rise time of digital pulses is reduced to the subnanosecond range, the skin effect becomes an important issue in high-speed digital systems. The *skin effect* is defined as: an effect characteristic of current distribution in a conductor at high frequencies by virtue of which the current density is greater near the surface of the conductor than in its interior. As the frequency, conductivity, and permeability of the conductor are increased, the current concentration is increased. This results in increasing resistance and decreasing internal inductance at frequencies for which this effect is significant. The conductor loss caused by the skin effect is the most significant contributor to losses in a microstrip line at high frequencies. It was shown in [159] that the maximum length of interconnections in an ULSI/VLSI package is limited by the skin effect. Therefore, the maximum length (upper-bound of wire pattern length plus via length) can no longer be ignored.

All the kinds of noise mentioned above increase delay or cause inadvertent logic transitions, and so they should be minimized through careful design. In addition to the performance issues just described, there are a number of routing requirements for MCMs (see [252]). Among them are:

- minimizing delay, given a priority weight and maximum/fixed delay assignment;
- equalizing delay for signal groups, with specified tolerance;
- producing minimum bends;
- handling stacked and unstacked vias, and controlling the number of vias;
- assigning layers constrained with the maximum number of vias allowed to a net;
- regulating lengths and delays to meet timing and noise margins;
- being easily extensible as technologies change;
- constraining routes to specified layers (e.g., designated power, ground, or signal layers);
- picking the right kind of via based on the layer change and technology used;
- specifying route ordering; and
- accepting preferred directions on specific layers or nets.

5.6.3 Routing Problem Formulation and Algorithms

To define the routing problem for an MCM, the conventional multilayer routing environment involving multiterminal nets will be used. Assume that a path goes from cell to cell rather than from grid point to grid point. Each plane of an MCM consists of a two-dimensional $m \times m$ grid (called a basic grid) of the plane, with 1×1 being the basic cell-grid size. Therefore, each cell contains at most one pin.

Formally, an instance of the routing problem is a six-tuple, $(k, m, S, T, \lambda, \sigma)$ (see Figure 5.18), where:

- k is the number of layers in the routing environment of the model, including a chip layer, a number of pin redistribution layers, and the signal distribution layers.

- m is the number of rows (or columns) of the tessellation.

- S is the set of grid pins (or contact pads) in the basic-grid of the chip layer (i.e., the top layer of MCM, where each pin in S is marked with O). The set is denoted by $S = \{(i, j) | i, j \in \{1, \ldots, m\}\}$, where (i, j) is the cell containing a terminal of S. Pins are labeled with $1, 2, \ldots, n$. Two pin configurations are

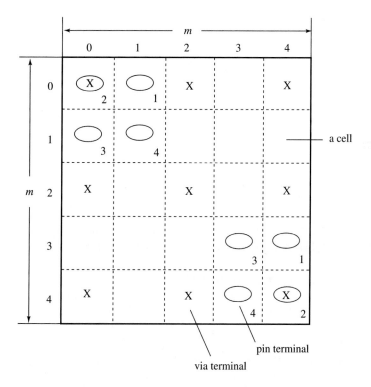

FIGURE 5.18
An instance of the pin redistribution problem, $O \in S$, $X \in T$.

used, square package with pins at the perimeter and a full two-dimensional array of pins (the densest configuration for a chip carrier).

- T is the set of grid points of the imposed uniform via-grid (marked with X in Figure 5.18), denoted by $T = \{(i, j)|i = 0 \ (\text{mod} \ \sigma) \text{ and } j = 0 \ (\text{mod} \ \sigma), i, j \in \{1, \ldots, m\}\}$. $|T| \geq |S|$ so that it is possible to redistribute all pins in S over the uniform via-grid. Note again that the uniform via-grid is used for pin redistribution.

- λ is the specified minimum spacing separating adjacent wires on the same layer. (Recall that intersecting wires must be placed on the same layer.)

- σ is the distance between two adjacent via-grid points in the pin redistribution layers, called the *separation*. After pin redistribution, signal distribution layers contain vias to be placed on grid points with a spacing of σ. The separation σ determines the minimum spacing for signal distribution layers in [22]. However, multiple wiring tracks are allowed between adjacent vias. Thus, in each signal distribution layer, $(\lfloor(\sigma - 2)/\lambda\rfloor + 1)$, signal lines (tracks) are allowed to run between adjacent via-grid points. Note that σ should be greater than or equal to λ so that the separation σ does not violate the minimum spacing in signal distribution layers.

Here the set of pins in S and the set of grid points in T are called *pin terminals* (or *source terminals*) and *via terminals* (or *target terminals*), respectively. A net is the set of electrically equivalent source terminals. The model does not assume that via holes are drilled through all the k layers. Thus a path on layer l can either change layers at any via terminal that is not occupied by any other net or cross over any empty via terminal. This allows the same xy location to be used as a via terminal by different layers, thus reflecting the availability of segmented vias in the MCM fabrication technology [342]. Each of the pin redistribution layers and signal distribution layers consists of a via-grid superimposed uniformly on the basic grid. The cell-grid size of the via-grid for the pin redistribution layers is $\sigma \times \sigma, \sigma > 1$. Note that the cell-grid size of the via-grid for the signal distribution layers is $\sigma \times \sigma$ or $\lambda \times \lambda$. Thus vias are only allowed in such via-grid points.

A single-layer routing assigns a net entirely to a specific layer, and an xy-plane-pair routing assigns a net to layers by splitting a net into one or more layers, to allow preferred wiring directions, horizontally or vertically, to layers.

THE PIN REDISTRIBUTION PROBLEM. In MCM routing, first, a pin redistribution, which aims to distribute pins uniformly using the pin redistribution layers, is done. The problem of pin redistribution is to assign $s \in S$ to the closest unique via-grid point $t \in T$ in the uniform grid. This is the pin redistribution problem (PR or PRP). As mentioned earlier, the pin redistribution layers have been used to provide a minimum spacing between signal wires in signal distribution layers in some MCMs [22]. Also, PRP has been used for engineering changes. The purpose of an engineering change is to correct design errors, enhance design performance, or modify the logic due to changing design specifications.

For the pin redistribution problem, the single-layer routing model is preferred. If we use the xy-plane-pair routing technique for pin redistribution instead, many vias between the xy-plane-pairs are introduced due to multiple bends in a net. Therefore, a planar routing is necessary and sufficient. Minimizing the number of layers is important because it helps to increase the processing yields and improve the package performance. Therefore, in PRP, the goal is to minimize the number of layers k such that $\sum_{i=1}^{k} |P_i| = n$, where P_i is a planar subset of n nets in the ith layer and n is the number of nets. As an output of PRP, an S-T assignment A is specified by a set of pairs such that $A = \{(s, t) | s \in S, t \in T\}$. Figure 5.18 is an instance of a PRP with eight pins in a 5×5 array.

One approach to the PRP is the concurrent maze router (CMR). For the balanced wire distribution of nets to utilize the routing area maximally, an algorithm should process all nets simultaneously. Thus, one idea is to allow the routing procedures to dictate the best choices by giving them equal priority. Consider a two-dimensional array, consisting of $m \times m$ identical cells. In each layer, process the $m \times m$ plane grid, cell by cell, from the upper-left corner to the lower-right corner in the array. Classify the cells into even and odd cells. Every iteration consists of two phases. In the first phase, scan all the even cells, expanding the corresponding nets into the adjacent odd cells; in the second phase, scan all the odd cells. The value $(i + j)$ is even for even cells and odd for odd cells. This ordering scheme is referred to as *even-odd ordering*. In each step of the concurrent expansion process, propagate the cells concurrently based on the ordering scheme and perform the following two tasks.

Phase 1. *Wave propagation.* During the wave propagation, the pairs of points that are closest together are connected first.

Phase 2. *Backtrace.* Once a source terminal (i.e., pin terminal) reaches any target terminal (i.e., via terminal), the two terminals are interconnected using the backtrace algorithm.

We continue this concurrent expansion process for all the nets until either the nets that have not yet been able to find their target terminals have been blocked by existing nets or all the nets have been assigned to target terminals in the current layer. Some source terminals still might be unassigned to target terminals in the current layer. In this case, they are brought to the next layer, where the same greedy strategy is applied again for them to be assigned to the closest unoccupied via terminals in the next layers. This process is repeated until every source terminal is assigned to its unique target terminal.

THE SIGNAL DISTRIBUTION PROBLEM. The redistributed pins are an input to wire the nets in the signal distribution layers. That is, the signal distribution layers are provided for distributing the redistributed chip I/O pins to the various wiring layers. This is the signal distribution problem (SD or SDP). To solve the

SDP, in each signal distribution layer, either single-layer routing or xy-plane-pair routing is performed based on the various wiring rules as required for specific design applications.

The major goal is to guarantee high circuit performance. Thus one good strategy is to assign the critical nets near the top layers so that the number of vias as well as via lengths required for those critical nets will be reduced. Since the routing procedures are being biased by the critical net data, some trade-off should be made to favor circuit performance compared with the processing cost. One effective strategy to solve the SDP is to use the single-layer routing (planar routing) technique exclusively for the critical nets. For the remaining noncritical nets as an input, xy-plane-pair routing is performed, producing a smaller number of layers than when using single-layer routing. The experimental results will show the success of the proposed approach.

If only single-layer routing is to be used, then only the stacked vias (see Figure 5.19 for the two via types) that connect the pins on the chip layer to the signal distribution layers are required. The via type is called the *primary via*. That is, by guaranteeing the minimum number of staircase vias (or avoiding them), the number of layers required for wiring entire nets will be greatly increased (thus, the number of stacked vias will also be increased). Otherwise, using xy-plane-pair routing, the routing model introduces many staircase vias per net between the two layers of each xy-plane-pair. This via type is the *secondary via*. The delay difference between the two via types in thin-film MCMs is minuscule, at less than 10 GHz [105]. However, in certain high-speed applications, the delay effect on the z-direction bends introduced by staircase vias cannot be ignored. Thus a stacked via may be preferred to a staircase via (as illustrated in Figure 5.19, one staircase via introduces two z-direction bends).

The cost of MCM routing can be summarized as cost $= w_k \times$ number of layers $+ w_{\text{strv}} \times$ number of staircase vias $+ w_{\text{strk}} \times$ number of stacked vias $+ w_l \times$ total wire length, where $w_k, w_{\text{strk}}, w_{\text{stkv}}$, and w_l are constants that control the relative importance of the number of layers, the numbers of staircase and stacked vias, and the wire length, respectively. Based on the above cost function, one can choose the best MCM routing strategy for each application.

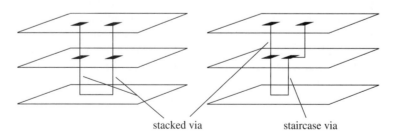

stacked via   staircase via

FIGURE 5.19
Two via types: a *stacked via* is a via where a net changes layers at the same grid points, while a *staircase via* is a via introduced when a net changes layers at the new position.

5.7 DISCUSSION

Earlier research work on single-layer detailed routing problem includes river routing (i.e., single-layer routing of two nets in a channel) and its extension [149, 210, 274], and single-layer routing involving only two nets (e.g., power and ground nets) [349, 396]. The problem of routing in an arbitrary planar graph with no vertex having degree greater than three (terminals can be anywhere in the graph) was shown to be NP-complete [290]. In [163, 284], the problem of detailed single-layer routing, with the restriction that each net has at most one bend, was studied. An efficient algorithm that finds a detailed single-layer routing around a rectangle (where terminals are on the boundary) and aims at minimizing the routing area can be found in [229]. Other research work related to single-layer routing problems can be found in [141, 216, 217, 243].

The single-row routing, first introduced for the backboard wiring, has been one of the fundamental routing methods for the multilayer high-density printed wiring boards (PWBs), due to topological fluidity, that is, the capability to defer detailed wire patterns until all connections have been considered. In the single-row routing, it is assumed that a multilayer board has fixed geometries. That is, the positions of pins and vias are restricted to nodes of a rectangular grid. The global routing process consists of two phases, via assignment and layering, that partition the total wiring requirements into single-row, single-layer routing problems.

In [363], the authors proposed a heuristic for the layering problem of multilayer PWB wiring, associated with single-row routing. The problem is restricted to the special case of street (routing regions above and below the row of terminals) capacities, up to two in each layer, and it is reduced to a problem in interval graphs by relaxing the original problem. The single-row, single-layer routing problem is, given k layers and t tracks per layer, to find k planar subsets of nets using the least number of layers. [117] presented a linear-time dynamic programming algorithm to solve the problem.

[60] solves the problem of sizing power/ground (p/g) nets in a VLSI chip composed of modules, where the nets are routed as trees in the channels between the modules. Constraints are developed in order to maintain proper logic levels and switching speed, to prevent electromigration, and to satisfy certain design rule requirements. The objective is to minimize the area of the p/g nets subject to these constraints. An optimization technique solves the problem more efficiently than the steepest descent method and Newton's method.

[219] described a bus router that is part of a custom IC mask layout system called CIPAR. The router is designed specifically to handle power and ground buses. It can route these nets completely on one metal layer. The router also automatically calculates and tapers the bus width based on current requirements specified in the input circuit description.

An algorithm is presented in [39] for obtaining a planar routing of two power nets in building-block layout. In contrast to other works, more than one pad for each of the power nets is allowed. First, conditions are established to guarantee a planar routing. The algorithm consists of three parts: a top-down terminal clustering, a bottom-up topological path routing, and a wire-width calculation procedure.

A new graph algorithm/ to route noncrossing VDD and GND trees on one layer is developed in [133]. The algorithm finds trees with as small an area as possible under metal migration and voltage drop constraints. Experimental results show that the power wire area is considerably smaller than a previously developed method for single-layer routing.

EXERCISES

5.1. Find a planar subset of a collection of two-terminal nets in a channel. Your algorithm should be more efficient than $\theta(n^2)$ time.

5.2. Design an efficient single-layer length minimization algorithm where there are no U's with vertical ceilings (i.e., all U's are either up-pushable or down-pushable).

5.3. Prove Theorem 5.5.

5.4. Give an instance of wire-length minimization where GREEDY performs $O(n^2)$ push operations.

5.5. Design an algorithm for routing two nets (e.g., power and ground). Consider two cases:
(a) each net has exactly one terminal on each module;
(b) a net may have any number of terminals on a module.

5.6. Consider two modules and a set of nets interconnecting them. Design an efficient algorithm for deciding if the nets are single-layer routable.

5.7. Design a bounded-delay, bounded-length routing algorithm (see Chapter 4) that produces planar routings.

5.8. For a placement of a set of modules and a two-terminal N, design the following. (In all cases below, analyze the time complexity of your algorithm.)

(a) An algorithm that finds a minimum-length routing interconnecting the terminals of net N.

(b) An algorithm that finds a minimum-bend routing interconnecting the terminals of N.

(c) An algorithm that finds a minimum-length routing interconnecting the terminals of N that also has the smallest number of bends.

(d) An algorithm that finds a routing interconnecting the terminals of N minimizing $\alpha l + \beta b$, where l is the length of the routing, b is the number of bends the routing has, and α and β are user-defined constants and are both between 0 and 1.

5.9. Given a circuit consisting of a set of modules and a collection of two-terminal nets, design an algorithm that decides if a single-layer placement and routing of the given circuit exists. Design an efficient algorithm that places the circuit and interconnects all nets.

5.10. Design an efficient algorithm that solves an arbitrary instance of the river-routing problem, a channel consisting of a set of nonintersecting two-terminals nets. Analyze the time complexity of your algorithm.

5.11. Consider a given single-layer detailed routing, involving an arbitrary number of modules and nets. Assume each net is x-monotone, that is, that each net is intersected by any vertical line at most once (see Figure E5.11a). Design an $O(n)$-time algorithm, where n is the input size, that is, the total number of segments (representing the nets and modules). You can spend $O(n \log n)$ time to preprocess the data.

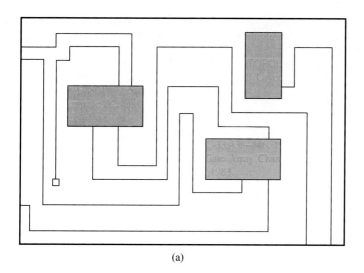

(a)

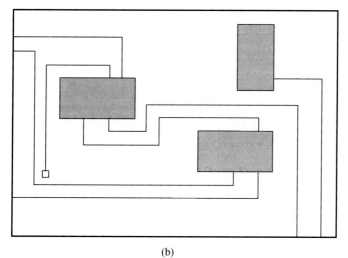

(b)

FIGURE E5.11
(a) An x-monotone layout. (b) A minimum-length layout.

COMPUTER EXERCISES

5.1. Consider a set of multiterminal nets on the boundary of a circle. Implement an algorithm for deciding if the nets are topologically realizable.

Input format.

1 2 2 3 3 4 3 3 2 4 4 1

(corresponding to Figure 5.1).

Output format. The output should show a circle and a YES or NO answer (in this example, NO).

5.2. Consider a switchbox involving a set of multiterminal nets. Implement MNP-ALG to find a maximum number (not maximum weight) of multiterminal nets.

Input format.

1 2 2 3 3 4 3 3 2 4 4 1

(corresponding to Figure 5.1).

Output format. The output should show a circle and the selected nets (as polygons). For example, nets 1, 2, and 3 of Figure 5.1 would be shown.

5.3. Implement greedy-routing in a square-grid switchbox.

Input format.

```
3 6              (* number of rows and columns, respectively, of the grid *)
T 0 2 0 0 5 0 (* terminals on the top row starting from NW corner,
                     0 means empty *)
B 1 0 0 2 0 5   (* there may be terminals on the corners of the switchbox *)
L 0 1 0
R 0 5 0
```

Output format. The output shows a complete routing in the grid or states that the problem is unsolvable (either the topological test fails or the detailed routing phase fails).

5.4. Consider a set of modules. Each module has a set of terminals on the upper side of its horizontal edge. Find a planar linear placement with small density, that is, one in which no two nets are allowed to intersect each other. First you have to decide if a solution exists.

Input format.

```
3                   (* number of modules *)
M1 6, 3 1;          (* module 1 occupies six grid points, at the third
                         grid point there is a terminal of net 1 *)
M2 4, 2 1, 5 4;
M3 7, 3 1, 2 4;
```

Output format. The output format is shown in Figure CE5.4. Draw the modules, show all nets and their routing, and report the density of your solution.

5.5. Implement a planar version of Lee's maze-running algorithm that interconnects k distinct two-terminal nets. The main goal is to route all nets. The second goal is to minimize the total length. The third goal is to minimize the number of bends.

Input format. The input specifies the grid size, the position of the terminals, and the position of the obstacles (the northeast corner is grid point $(1, 1)$).

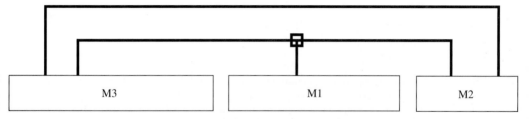

Number of tracks = 2

FIGURE CE5.4
Output format, a planar linear placement.

7 6;	(* size of the grid *)
2;	(* number of nets *)
5 5 , 2 2;	(* positions of source and target of the first net *)
1 5 , 6 5;	(* positions of source and target of the second net *)
3 1, 1 3,	(* positions of the obstacles *)
4 3, 5 3, 6 3, 3 4, 4 5.	

Output format. The output is shown in Figure CE5.5.

5.6. Implement a concurrent maze router. The input and output formats are as shown in the previous exercise.

5.7. Given a set of standard cells and a net list, implement an efficient algorithm for placing and (global, detailed, and over-the-cell) routing of all the nets. The number of rows is given, and the width of each row is also given. The goal is to route all nets minimizing the height of the chip. Note that in each cell, the same terminal appears on both the top side and the bottom side of the cell, and there are no terminals on the left and the right sides of the cell. The two middle tracks of each cell are used by power and ground nets, and the rest can be used for routing the signals.

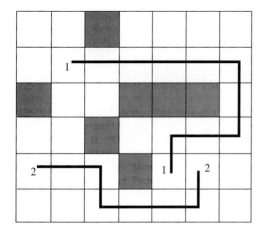

FIGURE CE5.5
Output, planar maze routing.

Input format.

3 7 200
1 3 0 4, ... ,
0 2 3 1 0 0

The above example contains a number of cells to be placed in three rows, each cell row has height 7, and the total number of columns is 200.

Two of the modules are shown, the first and the last. The first module occupies four columns: net 1 has a terminal at the first column, net 3 has a terminal at the second column, the third column is empty, and net 4 has a terminal at the fourth column. Note that two modules should be placed at least one unit apart.

Output format. Show a placement (containing the name of the modules) and a complete routing of the nets. Indicate the height of your routing (including the height of the cells and the total number of tracks used for routing). For example, your output can be similar to the one shown in Figure CE5.7.

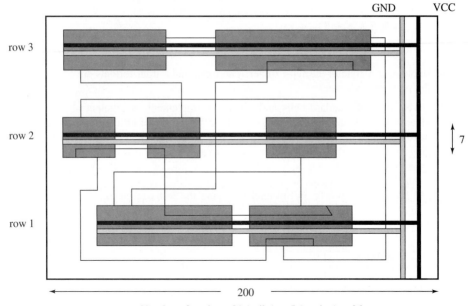

Number of tracks = 21 (cells) + 5 (routing) = 26

FIGURE CE5.7
Output format, placement and routing of standard cells.

CHAPTER

6

CELL GENERATION
AND PROGRAMMABLE
STRUCTURES

In VLSI design, a logic function is implemented by means of a circuit consisting of one or more basic cells, such as NAND or NOR gates. The implementation can be used as a library cell in the design phase. In CMOS circuits, it is possible to implement complex Boolean functions by means of NMOS and PMOS transistors. A cell is an interconnection of CMOS transistors. A CMOS cell is depicted in Figure 6.1. It consists of a row of PMOS transistors and a row of NMOS transistors corresponding to the PMOS and NMOS sides of the circuit, respectively.

The automatic generation of standard CMOS logic cells has been studied intensively during the last decade. The continuous progress in VLSI technology presents new challenges in developing efficient algorithms for the layout of standard CMOS logic cells.

The cell generation techniques are classified into random generation and regular style. A random cell generation technique does not exploit any particular structure to produce the cells. It uses a general technique, such as a hierarchical place-route algorithm (as discussed in Chapters 2–3). Random generation methods produce compact and (if desired) high-performance layouts. However, it takes a long time to design a cell. Regular generation techniques employ a predefined structure to design a cell. Cells generated in this manner occupy more area but can be designed faster. Traditional cell structures based on regularity are, for example, PLAs, ROMs, and RAMs. The disadvantage of the ROM-based cell is that it takes a lot of area, as it uses many redundant transistors. This chapter

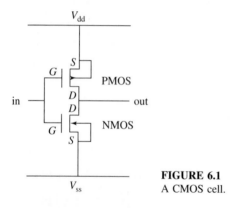

FIGURE 6.1
A CMOS cell.

discusses programmable architectures such as programmable logic arrays (PLAs), and regular structures used in cell generation, in particular, CMOS functional cell arrays and gate matrices.

6.1 PROGRAMMABLE LOGIC ARRAYS

A programmable logic array (PLA) provides a regular structure for implementing combinational and sequential logic functions. A PLA performs combinational functions of an input set to yield outputs. Additionally some of the outputs may be fed back to the inputs through registers, forming a finite-state machine as shown in Figure 6.2.

A PLA maps a set of Boolean functions in a canonical, two-level sum-of-product form into a geometrical structure. A PLA consists of an AND-plane and an OR-plane (see Figure 6.3a). For every input variable in the Boolean equations, there is an input signal to the AND-plane. The AND-plane produces a set of product terms by performing an AND operation. The OR-plane generates output signals by performing an OR operation on the product terms fed by the AND-plane. Reducing either the number of rows or the number of columns results in a more compact PLA. In practice, it is often the case that only 10% of the intersections are used (or "personalized") [120]. For such sparse PLAs, if the structure is implemented directly, a considerable amount of chip area will be wasted.

To reduce the area, two techniques have been developed, logic minimization for reducing the number of rows and PLA folding for reducing the number of

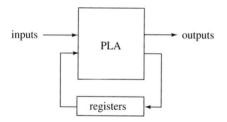

FIGURE 6.2
A PLA-based finite-state machine.

columns. Using logic minimization techniques [26, 27], the number of product terms can generally be reduced while still realizing the same set of Boolean functions.

For example, Figure 6.3b shows a PLA implementing the same functions as those of Figure 6.3a, but only four rows are used instead of six. This technique is the same as finding the minimal number of prime implicants for a set of Boolean equations. The minimization of product terms obtained by logic minimization has a direct impact on the physical area required since each product term

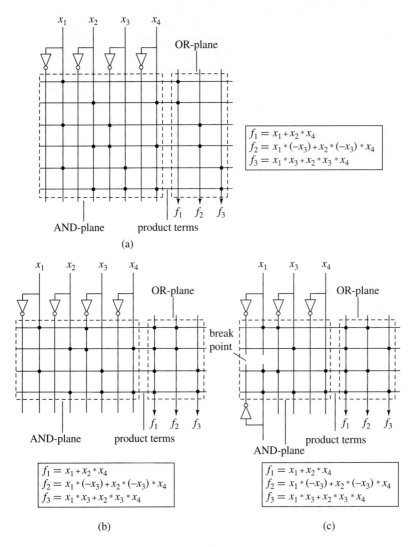

FIGURE 6.3
An example of a PLA. (a) A PLA stick diagram; (b) with logic minimization; (c) with folding.

is implemented as a row of the PLA. However, minimization of product terms is an NP-complete problem.

A logic minimization technique proposed in [26] follows. A PLA's *personality*, realizing a sum-of-product term of variables $x_1 \ldots x_n$, consists of a set of rows, each row representing a product term g. Each row is represented by a string $a_1 a_2 \ldots a_n$. $a_i = 1$ if the literal x_i appears in the term g. $a_i = 0$ if the literal $\overline{x_i}$ appears in the product term. Finally, $a_i = 2$ if neither x_i nor $\overline{x_i}$ appear in the term g. A *minimum personality* PLA corresponds to a minimum-area PLA. The first thing to do when trying to minimize a PLA's personality is to find redundant rows. Suppose that the inputs to the AND-plane are $x_1, \ldots, x_n$ and the outputs of the OR-plane are $y_1, \ldots, y_m$. Then each term represents a set of points in the n-dimensional Boolean cube. The points are input assignments for which the product term is 1 (true). For example, if $n = 4$, a row with label 1021 in the AND-plane represents the term $x_1 \overline{x_2} x_4$, which in turn represents the vertices of the four-cube with coordinates (1, 0, 0, 1) and (1, 0, 1, 1). To find out if a row r is redundant, consider every output y_i in which r has a 1 in the OR-plane, looking at the set S_i consisting of other rows that have 1 for output y_i. Let f be the term for the row r being tested for redundancy. Focus on a particular output for which this row has a 1, and let $g_1, \ldots, g_k$ be the terms of all other rows with 1 in that output. The question is whether $\overline{f} + g_1 + \cdots + g_k$ is true for every assignment of values to the input variables, that is, whether this expression is a tautology.

Testing whether a Boolean expression is a tautology is NP-complete. However, the following heuristic seems to be fairly effective. It is based on the idea that any expression $f(x_1, \ldots, x_n)$ can be written as $x_1 f_1(x_2, \ldots, x_n) + \overline{x_1} f_0(x_2, \ldots, x_n)$. If f is a sum of terms, as represented in a PLA personality, then f_1 is formed from f by taking the rows with 1 or 2 in the first column, and then removing the first column. f_0 is constructed similarly from the terms with 0 or 2 in the first column. Thus an n-column matrix is converted into two matrices of $(n - 1)$ columns each. The original matrix is a tautology if and only if both of the new matrices are. As columns are eliminated, it is often not necessary to extend the method recursively until the matrices of one column are reached. For example, if a matrix has a row with all 2's, then it is a tautology.

Figure 6.4 shows a sum of four terms, $\overline{x_2} \cdot \overline{x_3} + \overline{x_1} + x_2 + x_1 \cdot x_3$, represented in PLA personality form. It is split into two matrices, the upper matrix being the second and third columns of those rows with 0 or 2 in the first column and the

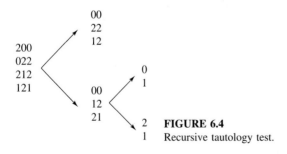

FIGURE 6.4
Recursive tautology test.

lower one being the same columns for rows with 1 or 2 in column one. The upper matrix contains a row 22, so it is a tautology. The lower one needs to be split again. Of the two matrices that result, the lower one has a row of all 2's, in this case, a single 2. The upper one has rows 0 and 1, which represents the expression $x_3 + \overline{x_3}$; it is also a tautology. Thus the original matrix is a tautology.

The algorithm for testing each row of the PLA personality, and eliminating it if it is redundant, follows:

> *procedure* PLA-redundancy-removal
> *for* each row r *do*
> *begin-1*
>   flag = REDUNDANT;
>   *for* each column c of the OR-plane in which row r has 1 *do*
>   *begin-2*
>   let R be the set of rows other than r with 1 in column c;
>   *if* the set of points covered by r is not a subset of the points
>   covered by the union of the members of R *then*
>   flag = NOT_REDUNDANT;
>   *end-2*;
>   *if* (flag = REDUNDANT) *then* delete r from PLA personality;
> *end-1*.

To illustrate the algorithm, consider the following PLA personality:

AND-plane	OR-plane
200	11
102	01
121	01
122	10

It is claimed that the second row is redundant. To test row 2, it is only necessary to consider column 2 of the OR-plane because row 2 has no other 1's in the OR-plane. The set R consists of rows 1 and 3, because these rows have 1 in the second column. Thus, test whether the complement of row 2, plus rows 1 and 3, is a tautology. The complement of the second row, which is 102 ($x_1\overline{x_2}$), consists of the two rows 022 and 212 ($\overline{x_1} + x_2$). After the tautology test is applied, it is concluded that only row 2 is redundant. Eliminate it to get a new PLA with fewer rows.

Next consider a different methodology for minimizing the area of a PLA, namely, PLA-folding. PLA-folding allows two or more signals to share a single row or column and thus reduces the total number of rows and columns. Figure 6.3c shows the result of folding a PLA by allowing signals $X1$ and $X2$ to share a single column. This column-folding reduces the total number of columns from 11 to 9 while keeping the number of rows unchanged. The folding of one signal pair may block another. The objective of PLA-folding is to find the maximum number of pairs that can be folded simultaneously.

The problem of optimally folding a PLA is known to be NP-complete [231]. Many algorithms and heuristics have been developed to solve this problem: the PLA decomposition technique [177], optimal bipartite folding [91, 288], branch-and-bound search [214], the graph-based algorithm [366], a best-first search heuristic [161], both row- and column-foldings [120], and simulated annealing techniques [388].

In the formulation of column-folding presented in [120], each PLA is represented by an undirected graph $G = (V, E)$, called a PLA *intersection graph*. Each node in this graph corresponds to a column of the PLA. There is an edge between nodes i and j if columns i and j of the PLA are both personalized along the same row (i.e., columns i and j cannot be folded on the same column). A pair of distinct columns p, q that can be folded in the same column forms an *unordered folding pair*. An *ordered folding pair* (p, q) specifies that the columns p and q are to be folded in such a way that p is above q. A set S forms ordered folding pairs (or *a folding set*) if each line of the PLA appears in at most one element of each pair in S. The PLA appears in at most one element of each pair in S, forming an ordered folding pair. Each ordered pair in S is represented in the intersection graph by a directed edge. If columns p and q have been folded with column p above column q, the directed edge from p to q is added. Thus a mixed graph is obtained, containing both directed and undirect edges, $M = (V, E, A)$, where A is the set of directed edges. Since each column is folded at most once, the directed edges in A form a matching (i.e., no two edges in A are incident on the same node). An alternating cycle C in the mixed graph is a cycle in which directed and undirected edges are traversed alternately. Based on the previous discussion the following theorem is easily proved.

Theorem 6.1. Given an intersection graph $G = (V, E)$ and an ordered folding set S, the set S is implementable if and only if the mixed graph $M = (V, E, A)$ obtained from G and S does not contain any alternating cycles.

A folding technique proposed in [366], a generalization of the algorithm proposed in [120], follows. When a PLA is folded, allow AND-plane inputs to arrive at either the top or bottom of the PLA. To save a column, two signals must be able to share a column. In Figure 6.3c, note that two inputs x_1 and x_2 are paired up at the same column. The row separating two signals sharing a column is called a *breakpoint*. Note that the breakpoints need not be the same for all pairs. Assume that complemented and uncomplemented wires of an input are placed in adjacent columns of the PLA. Thus, either a straight folding as in Figure 6.3c, or a twisted folding, pairing x_1 with $\overline{x_2}$ and x_2 with $\overline{x_1}$ in Figure 6.3c, can be performed.

All the pairs of columns that can be paired are considered, that is, columns that do not have 1's or 0's in the same row and have not yet been paired with another column. Consider pairing two columns C and D. A directed graph is maintained representing the constraints on the order of rows. Pair columns C and

D, with C above D, as a straight (not twisted) folding. Create two new nodes, p and q. Node p (q) represents the constraints that the term using C ($\overline{C}$) wires must be placed above the term using D ($\overline{D}$) wires. Arcs are formed based on the following rules:

1. If row r has 1 in column C, then create arc $r \rightarrow p$.
2. If row r has 1 in column D, then create arc $p \rightarrow r$.
3. If row r has 0 in column C, then create arc $r \rightarrow q$.
4. If row r has 0 in column D, then create arc $q \rightarrow r$.

If C and D are twisted, do essentially the same thing, exchanging the roles of p and q in **2** and **4**.

Next, test to see if pairing C and D introduces any cycles in the graph. If a cycle is introduced, then the rows cannot be properly permuted (i.e., there will be wire overlaps). In that case, remove the arcs just added and consider the next possible pairing, either the pairing of C and D in other arrangements or the pairing of other variables. At the end, a maximal set of pairings is obtained that results in an acyclic graph. The graph represents a partial order on the rows, and a total order consistent with the partial order is chosen, a process known as *topological sort*.

Let c be the number of columns and let w be the weight of the personality, that is, the total number of 0's and 1's. Since each 0 and 1 can account for only one arc of the graph under construction, the number of arcs is bounded by w. It is possible to test for the acyclicity of the graph in time proportional to the number of arcs by constructing a depth-first spanning forest and checking for the existence of backward arcs. There are $4\binom{c}{2}$ possible pairings to consider, that is, $\binom{c}{2}$ pairs of columns, which may be paired with either above or below, and may be paired straight or twisted. Thus the algorithm finds a feasible solution in $O(wc^2)$ time, where w is proportional to the number of rows and c is the number of columns.

The main advantage of a PLA approach to cell generation is that its regular grid pattern makes the generation task simple. Design time is short. However, there are several disadvantages. First, it requires a large area for most circuits and produces long signal lines, thus increasing circuit parasitics, which in turn degrades circuit performance. Moreover, the designer cannot control a PLA's I/O pin positions and aspect ratio. Hence, a PLA is not effective for realization of large circuits.

6.2 TRANSISTOR CHAINING

This section discusses the implementation of a logic function on an array of CMOS transistors. AND/OR gates in a CMOS functional cell correspond to series/parallel connections in the circuit diagram. One can obtain a series-parallel implementation in CMOS technology, in which the PMOS and NMOS sides are dual of each other.

Examples of the dual topology of the PMOS side and that of the NMOS side are shown in Figures 6.5a,b. In transistor chaining, transistors are ordered left to right and the corresponding connections are made. These connections are done

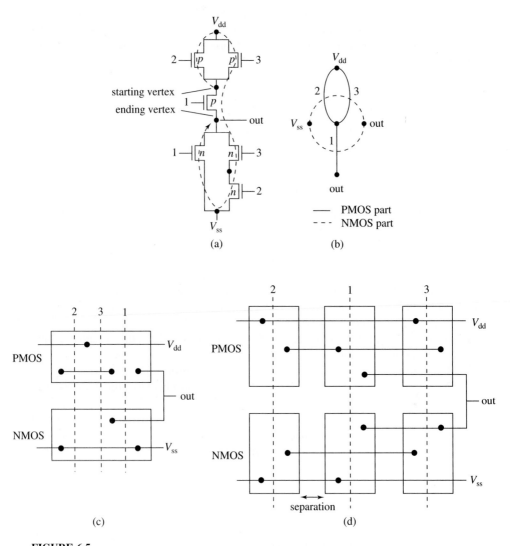

FIGURE 6.5
An example of transistor chaining. (a) Logic diagram and an Eulerian path; (b) a dual representation; (c) optimum area layout; (d) an alternative layout that is not optimal.

by sharing diffusion regions (Figure 6.5c) or by metal connection (Figure 6.5d). In general, the area of a functional cell is calculated as follows: area = width × height, where the height is constant and the width is (basic grid size) × (2 × (number of inputs) + 1 + number of separations). A separation is required when there is no connection between physically adjacent transistors, as illustrated in Figure 6.5d. In the transistor chaining problem, an optimal layout is obtained by minimizing the number of separations. In a graph every source/drain is represented by a vertex, and every transistor is represented by an edge connecting the vertices that represent the source and drain. Of interest for optimal layout of CMOS circuits is the property of the graph model that if two edges are adjacent in the graph, then the corresponding gates are placed in adjacent positions and the edges are connected by a common diffusion area. In order to find the minimum number of separations, it is necessary to find a set of minimum-sized paths that correspond to chains of transistors in the array.

A graph-based algorithm [365] is used to minimize the array size. First consider the idea in an NMOS circuit. If there exists an Eulerian path (i.e., a path that contains each edge exactly once), then the transistors corresponding to edges of an Eulerian path can be chained by diffusion areas. A graph having an Eulerian path is called an Eulerian graph. As illustrated in Figure 6.5c, such a set results in a minimum width layout. If an Eulerian path does not exist, then the graph can be decomposed into several subgraphs, each of them Eulerian. Each Eulerian path is a chain of transistors that is separated by a separation area. Therefore, the problem of finding an optimal layout is equivalent to decomposing the graph into a minimum number of disjoint Eulerian subgraphs that cover the graph.

It is well known that a graph with at most two of its vertices having odd degrees (and the rest of the vertices having even degrees) is Eulerian, and vice versa. A heuristic algorithm introduces pseudoedges (transistors) to obtain an Eulerian graph. An Eulerian path dictates an ordering of transistors. Then the pseudotransistors are removed, producing spacings. This heuristic algorithm does not necessarily produce an optimal layout.

An alternative algorithm is as follows. Start traversing the graph from an odd-degree vertex. If while traversing the graph, a vertex is reached that cannot be exited (note that this vertex must also be an odd-degree vertex), then remove all visited edges. The order in which the edges are visited dictates an ordering of the corresponding transistors. Repeat the same procedure starting from another odd-degree vertex. This task is repeated for s times, where $s = $ (number of vertices)$/2$ -1. At this stage, all edges have been visited. The number of separations is equal to s, being the minimum possible.

Next the transistor chaining problem in CMOS circuits is discussed. A linear time algorithm [222] has been proposed to obtain a minimum width cell layout of a static CMOS circuit. The problem is reduced to finding a minimum number of edge-disjoint dual paths that cover the pair of graphs. A dual path is one that is a path in both graphs, the NMOS part and the PMOS part. An $O(l)$ time heuristic is presented in [222], where l is the number of literals in the given input prefix logic expression. The algorithm utilizes the decomposition tree (DT) structure of the dual multigraphs (see Figure 6.6). A *two-terminal series-parallel multigraph* (2T-SPMG) M is defined as follows. An edge is a 2T-SPMG (basic graph). Let M_1 and M_2 be 2T-SPMGs, where u_i and v_i are the two terminals of M_i, for $i = 1, 2$. Then the graph obtained by merging v_1 and u_2, denoted by $M_1 * M_2$, and the graph obtained by merging the pair u_1 and u_2 and v_1 and v_2, denoted by $M_1 + M_2$, are 2T-SPMGs.

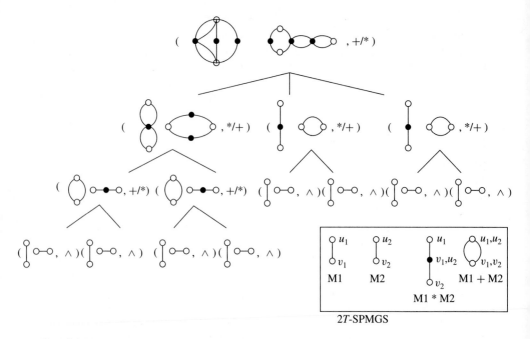

FIGURE 6.6
A decomposition tree (DT) of a dual graph pair.

In order to process a DT in a bottom-up fashion, its components are classi-fied into types and composition rules are established that order and orient those components so as to minimize the number of dual paths in a dual path cover.

Previous works require that the n-part and the p-part of the circuit graph be mutual duals and moreover be a series-parallel graph. [383] relaxed any con-straining conditions on the circuit graph and provided a general algorithm that can handle arbitrary graphs optimally. This is done by allowing a p-type and an n-type transistor to share a column (i.e., have the same position in a chain) not only when they are complementary, but also when they have a source or drain part in common.

This method shows that the functional cell approach can reduce the space of a conventional NAND gate realization considerably. However, the transistor chaining technique may produce a large number of tracks (increased height), more specifically, $O(n)$ tracks (or rows), where n is the number of transistors. An important second objective is to find the minimum number of tracks, or to trade off width (the number of separations) and height.

6.3   WEINBERGER ARRAYS AND GATE MATRIX LAYOUT

Weinberger [379] introduced one of the first methods for regular-style layout. The idea is to have a set of columns and rows, with each row corresponding to a gate

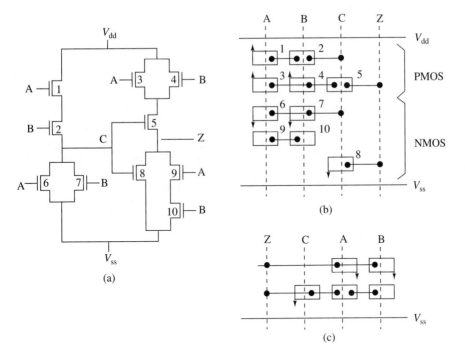

FIGURE 6.7
An example of gate matrix layout. (a) A CMOS transistor circuit; (b) an example of column permutation; (c) optimal area layout for the NMOS part.

signal and the columns responsible for the realization of internal signals (e.g., transistor to V_{dd} connection). Weinberger array techniques have resulted in other regular-style layouts, one being gate matrices.

Gate matrix layout was first introduced by Lopez and Law [228] as a systematic approach to cell generation. The gate matrix layout problem is to find a layout that uses a minimum number of rows by permuting columns. The structure of gate matrix layout is as follows (see Figure 6.7):

- All p-transistors are placed in the top half of the matrix and all n-transistors in the lower half.
- A vertical polysilicon wire corresponding to an input is placed in every column. Polysilicon is of constant width and pitch; thus, the number of columns is fixed.
- All transistors using the same gate signal are constructed along the same column.
- A transistor is formed at the intersection of a polysilicon column and a diffusion row.
- Transistors placed in adjacent columns of the same row are connected using shared diffusion. In addition, short vertical diffusion stripes are sometimes

necessary to connect drains and sources of transistors to metal lines on different rows (e.g., see Figure 6.8a).

- Transistors separated by one or more polysilicon columns are connected by metal runners. Connections to V_{ss} (ground) may be placed on a second layer of metal.

The advantage of the gate matrix is that it is a systematic layout style that can be formally described, and that symbolic layout can be easily updated for new design rules. Its disadvantages are its inflexibility of pin assignment, its long average wire length for large circuits, and its requirement of two metal layers.

The size of a gate matrix layout is proportional to the product of the number of columns and rows. To minimize a gate matrix layout area, the number of rows must be reduced, since the number of columns is fixed to the number of nets in the circuit schematic. Because a row can be shared by more than one net, the number of rows depends heavily on both the column ordering and the net

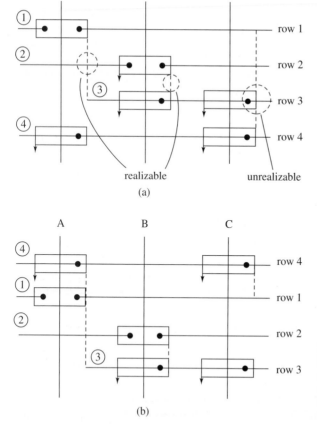

FIGURE 6.8
Permutations of the rows to realize all vertical diffusion runs. (a) A layout containing unrealizable placement; (b) a realizable layout.

assignment to rows. Figure 6.7 shows how column ordering affects the size of the layout area.

Consider a set of horizontal line segments. A graph with vertices corresponding to the segments and edges indicating that the corresponding intervals share a point on the horizontal axis is called an *interval graph*.

Wing et al. [384] solved the gate matrix layout problem using a graph-based technique. Gate matrix layout is represented as an interval graph, $I(L) = (V, E)$, where V is the set of nets and E is the set of edges connecting vertices whose nets overlap in one or more columns. The number of rows used by a layout is equal to the size of the maximum dominant clique of $I(L)$. A clique of a graph G is a complete subgraph of G. A dominant clique of G is a clique that is not a proper subgraph of any other clique of G. Wing et al. transformed the gate matrix problem into the problem of finding an interval graph in which the maximum clique size is minimized. From an interval graph, the optimal column ordering can be obtained by applying the algorithm of [265]. The left-edge algorithm can be used to assign nets to rows.

Consider a transistor whose drain and source are connected to nets n and m, respectively. In a layout, it may not always be possible to place two nets on the same row. If net n is on the same row as the transistor while m is on a different row, a short vertical diffusion run will be required to connect m to the source.

In formulating the problem, only the n-transistors will be considered since the p-part of the circuit is determined once the topology of the n-part is known. However, a good solution for n-part may be a very bad solution for p-part. Define the set $G = \{g_i | g_i$ is a distinct transistor gate or an output terminal$\}$. The polysilicon strips will be referred to as *gate lines*. Let $C = \{c_i\}$ ($R = \{r_i\}$) be the set of columns (rows) of the gate matrix. In a layout, let $f : G \rightarrow C$ be a function that assigns the gates to the columns, the *gate assignment function*. In a circuit, denote the set of nets (nodes) by $N = \{n_i\}$. Every transistor (channel) is connected between two nodes (nets) or between a node and ground. Each net will be realized by a segment of horizontal metal line that is connected to a drain, source, or gate of the transistors of the node. In a layout, define a *net assignment function* $h : N \rightarrow R$ as the function that assigns the nets to the rows. For each pair (f, h) of functions, there is a layout $L(f, h)$.

In a given layout $L(f, h)$, for each pair of nets (n, m) such that $h(n) \neq h(m)$, define the set of diffusion runs $D(n, m) = \{g_k | h(n) \neq h(m)$ and there exists a vertical diffusion run between net n and m near gate line $k\}$. $D(n, m)$ is *realizable* if each of its vertical diffusion runs can be placed such that it does not conflict with (overlap) any transistor in the same column on a row between row $h(n)$ and row $h(m)$. A layout is not realizable if a vertical diffusion line must cross over transistors or makes contact in the same column. An unrealizable layout can be turned into a realizable layout in one of two ways, by permuting rows and thus removing obstacles or by increasing the space between two columns.

Then the problem can be restated as follows. Given a set of transistors T together with the set of distinct gates G and the set of nets N, find a pair of functions $f : G \rightarrow C$ and $h : N \rightarrow R$ such that in the layout $L(f, h)$ the number of rows, card$[R]$, is minimum, and $L(f, h)$ is realizable.

First attempt to find a layout that results in a minimum number of rows. If it is realizable, the problem is solved; otherwise, find a permutation of the rows to obtain a realizable layout, keeping the number of rows unchanged. If none exists, enlarge the spacing of some pair of polysilicon strips to allow the vertical diffusion runs to be placed between them without overlapping any transistors, as shown in Figure 6.8b.

Associated with each net n_i is a net-gates set $X(n_i) = \{g_j | n_i$ is connected to the drain, source, or gate of a transistor assigned to $g_j\}$. For example, in Figure 6.8, $X(1) = \{A\}$. Similarly, a gate-nets set $Y(g_j) = \{n_i | g_j \in X(n_i)\}$ is associated with each gate g_j. For example, in Figure 6.8, $Y(A) = \{1, 4\}$.

The gate-nets sets are represented by a *connection graph* $H = (V, E)$ with the vertex set $V = \{v_i | n_i \in N\}$ and the edge set $E = \{\langle v_i, v_j \rangle |$ such that n_i and $n_j \in Y(g_k)$ for some $k\}$. The connection graph describes how the nets are connected to the gates. In an abstract representation of a layout, the nets can be regarded as intervals that overlap one another. Series-connected transistors are treated as a group and one net is assigned to represent the group.

We define an interval graph associated with a layout $L(f, h)$ as $I(L) = (V, B)$ where V is the same vertex set as in H and the edge set $B = \{\langle v_i, v_j \rangle | n_i$ and n_j overlap in $L\}$. Figure 6.9 shows an example.

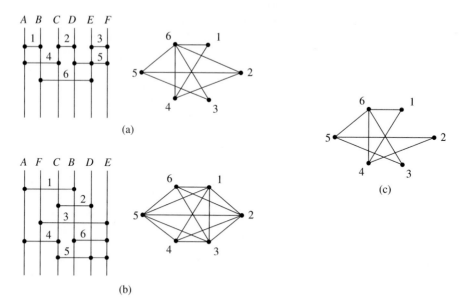

(a)

(b)

(c)

FIGURE 6.9
(a) and (b) Two interval graph representations of two gate matrix layouts. (c) The connection graph of the circuit.

The following are self-evident:

- Each layout L realizes a connection graph H and the interval graph $I(L)$ contains H as a subgraph.
- The number of rows in a layout L equals the size of the largest dominant clique of $I(L)$. (A dominant clique is a clique that is not a proper subset of another clique. The size of a clique is the number of vertices it contains.)
- A lower bound on the minimum number of rows required in any layout is equal to the maximum number of nets connected to any gate.

The problem can be restated as: Given a connection graph H, find an interval graph I which contains H as a subgraph such that the size of the largest dominant clique of I is minimized. A heuristic algorithm is given in [384] to find an interval graph I such that the size of the dominant clique is minimally increased as I is constructed.

After attempting to find a layout with the minimum number of rows, find a permutation of the rows to obtain a realizable layout. Consider two nets n_i and n_j for which the set of vertical diffusion runs $D(i, j)$ is not empty. Since diffusion runs should be as short as possible, the two rows to which the nets are assigned, $h(n_i)$ and $h(n_j)$, should be adjacent. However, it is apparent that if there exists a third net n_k for which neither $D(i, k)$ nor $D(j, k)$ is empty, the neighboring requirement is cyclic. But if the diffusion runs induced by h are all realizable, then h is acceptable as a solution.

A greedy algorithm for constructing a sequence, node by node, proceeds as follows. Each node is added to the partial sequence, ensuring all diffusion runs induced by the partial sequence are realized. When no such node can be found, the spacing between two polysilicon lines is enlarged to make room for the vertical diffusion run that cannot be realized otherwise.

An efficient algorithm for bounding the number of rows in a gate matrix for a given class of circuits, the AND-OR circuits or series-parallel circuits, follows. A graph $G = (V, E)$ has a $|V_3|$-separator if its vertices can be partitioned into three sets V_1, V_2, and V_3, where $|V_1|$ and $|V_2|$ are between $\left(\frac{1}{3}\right)|V|$ and $\left(\frac{2}{3}\right)|V|$, respectively, and the removal of V_3 disconnects V_1 and V_2.

It is well known that trees have a one-separator. First, choose an arbitrary node as the root of the tree. Traverse down the tree and check the number of nodes of each subtree. If one of the children has nodes between $\frac{1}{3}|V|$ and $\frac{2}{3}|V|$, then the root itself is the one-separator. Otherwise, recursively traverse the path of the subtree whose number of nodes is greater than $\frac{2}{3}|V|$. A separator will be found (proof by a simple case analysis). Consider a series-parallel graph $G = (V, E)$, where $|E| = n$. Now, to transform G into a tree, find a maximal-size tree by deleting the smallest number of edges from G. However, the worst-case example (such that there are $n - 1$ edges connecting two vertices) shows that $n - 1$ edges must be removed to have a tree.

The hierarchical tree representing the construction of a series-parallel graph is called a series-parallel tree (Figure 6.10). To construct a tree $BDT = (V, E)$,

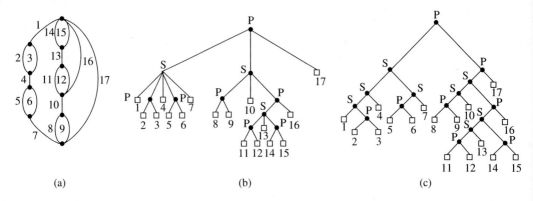

FIGURE 6.10

Examples of series-parallel trees. (a) A series-parallel graph; (b) a series-parallel tree; (c) a series-parallel binary tree.

called a binary decomposition tree, associated with the series-parallel tree T, let T_1 and T_2 be the two subgraphs obtained after separating T. The left child and the right child of BDT are associated with T_1 and T_2, respectively. The tree BDT is constructed recursively in this fashion.

Lemma 6.1. Any planar graph $G = (V, E)$ that has a k-separator can be laid out, along with its dual G^d, in a gate matrix style, using $O(k \log n)$ tracks (rows).

Proof. Note that BDT contains $\log_{3/2} n$ levels, since the number of nodes of any subgraph of T is at most $\frac{2}{3}$ of its parent graph in T. There are at most k vertices (separators) between T_1 and T_2, and thus k rows are required for connections at each level. Since there are $O(\log n)$ levels, connections occupy a total of $O(k \log n)$ tracks (rows). Analogously, G^d requires $O(k \log n)$ tracks.

It can be shown that the series-parallel graphs, representing AND-OR circuits, have two-separators. (This can be proven by considering the corresponding series-parallel tree.)

Theorem 6.2. Any (AND-OR) CMOS cell with n gates can be laid out, in gate matrix style, using at most $O(\log n)$ rows in $O(n \log n)$ time.

Proof. Since series-parallel graphs have two-separators, any CMOS cell with n gates can be laid out using $O(\log n)$ rows. The total time required for constructing BDT is $T(n) \leq 2T(\frac{2}{3}n) + O(n)$, or $T(n) = O(n \log n)$.

It can be shown that the above upper bound on the number of rows cannot be improved by showing instances of CMOS cells, which require $\Omega(\log n)$ rows [345]. The assumption that the permutation of both the PMOS and NMOS sides is the same is not true in practice. Thus, [261] proposed a heuristic that considers both PMOS and NMOS simultaneously, thereby improving the previous results. Previous works that use the problem formulation described above do not take into account the fact that the mapping from logic representation to circuit representation is one to many. In other words, circuits with different transistor orderings, and thus different net lists, are logically equivalent. Taking advantage of the flexibility offered by transistor reordering can result in layouts with significantly reduced areas. Therefore, [336] proposed a general transistor reordering algorithm to determine the gate sequence and net list, which optimize the layout of the gate matrix.

6.4 OTHER CMOS CELL LAYOUT GENERATION TECHNIQUES

This section reviews several existing cell generation techniques. TOPOLOGIZER [188] is an expert system for CMOS cell layout. It uses rules specified by an expert designer to produce a symbolic layout from a transistor description and a description of the environment in which the cells reside. The placement rules include moving transistors and rotating transistors. The routing expert consists of a prerouter and a refinement router. This procedure produces a rough routing by assigning a unique track to each pair of terminals to be connected. The refinement router then improves the rough routing by applying a set of rules including U-turn elimination and row sharing.

Sc2 [137] accepts a logic description and automatically generates a layout. It consists of two subtasks, transistor placement and intercell routing. Placement starts with pairing the p-type transistors with the n-type transistors. Then the transistor pairs are oriented to maximize the number of adjacent sharing. However, optimal results in terms of minimum diffusion are not guaranteed. The goal of the routing step is to reduce the total wire length. Several routing methods are employed, starting with abutting the transistors, then routing over the transistors, and finishing with channel routing. It permutes the transistors to facilitate more diffusion adjacencies under the constraints that it does not change the circuit's functionality.

GENAC [267] proposes a hierarchical method for the placement of the transistors. It minimizes the number of diffusion breaks as well as the routing density. GENAC also does transistor-folding and routing over transistors.

Layout Synthesizer [45] uses a graph-based placement algorithm to simultaneously maximize diffusion sharing and minimize the wiring area. It does not perform transistor-folding nor does it route over the transistors.

LiB [148] consists of the following nine steps:

1. **Clustering and pairing.** In order to handle large circuits, LiB partitions a circuit into several groups of strongly connected transistors. PMOS and NMOS are good candidates to be paired if they are both within the same group and controlled by the same gate signal.

2. **Chain formation.** In order to achieve a minimal size layout, it is necessary to abut as many drain/source connections as possible. LiB uses an optimal transistor chaining algorithm [155] (a branch-and-bound search method is used

to exhaustively search for an optimal routing) to minimize the number of chains needed.

3. **Chain placement.** LiB determines the position and orientation for each chain in a linear array such that the connection requirement (measured as chain density) is minimized. If the number of chains is fewer than five, an exhaustive search method is used to find out the best placement; otherwise, a modified min-cut placement algorithm is applied.

4. **Routing region modeling.** In order to effectively utilize the space and reduce the routing complexity, a cell is divided into five routing regions, as shown in Figure 6.11. Among them, regions p and n are handled by subtask 6 while regions U, M, and L are individually routed using a general purpose channel router in subtask 8.

5. **Large transistor-folding.** The transistor size specification is considered during this stage. For each transistor pair, the sum of the routing space and the diffusion space needed should not exceed the cell height. LiB estimates the routing area needed for each pair according to the local net density and design rules. If the sum of the routing area and diffusion space exceeds the cell height, the algorithm folds that pair into multiple columns.

6. **Routing on the diffusion islands.** In this layout style, the PMOS and NMOS transistor rows are also used as routing regions for the first metal wires. In this stage, LiB routes as many net segments on the PMOS transistor row and the NMOS transistor row as possible.

7. **Net assignment.** LiB determines which net goes to which region such that the routing area and the number of vias needed are minimized. A bipartite graph model for this problem is proposed.

8. **Detailed routing.** A general channel/switchbox router is used.

9. **Compaction and geometry transfer.** Thus far, LiB has generated a design-rule-independent symbolic layout. LiB feeds the result into Symbad/OED, a commercial compactor, to generate a mask-level layout.

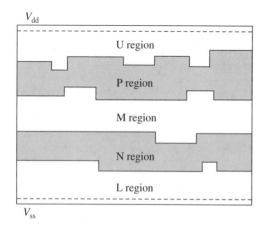

FIGURE 6.11
Routing regions.

The one-dimensional CMOS layout style consists of rows of p- or n-transistors. Their order is determined by maximizing the diffusion abutment or optimizing the routing. Recently, [99] proposed linear time algorithms that solve the two multirow layout problems optimally. These two layout problems are presented here:

- **Transistor orientation.** If the right net of the transistor in position (i, j) is the same as the left net of the transistor in position $(i, j + 1)$, the corresponding terminals can abut. Otherwise, a space must be inserted between the two terminals, thus increasing the layout width. The goal is to orient the transistors in such a way that the number of space insertions is minimized.
- **Symbolic layout optimization.** Another problem that arises in layout generation occurs during symbolic-to-geometric mapping (also see Chapter 7). This process includes the mapping of each symbolic object to geometric shapes and determining their absolute locations, subject to ground rule constraints. Due to rapid changes in technology, contact-to-contact spacing rules may be larger than gate-to-gate spacing rules. Since the gate-to-gate minimum distance is the smallest ground rule constraint, it dictates the layout grid. The following two problems tackle the issue of modifying a symbolic layout such that ground rule violations are avoided, and the amount of added space is minimized.

    Vertical contact relocation. One way to solve the ground rule problem is to shift a contact vertically to upper or lower vacant tracks, if possible. If two contacts reside on the same track in two consecutive columns, shifting one of them to another track will resolve a conflict, and thus will avoid an extra spacing.

    Horizontal contact offsetting. Another way to handle the contact adjacency problem is by moving the contacts off-grid. In some technologies, offsetting the left contact leftward and the right one rightward yields a legal distance between them. This offsetting is large enough to avoid a ground rule violation, and small enough to keep the contact connected to the gate below it. The problem is then to decide about contact-offsetting such that the unresolved contact adjacencies can be handled with a minimal number of column insertions.

A dynamic programming approach to one-dimensional cell placement was presented in [275]. Efficient search techniques (depth-first) are employed to bound the solution space.

6.5 CMOS CELL LAYOUT STYLES CONSIDERING PERFORMANCE ISSUES

Timing issues should be incorporated in designing cells to optimize circuit performance. In the case of custom logic layouts with two-level metal layers, no structured methods have been reported. The future generation of CMOS technologies with reduced supply voltage may require both p-type and n-type polysilicon

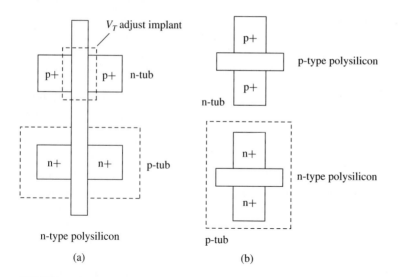

FIGURE 6.12
Two layout structures for the inverter circuit. (a) Twin-tub CMOS processing; (b) separated polysilicon gates.

gate materials to better control threshold voltages. In such a case, the conventional layout style (e.g., a contiguous n-type polysilicon feature has been used for crossing both p^+ and n^+ diffusion areas as shown in Figure 6.12a) would be vulnerable to the interdiffusion problems (i.e., inaccuracy due to the diffusion process) at finer design rules. A more desirable layout structure may consist of separate polysilicon features as shown in Figure 6.12b. With this approach, the PMOS and NMOS transistors can be optimized independently without any interdiffusion problems.

A new layout method for high-speed VLSI circuits in single-polysilicon and double-metal MOS technology, the metal-metal matrix (M^3) layout, has been proposed in [115, 174, 175]. With an emphasis on the speed of performance, M^3 makes the maximal use of metal interconnections while restricting delay-consuming polysilicon or polycide only to form MOS transistor gates. M^3 uses a metal-metal matrix structure to restrict the use of polysilicon to form transistors or to connect the same type of transistors. A minimal amount of polysilicon suppresses parasitic RC delays. In the M^3 style, all signals run vertically with the upper level metal lines to form the signal columns. Interspersed between these metal lines are diffusion columns. The lower-level metal lines are used to horizontally connect the signal columns to the gates of MOSFETs, to connect the sources and drains of MOSFETs, and to run power and ground busses. These horizontal nets can be folded to reduce the layout area. However, when the layouts suffer from high parasitic resistances and capacitances due to long diffusion runners, the roles of both metal layers can be exchanged so that long diffusion runners can be replaced with metal.

The new MOS VLSI layout structure for two-level metal technology takes advantage of the layout trend, chip speed improvement and amenability to existing automatic layout tools. However, there remain some open questions related to the increased number of contacts, processing yields, and cost/performance evaluation.

Gate signals running from top to bottom in polysilicon result in a high capacitive load and degraded performance. Also, due to the fixed number of gate lines, the aspect ratio cannot be customized. Furthermore, a gate matrix uses only a single layer of metal and no provision has been made to utilize an additional layer of metal efficiently. M^3 is designed to utilize two layers of metal to reduce delays in layouts. However, no area reduction over a gate matrix has been realized.

Therefore, [143] proposed a new layout style called the *flexible transistor matrix* (FTM) for large-scale CMOS module generation. The FTM uses two layers of metal. Compared to a gate matrix, the FTM can generate significantly better results. In addition, the algorithm can control the aspect ratio, I/O pin positions, and different transistor sizes.

The FTM also allows the use of diffusion and polysilicon for secondary interconnects. In the FTM, the transistor area is shared with the routing area to achieve more compact layouts. This is done by overlapping transistors with two metal layers. The FTM uses both polysilicon and the top layer of metal to route signals across the p and n planes. This reduces the lower bound on the width of the layout and allows for a wider range of aspect ratios.

The columns consist of signal nets and transistors, whereas the rows consist of interconnects between terminals of transistors and signal nets (see Figure 6.13). Signal nets run vertically in the top layer of metal (metal 2) and interconnects run horizontally in the bottom layer of metal (metal 1). All p-diffusion (and n-diffusion) and their corresponding interconnects have to be placed anywhere in the matrix.

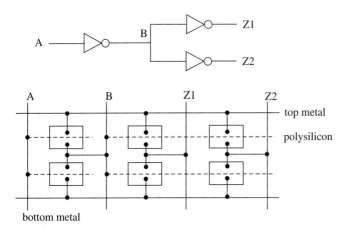

FIGURE 6.13
CMOS inverter circuit layout for single-polysilicon double-metal technology.

[184] proposed a performance-driven approach to module generation for static combinational and a restricted class of sequential CMOS logic circuits. Their flexible layout style provides a foundation for synthesizing high-speed circuits. In this style:

- the use of highly parasitic interconnect materials, such as diffusion and polysilicon, is minimized;

- transistor sizing and reordering of devices are supported; the transistor channel lengths are fixed at the minimum imposed by the design rules, and only the widths are varied;

- metal layers are used as the primary interconnection media; and

- wires at different levels are prevented from running overlapped for a non-negligible distance, thus avoiding the creation of unwanted capacitors that can cause extra delay or cross-talk problems.

The layout style used in [184] is somewhat similar to M^3, sharing all the advantages of M^3. Further improvement in circuit speed over M^3 is achieved using diffusions only when it is necessary (i.e., for creating transistors). A layout is a two-dimensional arrangement of gates; it is composed of vertically stacked rows, each of which contains horizontally aligned gates. The number of rows used is specified by the designer. A static gate consists of a pull-up structure and a pull-down structure. The pull-up structure is constructed by arranging transistors on one column of p-diffusion in the pull-up plane of a row, and making connections between the transistors, either with the first-level metal (or metal 1) wires and diffusion contacts or by abutting the source and drain of adjacent transistors. The pull-down structure can contain n-type transistors only, and is constructed in the pull-down plane in the same way. A CMOS pass gate can be considered as being composed of a pull-up transistor and a pull-down transistor. Hence, its implementations are very similar to that of a CMOS inverter except that the power connections are replaced by a connection to a signal net.

The gates are grouped into pairs and attached back-to-back so that they can share the vertical metal 1 segments connected to power and grounds as well as polysilicon wires for transistor gate terminals. This is the only occasion when polysilicon is used for routing, and the rest of the horizontal interconnections are done in the second-level metal (or metal 2). Polysilicon and metal-two wires never run vertically until the layout is compacted. Metal 1 wires run vertically when used for inter-row routing and intrarow routing. Another usage of the first-level metal is for intragate connections.

[227] presented a doughnut layout style for CMOS gates that improves the switching speed when compared to the standard layout style. In such a design, the output diffusion is surrounded by the transistor channel and gate. This significantly reduces the area and length of perimeter of the output diffusion while permitting relatively large W/L ratios: the factor W/L is the effective width-to-length ratio of the p- or n-channel for the gate and represents the primary variables available to the designer for controlling switching speed. The designer can im-

prove the switching speed of a gate by increasing the ratio, $S = (W/L)/C_{out}$, where C_{out} is the capacitance at the output node due to layout self-loading. The self-loading capacitance is dependent on the drain diffusion and can be influenced significantly by the layout style. This improvement is also obtained by decreasing the self-loading output capacitance for a given W/L ratio of the transistor channels.

6.6 DISCUSSION

A current trend in cell generation is to contribute to area-efficient design methodologies and tools while solving the delay problem. This chapter presented various algorithms that have been developed to minimize the layout area and/or maximize the performance in the systematic cell generation techniques.

Several automatic CMOS cell layout generators have been introduced. Recently, Feldman et al. [99] presented a linear-time algorithm that solves multirow layout problems. An interval graph-based algorithm has been applied to the gate matrix layout by Wing et al. [384] and to PLAs by Yu and Wing [402].

The performance impact of transistors has been observed widely by VLSI designers, and much effort has been devoted to solve the delay problem. Several CMOS cell layout styles emphasize further performance issues: the metal-metal matrix (M^3) layout style [174], the doughnut layout style [227] that improves switching speed compared to the standard layout style, the flexible transistor matrix (FTM) [143] that generates significantly better results compared to the gate matrix style, and finally a performance-driven layout style [184] that improves circuit speed over M^3.

Two types of design techniques that are expected to dominate a large portion of the IC industry in the coming year are design for testability and design for manufacturability. Design for testability is a relatively mature concept, whereas design for manufacturability is quite a recent one. The objective is to assure that the design is manufacturable with a high yield, starting at an early stage of the design process. Therefore, taking timing issues into consideration in the CMOS cell generation step is crucial in the early design process.

EXERCISES

6.1. Give an example where using two PLAs requires less area than using one PLA.

6.2. Realize the following set of functions using a PLA with
(a) logic minimization alone,
(b) folding alone,
(c) logic minimization and folding.

$$F_1 = x_1 + x_2 x_3 x_4$$
$$F_2 = x_1 x_2 x_4 + x_2 x_3 x_4 x_5$$
$$F_3 = x_1 x_3 + x_2 x_4 + x_1 x_2$$
$$F_4 = x_3 x_4 x_5 + x_2 x_3$$

6.3. Use the recursive tautology test algorithm to decide if each of the following is a tautology:

(a) $x_1\overline{x_2x_3} + x_2 + \overline{x_1}x_3$
(b) $x_1\overline{x_2x_3} + x_2x_3 + \overline{x_1}$
(c) $x_1\overline{x_3} + x_2x_3 + \overline{x_1x_2}$

6.4. Design an efficient algorithm for obtaining the minimum number of Eulerian paths covering a given graph (each edge of the graph should be contained in exactly one path). Employ your algorithm to obtain a minimum-width layout of a general NMOS circuit.

6.5. Prove that series-parallel graphs are two (vertex) separable.

6.6. Use the result of the previous exercise to design an algorithm for generating a gate-matrix layout of a series-parallel cell.

6.7. Are trees one edge separable? Why?

6.8. Show that minimizing the number of separations in the transistor chaining model may require a large number of tracks. Show that minimizing the number of tracks in the transistor chaining model may require a large number of separations.

6.9. Design a transistor chaining algorithm that minimizes both the number of separations and the number of tracks. Analyze the time complexity of your algorithm.

COMPUTER EXERCISES

6.1. Implement a PLA column-folding algorithm.

Input format. The input is a set of equations

$F1 = 1 + 2 - 34$ (* $x3$ is complemented *)
$F2 = 124 + -2345$ (* $x2$ is complemented *)

Output format. The output is a drawing of the original PLA (before folding) and a drawing of the PLA after folding (as shown in Figure 6.3c).

6.2. Implement a transistor-chaining algorithm for CMOS cells.

Input format. The input is a series-parallel network

$F = 124 + +15 + -2345$ (* $x2$ is complemented *)

Output format. The output is a drawing of a minimum width layout (as shown in Figure 6.10c).

6.3. You are given a set of Boolean functions. Since using one PLA to implement all the functions takes too much area, you are to implement them using a set of PLAs; additional routing is necessary to interconnect the PLAs. Analyze the problem and design an algorithm for performing this task. Then implement your algorithm.

Input format. The input is a series-parallel network

$F1 = 124 + +15 + -2345$ (* $x2$ is complemented *)
$F2 = 12 + +135 + 45$

Output format. The output is a drawing of each PLA (as shown in Figure 6.3c) and an estimate of the routing area for interconnecting the PLAs. Explain how you estimated the routing area.

CHAPTER

7

COMPACTION

Throughout this book, a number of VLSI physical design problems and various methodologies for solving them have been presented. In most cases, algorithms have been considered in a discrete framework called abstract layouts. Layout techniques have been described in a grid environment. In representation, transistors, wires, and vias are represented by point and line segments in the plane (or vertices and edges of a graph). That is, they are modeled as symbols. Such representation is called a symbolic layout. A symbolic layout can be used to describe both leaf cells that contain logic elements and routing cells that contain the intercell wires.

Various languages for describing symbolic layouts have been proposed. After a layout is obtained it is necessary to transform it into a physical layout, where elements satisfy given physical design rules (see Chapter 1). These rules dictate various minimum separation, width, and shape rules (e.g., a via is represented by a square). The resulting physical layout is then compacted. Finally, the compacted layout is tested to determine whether there are any violations of physical design rules. Since routing cells can be regarded as part of a symbolic layout, compaction techniques can be applied to various routing problems to reduce routing areas. Figure 7.1a shows the symbolic layout of a routing cell, and the corresponding physical layout is shown in Figure 7.1b. The input to a compactor is of the form shown in Figure 7.1b.

A larger example is given in Figure 7.2. The example is from [218]. The design of a carry-chain adder is shown in Figure 7.2a, and its symbolic layout is given in Figure 7.2b. Figure 7.2c shows the corresponding physical layout. We

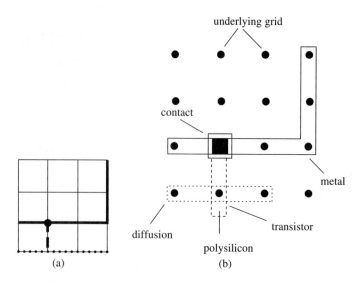

FIGURE 7.1
A symbolic layout and the corresponding physical layout. (a) Symbolic layout; (b) physical layout.

shall use this example for demonstrating some of the compaction algorithms. The constraints to be generated are related to the layout primitives.

There are two classes of compaction techniques. In *one-dimensional* (1D) *compaction*, the given physical layout is compacted in alternating directions, the *x*-direction followed by the *y*-direction. In *two-dimensional* (2D) *compaction*, the layout is compacted in the *x*- and *y*-directions simultaneously. Certainly, 2D compaction produces more compact layouts than 1D compaction does. However, a denser layout will be obtained. Most versions of 2D compaction problems are difficult, whereas several versions of the 1D compaction problem can be solved optimally in polynomial time. A class of 1D compaction techniques that performs compaction in one direction while incorporating heuristics to perform compaction in the other direction has been proposed and is called $1\frac{1}{2}D$ *compaction*.

This chapter will describe several 1D and 2D compaction algorithms and discuss their performance. Several $1\frac{1}{2}D$ compaction techniques will be mentioned in the discussion section.

7.1 1D COMPACTION

Consider for example the physical layout shown in Figure 7.2c. 1D compactors try to minimize the total area corresponding to a given physical layout by performing a sequence of *x*- and *y*-direction compactions. Each compaction step is done independently of the other steps. There are two classes of algorithms, the compression-ridge techniques and the graph-based techniques.

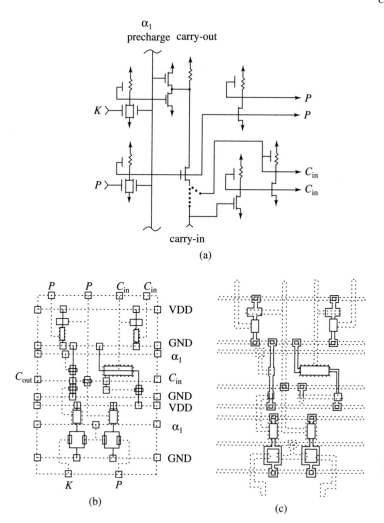

FIGURE 7.2
A carry-chain circuit.

7.1.1 Compression-Ridge Techniques

Here, compaction in the x-direction is discussed. The same technique can be employed to perform a y-direction compaction. Consider a physical layout. The width of the layout can be reduced by w units if an empty vertical slice of width w can be found going through the entire layout. See Figure 7.3. The original layout is shown in Figure 7.3a (the shaded regions are not part of the layout). The compressed layout is shown in Figure 7.3b. The slice does not have to be in one piece. In the figure, the slice consists of the three pieces of shaded regions. The compression-ridge notion was introduced in [6] along with an algorithm for

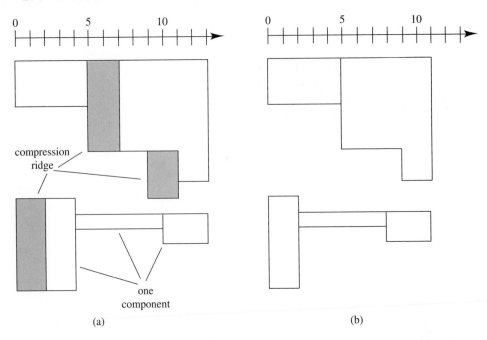

FIGURE 7.3
The compression-ridge method. (a) Before compaction; (b) after compaction.

obtaining compression ridges. The basic concept is to find vertical slices and to compact the layout as shown by the slice.

Formally, a compression ridge is a set of rectangles whose removal is equivalent to making its left boundary adjacent to its right boundary. Certainly, not every set of rectangles is a legal set of compression ridges. The union of the projection of the ridges on the y-axis as in Figure 7.3 (or x-axis, if a compaction in the y-direction is being performed) should cover the entire height of the layout without gaps. Note that if there are gaps, a compression (compaction) cannot be performed.

Dai and Kuh [78] later improved on the technique by introducing an efficient algorithm for finding the vertical slices, that is, the compression ridges. Their technique is based on the following graph representation of the layout, in particular of the empty regions. Empty regions are partitioned into a collection of rectangles. The top side of the layout is considered an empty rectangle and is denoted by s. The bottom side of the layout is also considered to be an empty rectangle and is denoted by t. A vertical dependency graph $G = (V, E)$ is constructed as follows. Each vertex is associated with an empty rectangle. There is an edge of weight w from a rectangle r_a to a rectangle r_b, if r_a is above r_b and r_a and r_b share a horizontal edge of length w. Also, there is an edge of weight ∞ from r_b to r_a (the edge drawn upward in the figure). An example is shown in Figure 7.4, where weights correspond to the capacity of the edges. Note that

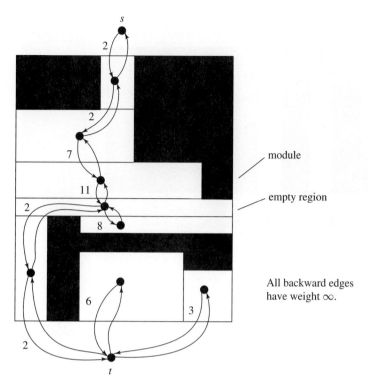

module

empty region

All backward edges
have weight ∞.

FIGURE 7.4
The flow-based compression-ridge method.

a path in G corresponds to a compression region. The amount of reduction in length is given by the minimum-weight edge in the path. The maximum flow in Figure 7.4 corresponds to the compression ridge of Figure 7.3. Thus our goal is to find a set of paths or, equivalently, a maximum flow from s to t. This problem has been studied extensively in the literature. See [199] (or any book on combinational optimization) for maximum-flow algorithms.

Consider Figure 7.5a. To find a compression ridge, edges with weight ∞ are used, as shown. The compacted layout is shown in 7.5b. Note that these ∞-weight regions are not removed from the layout and that is why their weight (capacity) is set to ∞.

The main advantage of the compression-ridge technique is its relative simplicity. Also, it is possible to do a little x-compaction (i.e, one ridge) followed by a little y-compaction (i.e., it is not necessary to do a complete x- or y-compaction), and this makes the technique (if it is done right) closer to a real 2D compactor. Note that doing it right means coming up with a good order. Recall from previous chapters that coming up with a good order is, in most problems, not easy. The main disadvantage of the compression-ridge method is that even the 1D compaction problem cannot be solved optimally.

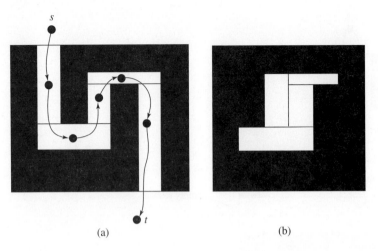

FIGURE 7.5
An application of compression ridge with ∞ edges.

7.1.2 Graph-based Techniques

To obtain a layout with minimum width, the circuit elements and vertical wires are moved horizontally to the leftmost possible position (assuming an imaginary left boundary), while horizontal interconnections are stretched accordingly. Consider a physical layout $\mathcal{F}$ (see Figure 7.2c). A directed acyclic graph (DAG) G, called the constraint graph, is associated with $\mathcal{F}$. Each maximal geometrically connected set of circuit elements and vertical interconnections in the symbolic layout forms a group. The left and right (imaginary) boundaries form groups V_0 and V_n, respectively. Other groups are denoted by $V_1, \ldots, V_{n-1}$. The algorithm to be described here was proposed in [218].

Figure 7.6a shows the vertical groups of the circuit in Figure 7.2. Vertical groups denote vertices of G, and spacing requirements among the groups are represented as the edges of G. Let $X(V_i)$ denote the x-coordinate of the centerline of group V_i, for all i. If the design rules require a minimum spacing ξ between groups V_i and V_j, then

$$X(V_i) + \xi \leq X(V_j),$$

where V_i lies to the left of V_j. A directed edge from V_i to V_j, with weight ξ, is inserted. Note that for two groups that are not visible to each other, an edge does not need to be inserted. (Recall that two groups are visible if a horizontal line segment can be drawn crossing only the two groups and not any other groups.) It can be shown that the number of visible edges is linear in the number of segments [313].

After design-rule edges have been added, user-specified constraints are translated into edges of G. Consider groups V_i and V_j. If the user requests that V_j be at least α units to the right of V_i, then the type of inequality is the same as described earlier. That is, a directed edge from V_i to V_j, with weight α, is inserted. If the

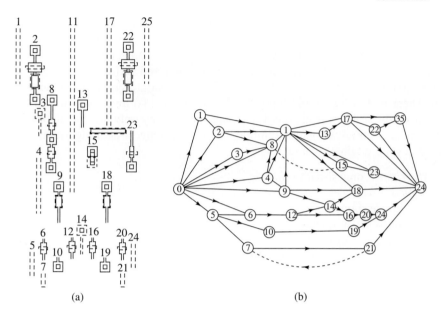

FIGURE 7.6
Modeling a compaction problem. (a) Vertical elements of a layout; (b) the constraint graph.

user specifies that V_j should be not more than β units to the right of V_i, then the constraint is specified by

$$X(V_j) - \beta \le X(V_i).$$

A directed edge from V_j to V_i with weight $-\beta$ is inserted. If the user requires V_j to be exactly γ units to the right of V_i, the inequality becomes

$$X(V_i) + \gamma = X(V_j).$$

This inequality is equivalent to the following two inequalities:

$$X(V_i) + \gamma \le X(V_j), \quad \text{and}$$

$$X(V_j) - \gamma \le X(V_i).$$

Two edges are added to G, one from V_i to V_j with weight γ and one from V_j to V_i with weight $-\gamma$. The constraint graph of the physical layout in Figure 7.6a is shown in Figure 7.6b. The graph contains both design-rule constraints and user-specified constraints (the weights are not shown). If G has multiple edges, all of them are deleted except the one with the largest weight, as it implies all other constraints. By construction, self-loops do not appear in G (edges with negative weights are shown as dashed lines).

Given G, set $X(V_0) = 0$. That is, the left boundary is set to x-coordinate zero. A solution to the 1D compaction problem is an x-coordinate for each V_i, such that all constraints are satisfied. An optimal solution is one with the x-coordinate of the last group being at its minimum possible value. Note that, in general, G may contain conflicting constraints, and a solution may not exist. For example, the two constraints $X(V_1) + 1 \leq X(V_2)$ and $X(V_2) + 2 \leq X(V_1)$ are conflicting.

Let E_r and E_l denote the set of right-directed and left-directed edges of G, respectively. Define $G_r = (V, E_r)$, a single-source single-sink directed acyclic graph. There is, thus, a natural topological order in G_r. To find a solution to the 1D compaction problem if the constraints are specified by only G_r, the algorithm for obtaining a minimum-width solution, the classic longest-path algorithm (for directed acyclic graphs), is used.

> *Algorithm* Longest-path (G_r)
> *begin-1*
> create a queue;
> push V_0 on queue;
> *for* $i = 1$ to n *do*
>   in-degree(V_i) = in-degree of V_i in G_r;
> *while* queue not empty *do*
> *begin-2*
>   pop queue's node V_i;
>   *for* each edge (V_i, V_j) in G_r *do*
>   *begin-3*
> $X(V_j) = \max (X(V_j), X(V_i) + w_{ij})$;
>   (* w_{ij} is the edge weight *);
>   in-degree(V_j) = in-degree$(V_j) - 1$;
>   *if* in-degree$(V_j) = 0$ *then* push(V_j) into queue;
>   *end-3;*
> *end-2;*
> *end-1.*

The algorithm longest-path runs in $O(|E_r|+|V|)$ time. A solution, satisfying all constraints in G_r, is obtained. Next, left edges are examined one by one. If the constraint imposed by a left edge is satisfied, then the next is examined. Otherwise, the layout is modified by the least amount to satisfy the constraint. Specifically, consider a left edge (V_i, V_j) and let $X(V_i)$ and $X(V_j)$ denote the current positions of the corresponding vertices. If $X(V_i) + w_{ij} > X(V_j)$, $X(V_j)$ will be set to $X(V_i) + w_{ij}$. Note that w_{ij} of a left edge is typically a negative value.

All left edges are examined as described. Then, if all constraints are satisfied, the procedure terminates. Otherwise, another pass of the longest-path algorithm is applied. This iteration proceeds until all constraints are satisfied. It can be shown that the number of iterations is at most $|E_l| + 1$ if there are no conflicting constraints. At that time, an optimal solution is obtained, assuming a solution exists. An example demonstrating the iteration process is shown in Figure 7.7. The input and first application of the longest-path algorithm are shown in Figure 7.7a–b. Then $X(V_2)$ and $X(V_3)$ are increased after examining the two negative edges in Figure 7.7c. The result of the second call to algorithm longest-path is shown in Figure 7.7d. Left edges are used to examine $X(V_4)$ (Figure 7.7e). Algorithm

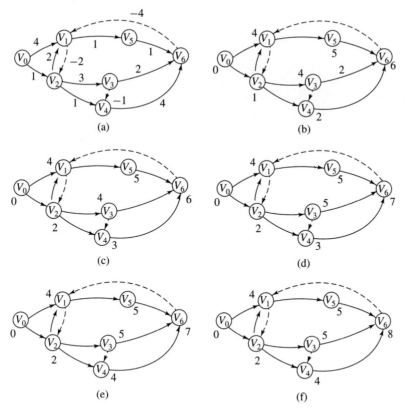

FIGURE 7.7
Demonstration of the graph-based 1D compaction.

longest-path is then applied for the third time, and left edges are checked one more time to verify all the constraints. Thus the result in Figure 7.7f is indeed the minimum-width solution.

7.1.3 Wire-Length Minimization

In the compacted layout derived with a longest-path algorithm, all objects are pushed toward one side in the direction of compaction, often resulting in unnecessarily long wires. This effect can be dramatically demonstrated in the application of compaction algorithms to channel routing. See Figure 7.8a. (The direction of compaction is toward the bottom.) After straightening out the wires, a smoother layout with shorter wire length results, as shown in Figure 7.8b.

The longest-path solution is just one of many minimum-width solutions to the constraint equations. All of these solutions yield the same positions for those objects on the critical path from the source to the sink but have different positions for those objects not on the critical paths. The freedom to move objects not on the critical path allows the compactor to optimize the layout to achieve the secondary

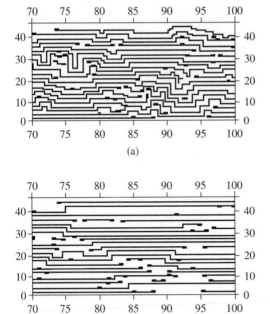

(a)

(b)

FIGURE 7.8
(a) Compacted horizontal layer features.
(b) Straight version of (a).

goal of wire-length minimization in addition to the primary goal of minimum silicon area. The minimization of wire lengths [310] leads to the shortening of slack wires, which has the following beneficiary effects.

- Shorter wires carry less parasitic effects and thus make faster circuits.
- Better compaction may be achieved in the orthogonal direction, since shorter wires present less blockage.

Different masks are assigned different unit weights, c, to reflect parasitic strengths. For example, because of its high capacitance, a diffusion mask is given more weight than a polysilicon or a metal mask. The cost of a horizontal wire segment, $[x_i, x_j]$, is defined as the unit weight times the area covered: $c \times$ width $\times (x_j - x_i) = c_{ij}(x_j - x_i)$, where width is the wire width and $x_i = X(v_i)$. For polygon shapes, there are two ways to define the cost function.

- **Polygon perimeter times weight.** In this case, the same formula for wires can be applied to the horizontal edges of polygon shapes with width $= \lambda_w/2$, where λ_w is the minimum wire width. (Half is used because the number of horizontal edges is always even, and both the top and bottom edges contribute equally.) For the polygon in Figure 7.9a, the length of horizontal perimeters is $pl = (x_3 - x_1) + (x_3 - x_2) + (x_4 - x_2) + (x_4 - x_1)$, and the cost function is proportional to the shaded area, which is $pl \times \lambda/2$.
- **Polygon area times weight.** In this case, the polygon is sliced horizontally. The slicing can be done by sorting and scanning the horizontal polygon edges

vertically. The area of a polygon is then the sum of the areas of rectangle slices. For the polygon in Figure 7.9b, three horizontal slices are cut out, and the total area is $(x_3 - x_1)(y_2 - y_1) + (x_2 - x_1)(y_3 - y_2) + (x_4 - x_1)(y_4 - y_3)$.

The choice of using the polygon perimeter or the polygon area method for determining the cost function affects the final solution. For example, Figure 7.9c is minimum in the diffusion perimeter. Figure 7.9c may be better in terms of capacitance, but Figure 7.9d is better in resistance.

For compaction in the x-direction, the cost function for the wire-length minimization is

$$L = \sum_{ij} c_{ij}(x_j - x_i) = \sum_i C_i x_i, \quad C_i \equiv \sum_j (c_{ji} - c_{ij}), \qquad (7.1)$$

where the summation is taken over all horizontal wires and horizontal polygon edges. The wire-length minimization problem is to minimize L subject to the design rule constraints and the minimum cell width. This is a linear programming (LP) problem.

The following heuristic improvements can be made. In operations research, the two notions local-slack and global-slack are used to describe the possible delay of activities. The *local-slack* is the period of time in which one activity may be delayed without delaying any of the other activities. In this context, local-slack means the distance along which an element may be shifted against the direction of compaction without interfering with another element. The *global-slack*

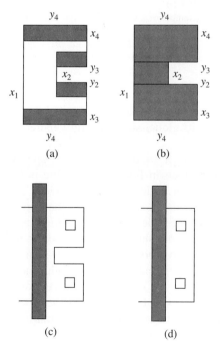

(a)

(b)

(c)

(d)

FIGURE 7.9
(a) Polygon perimeter; (b) polygon area;
(c) minimum diffusion area; (d) minimum diffusion perimeter.

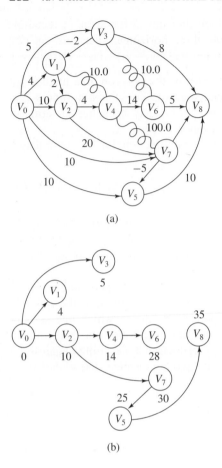

(a)

(b)

FIGURE 7.10
(a) An example of a constraint graph. The normal arcs are constraints, while the spring arcs are wire length costs. (b) The minimum size solution and the longest-path tree.

is the difference between the maximum coordinate of an element and its minimum coordinate. Positioning an element according to its global-slack generally increases the coordinates of other elements.

For example, consider the constraint graph of Figure 7.10a. The longest-path solution with a minimum cell size 35 is given in Figure 7.10b. The union of longest paths from the source node to all other nodes is also displayed. The longest path, $V_0 V_2 V_7 V_5 V_8$, from the source to the sink node V_8 defines the critical path. For the longest-path solution, the positions of nodes not on the critical path are biased toward the source. In other words, there are slacks for nodes not on the critical path: V_1, V_3, V_4, and V_6. V_1, V_3, and V_6 have positive local-slacks, since these vertices may be shifted to the $+x$-direction without interfering with others. V_4 has zero local-slack, because it is constrained by node V_6. However, once V_6 shifts to the right, V_4 gets free space to move in. In such a case, it is said that V_4 has positive global-slack. The reduction operates in two phases, first local improvements, and then group improvements.

Next, local improvements are made. In the cost function, $L = \sum_i C_i x_i$, C_i is the unit cost function, that is, the cost of moving a vertex V_i to the right by one unit. If C_i is negative, the vertex can be moved to the right as far as it can go (i.e., to reduce its local-slack to zero). If the vertices could be topologically sorted and this process could run from the high vertices back to the low vertices, it would finish in one pass. However, this cannot be done completely because of the cycles in the graph. Thus, the process repeats until nothing moves on a pass through the vertices. An algorithm for local improvement can be described as follows:

Procedure Iterative-improvement-for-wire-length
> *begin-1*
>> topologically sort the nodes in the order of decreasing x-coordinates;
>> *repeat*
>> *begin-2*
>>> Nothing-Moved = Yes;
>>> *for* each x_i *do*
>>> *begin-3*
>>>> *if* $C_i \leq 0$, *then*
>>>> *begin-4*
>>>>> let δx be a large positive integer;
>>>>> *for* each arc (V_i, V_j) *do*
>>>>> if $\delta x > (x_j - x_i - w_{ij})$ then set $\delta x = x_j - x_i - w_{ij}$;
>>>>> let $x_i = x_i + \delta x$;   (* move x_i to the right by δx *);
>>>> *end-4;*
>>> *end-3;*
>>> *if* $\delta x > 0$ *then* Nothing-Moved = No;   (* for each x_i *);
>> *end-2;*
>> *until* Nothing-Moved = Yes;
> *end-1.*

Note that δx is the local-slack of node V_i.

To illustrate the algorithm, consider the constraint graph example in Figure 7.10a. The longest-path solution in Figure 7.10b indicates that V_1, V_3, V_4, and V_6 have slacks. The costs of moving these nodes are $C_1 = -10.0$, $C_3 = -10.0$, $C_4 = -90.0$, and $C_6 = 10.0$. Since only the first three are negative, procedure Iterative-improvement-for-wire-length will try to move these nodes to the right in the order of V_4, V_3, V_1. V_4 cannot move because of the arc (V_4, V_6), while V_3 can advance one unit to the right until it is constrained by the arc (V_3, V_1). Then V_1 advances four units to the right until it is constrained by the arc (V_1, V_2). Figure 7.11 shows the result of the iterative improvements.

After local improvement is done, there are no local-slacks left for wire-length reduction. However, the global-slacks may still be available for the wire-length minimization. Schiele [310] and Kingsley [185] propose heuristic methods to cluster several nodes into a group and move the whole group. These heuristic methods have drawbacks in that they cannot stretch one wire in order to reduce another wire that is connected to it.

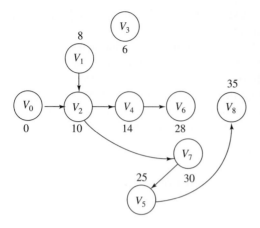

FIGURE 7.11
Result of the iterative improvement.

Global optimization was studied in [205, 224]. The wire-length minimization (formulated as an LP) can be solved with the well-known Simplex method. The Simplex method starts with an initial solution that is not the optimum, and uses iterative steps of pivoting to derive an optimal solution. A basic feasible solution for n independent variables $x_1, x_2, \ldots, x_n$ (the position of source node x_0 is a free parameter) is defined by an independent set of n constraint equations called the basis. For a compaction problem, the constraint equations in the basis take the form $x_i - x_j - w_{ij} = 0$. Let the slack variables be $s_{ij} = x_i - x_j - w_{ij}$. In other words, each equation in the basis corresponds to a tight arc with zero slack. It can be shown that these n tight arcs form a spanning tree on G. On the other hand, any spanning tree of tight arcs on G is a basic solution, so the longest-path tree in the compaction problem is a basic feasible solution that can be used to start the Simplex method.

The key step of the algorithm Simplex is pivoting from one basic feasible solution to another one with less wiring cost. In other words, the spanning tree corresponding to the current feasible solution is replaced by another spanning tree of a lower cost through an elemental tree transformation. The two trees before and after such a tree transformation differ in only two arcs. One arc, say $e_{\text{in}} = (V_i, V_j)$, in the old tree is removed, while a new arc, e_{out}, is added to form the new tree. Examine these steps one at a time. First remove arc (V_i, V_j). This partitions the old tree into two subtrees, T_1 and T_2. Let T_1 be the subtree that contains the source node. Since both subtrees remain tight during the pivoting, nodes in T_1 must be stationary while nodes in T_2 need to move together as a whole. Let δx be the maximal allowable move for T_2 in the direction of separating V_i and V_j, and dir be the move direction: $dir = +1$ if $V_j \in T_2$ and $dir = -1$ otherwise. If all the nodes in T_2 move by δx, then the new tree can be formed by picking a new tight arc between the two subtrees. Of course, this tree transformation will be done only if it is profitable. The cost difference between the old tree and the new tree is $C(T_2)\delta x$, and the profitability is defined by a negative unit cost change,

$C(T_2)\frac{\delta x}{|\delta x|} = dir \times C(T_2) < 0$. $C(T)$ is the unit wire-length cost function for moving a tree.

$$C(T) = \sum_{i \in T_i} C_i.$$

When pivoting with respect to all tree arcs produces no decrease in the wiring cost, the Simplex algorithm stops, and an optimal solution has been found. A formal description of Simplex follows.

> *Procedure* Simplex
> *begin-1*
>> mark all the arcs of the spanning tree as not visited;
>> *while* there is any arc marked as not visited *do*
>> *begin-2*
>>> take a not visited arc $e_{in} = (V_i, V_j)$;
>>> partition the spanning tree into two subtrees $\{T_1, T_2 | V_0 \in T_1\}$;
>>> let *dir* be the move direction of T_2;
>>> let δx be the maximal allowable move;
>>> let $e_{out} = e_{in}$;
>>> *if dir* $\times$ $C(T_2) < 0$
>>> *begin-3*
>>>> *if* $\delta x \neq 0$ *then*
>>>> *begin-4*
>>>>> move nodes in T by δx;
>>>>> set the new tight arc = e_{out};   (* nondegenerate case *);
>>>> *end-4*
>>>> *if* $\delta x = 0$ *then* choose any tight arc between
>>>>>>> T_1 and $T_2 = e_{out}$;   (* degenerate case *);
>>> *end-3*
>>> *if* $e_{out} = e_{in}$, *then* mark e_{in} as visited;
>>> *if* $e_{out} \neq e_{in}$, *then*
>>> *begin-5*
>>>> replace e_{in} with e_{out};
>>>> mark e_{out} as visited;
>>>> mark the rest of the tree arcs as not visited;
>>> *end-5;*
>> *end-2;*
> *end-1.*

For the example in Figure 7.12, Simplex starts with the longest-path tree solution in Figure 7.10b. The first arc (V_0, V_1) is a suitable candidate for pivoting, since the removal of (V_0, V_1) creates two subtrees $T_2 = \{V_1\}$ and $T_1 = V - T_2$, and $C(T_2) = -10$. So V_1 may be shifted to the right as close as possible to V_2. This pivoting generates the solution for Figure 7.12a. Next, examine (V_0, V_2). This is not a good candidate, since the cost of subtree $T_2 = \{V_2, V_4, V_5, V_6, V_7, V_8\}$ is

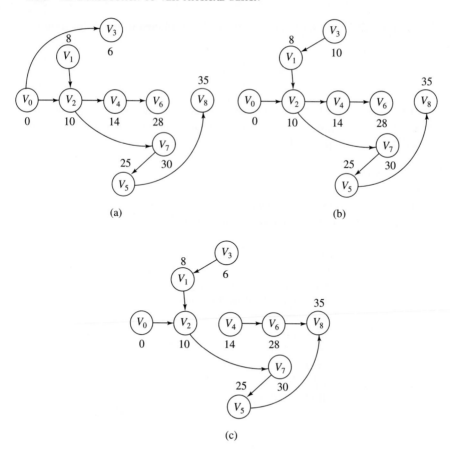

FIGURE 7.12
Illustration of Simplex.

$10 + 10 = 20$. So move on to the next arc (V_0, V_3), which is a suitable candidate for pivoting, since $C(\{V_3\}) = -10$. With this pivot, the spanning tree solution shown in Figure 7.12b is produced. Finally, a pivoting from arc (V_2, V_4) to arc (V_6, V_8) is performed, since $C(\{V_4, V_6\}) = -90 + 10 = -80$. This then produces the optimum solution as shown in Figure 7.12c, because no further pivoting can reduce the wire length cost.

7.1.4   Compaction with Automatic Jogs

Jog points are locations at which a wire can be bent and then continued a short distance away. Such a bending may result in a more area-efficient layout, as shown by the example in Figure 7.13. (A higher area efficiency cannot be guaranteed unconditionally since the bent wire must occupy some space in the direction of the bend.)

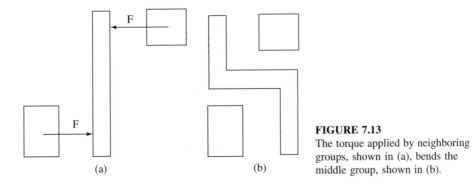

FIGURE 7.13
The torque applied by neighboring groups, shown in (a), bends the middle group, shown in (b).

Although jog points may be put into the symbolic layout plan manually, it would be much simpler if the compactor could automatically generate jogs at any stage of the compaction operation. See [152].

FORCE-DIRECTED JOG STRATEGY. The longest path information derived during the initial layout compaction operation may be used to determine all the possible jog points at once. Specifically, jog points may be included in a group where the longest path goes into and comes out from two different intervals of the group. In essence, the two segments of the longest path may be viewed as force vectors that exert a torque on the group. See Figure 7.13. The jog point that allows the group to be torn apart by the torque is usually introduced at the middle point of the wire. This midpoint policy is not a good choice when there are multiple wires running between two objects.

CONTOUR COMPACTION STRATEGY. [84, 397] developed a contour compaction strategy for automatic jog generation. Suppose that the compaction is to the left; objects are pushed as far left as possible. Then the wires will bend around the contour to form a river routing pattern. See Figures 7.14 and 7.8.

This type of contour compaction algorithm works well with the channel routing problem. The area saving is typically 10–20%. Given a channel routing problem, [84] generated a solution using a channel routing algorithm (see Chapter 3) and applied the following transformation to the horizontal layer features.

Step 1. Initialize the contour to be the channel's bottom edge.

Step 2. Initialize the current track to be the bottom track.

Step 3. Route each feature assigned to the current track so that its bottom edge is exactly the minimum separation distance above the contour. Note that following this contour may introduce jogs.

Step 4. Update the contour based upon the top edges of all features currently routed.

Step 5. If one or more tracks have not yet been processed, set the current track to the bottom-most unprocessed track and go to step 3.

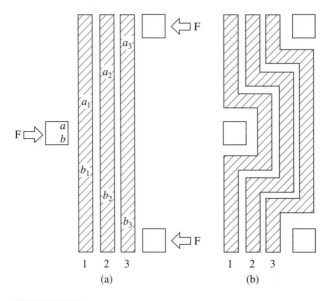

FIGURE 7.14
(a) Jog points due to convex corners. (b) Jogs after compaction.

Step 6. Since all the tracks have been processed, the channel height is the maximum contour value plus the minimum separation distance.

JOG INTRODUCTION ON NONCRITICAL PATHS. The above two algorithms focus on jog generation to minimize the layout size, and, hence, jogs are created mainly to reduce critical path lengths. Jogs on noncritical paths can be equally important for producing a high-quality layout with less parasitic capacitance and good performance. A good automatic jog strategy for a layout compactor should take both into account. In other words, jogs need to be introduced not only at places where the cell size (span) can shrink, but also at places where overall wire lengths and parasitic capacitance can be reduced.

Assume that the compaction direction is along the x-axis. (Analysis of y-compaction can be done in the same way.) The compaction process can be modeled as follows:

1. A pair of enormous push forces along the x-direction is exerted on the cell bounding box to minimize the cell size.
2. Pulling forces along the x-direction are exerted at two ends of the horizontal wires to shorten the wire lengths. The sizes of these pulling forces are determined by parasitic resistance or capacitance of the horizontal wires. See Figure 7.15.

For compaction of the leaf cell, this is more important, since polysilicon and diffusion carry more parasitic capacitance/resistance than metal.

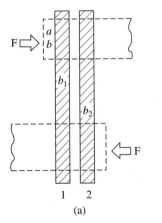
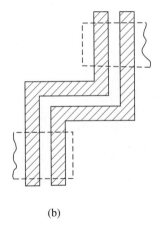

FIGURE 7.15
(a) Jog points due to concave corners. (b) Jogs after compaction.

AN EXAMPLE OF A STANDARD CELL LAYOUT. A six-way NAND CMOS circuit example is shown in Figure 7.16. In this layout, diffusion, polysilicon, and metal masks are drawn as dashed, dotted, and solid lines, respectively. The two rows of transistors are defined by the intersections of the diffusion and polysilicon masks. Parasitic weights, 1000, 100, and 1, are assigned to the diffusion, polysilicon, and metal layers, respectively. The loose layout, Figure 7.16a, is compacted first in the y-direction. The diffusion boundaries are smoothed out, while metal wires are constrained on the grid to allow more horizontal feed through channels, as shown in Figure 7.16b. The layout is then compacted in the x-direction. During this pass, the critical paths go through the lower row of transistors (PFET), while there are some slack spaces in the upper row of transistors (NFET). The series path in the NFET row is also the bottleneck of the circuit delay time. After the x-compaction (Figure 7.16c), NFET transistors are packed very tightly with the help of many jogs placed on the polysilicon wires. Some of these jogs occur on noncritical paths.

7.1.5 Grid Constraints

Grid constraints, which require selected objects to be placed on specified grids, will be considered here for two reasons.

- The pin positions of cells/macros are often required to be placed on the wiring grids, so that grid-based routers [134, 200] can wire them up. Even though significant progress has been made on the gridless routing problem [46], most industrial wiring tools in practical use today are still grid-based.
- Cells/macros need to be designed in such a way to support feed-through channels. As more circuits are packed into one chip, wiring them up becomes very hard, and it is necessary to feed global wires directly through

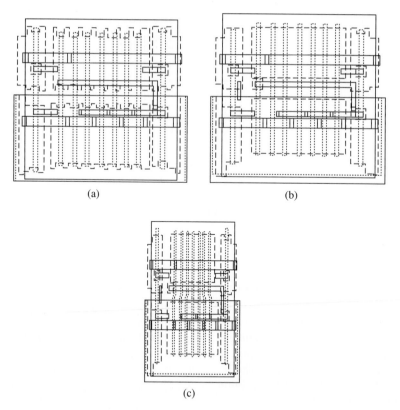

FIGURE 7.16

A six-way NAND circuit (a) before compaction; (b) after y-compaction; (c) after y- and x-compaction.

cells/macros when wiring channels are congested. The sea-of-gate layout style represents an extreme case where the wiring channels are completely eliminated, and all the wires go through the interior of cells/macros. Therefore, it is desired to keep as many feed-through channels inside cells/macros as possible.

In Figure 7.17, vertical wires A and C are located at $x = 0$ and $x = 3$, respectively, and wire B runs between wires A and C. Let the minimum spacing between wires be 1. If wire B is placed at the midpoint, $x = 1.5$ (Figure 7.17a), then no global wire is allowed to go between either A and B or B and C, and the porosity is zero. If wire B is placed at $x = 1$ (Figure 7.17b), then one global wire can squeeze through between B and C, and the porosity is increased to one. So, if the internal wires of a cell/macro can all be placed on grids, a larger porosity can be achieved. The compaction problem with grid constraints turns out to be a mixed integer problem. Even though a general mixed integer problem is NP-hard, this particular case admits a polynomial-time algorithm. Compaction algorithms have been given in [202, 203].

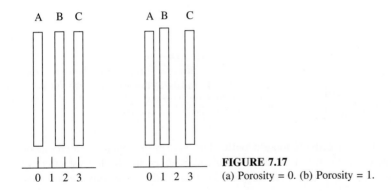

A B C   A B   C

0 1 2 3   0 1 2 3

FIGURE 7.17
(a) Porosity = 0. (b) Porosity = 1.

With mixed grid constraints, partition the nodes of V into two sets, $\{V_g\}$ and $V - \{V_g\}$, according to whether the node is constrained to the grid or not. Those nodes in $\{V_g\}$ are called *grid nodes*. Pictorially, the grid nodes will be represented by squares and nongrid nodes by circles. The grid constraint then takes the form

$$\text{for } V_j, \quad V_i \in \{V_g\}, \quad X(V_j) - X(V_i) = kd,$$

where d is the grid constant and k can be any integer. In the case where all nodes are grid nodes, the compaction problem can be solved by simply rounding up the weights of all edges to the unit of d and finding the longest-path solution. In the case where only some nodes are grid nodes, there is a mixed integer problem, the solution of which is not trivial.

In the case where the source node is on the grid, the longest-path method can be modified to treat grid constraints as follows. Consider a path p from the source V_0 (a grid node) to any node V_i. Suppose that along this path (excluding two endnodes) there are k grid nodes $U_1, U_2, \ldots, U_k$, which divide p into $k + 1$ segments, $p_1, p_2, \ldots, p_{k+1}$. Let $U_0 = V_0$, and $U_{k+1} = V_i$. The first k segments span paths from grid node to grid node, while the last segment spans a path from a grid node to V_i, which may or may not be a grid node. The separation between U_j and U_{j-1} is greater than or equal to LENGTH(p_j), where LENGTH denotes the usual path length, namely, the sum of weights along the path. For the first k path segments, both ends of each segment are grid nodes, and, therefore, for $1 \leq j \leq k$, $X(U_j) - X(U_{j-1}) \geq [\text{LENGTH}(p_j)]$, where $[x]$ is the round-up function of x, that is, $[x] = kd$ if $(k - 1)d < x \leq kd$. Then $X(V_i)$ must be greater than or equal to the effective path length defined in the following:

EFFECTIVE LENGTH$(p) =$

$$\sum_{j=1}^{k+1} [\text{LENGTH}(p_j)] \qquad\qquad \text{if } V_i \in \{V_g\}.$$

$$\sum_{j=1}^{k} [\text{LENGTH}(p_j)] + \text{LENGTH}(p_{k+1}) \quad \text{if } V_i \notin \{V_g\}.$$

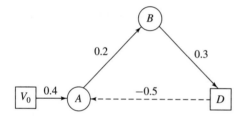

FIGURE 7.18
Assume that $d = 1$. The effectively longest path to D is V_0ABD, while the effectively longest path to A and B are V_0ABDA and V_0ABDAB, respectively.

Theorem 7.1. **Effectively longest path.** For the case in which the source node is a grid node, $X(V_i)$ in the minimum solution is given by the longest effective path length from the source node to node V_i in G, if the effectively longest path exists.

Such effectively longest paths may traverse through the same node twice (via a cycle). As an example, see Figure 7.18 in which the effectively longest path from V_0 to A is V_0ABDA.

Next, consider a pair of grid nodes with an equal constraint s between them. If s is less than the grid size d, then it is obvious that no solution on the grid can satisfy the constraint s, yet the cycle, which goes around these two nodes, has a zero cycle length. Therefore, even if there is no positive cycle in the graph, there may still be no effectively longest-path solution. In investigating this seemingly mysterious case, it is necessary to define an *effective cycle length* for a cycle. In the case that a cycle does not contain any grid node, the effective cycle length is clearly the cycle length in the ordinary sense. But if a cycle does contain grid nodes $U_1, U_2, \ldots, U_k(k > 1)$ that divide the cycle into k path segments $p_1, p_2, \ldots, p_k$, then the effective path length along that cycle path $\Sigma_{i=1}^{k}[\text{LENGTH}(p_i)]$ is defined as the effective cycle length. This definition is independent of the grid node selected for the starting node of the cycle path. A cycle with a positive effective cycle length is then called an effectively positive cycle.

Observe that if a constraint graph does not have any effectively positive elementary cycles, some composite cycles may still have positive effective length. For example, the graph in Figure 7.19 contains two elementary cycles with zero effective length, while the union of two cycles yields a composite cycle of effective length one.

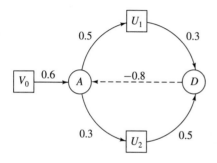

FIGURE 7.19
Assume that $d = 1$. Both elementary cycles U_1DAU_1 and U_2DAU_2 have an effective cycle length of zero, while the composite cycle $U_1DAU_2DAU_1$ has an effective cycle length of zero also.

The longest-path algorithm can be generalized to find the effectively longest path. See [202, 203].

Theorem 7.2.   **Existence theorem**. For the case in which the source node is a grid node, the necessary and sufficient condition for the existence of the effectively longest path from the source node to any node is that there is no effectively positive cycle on any path to that node.

7.2   2D COMPACTION

It can be shown that performing a 2D compaction is, in general, much better than performing two 1D compactions. An example is shown in Figure 7.20.

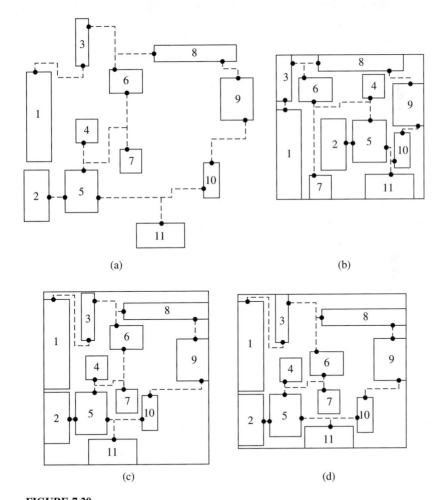

(a)   (b)

(c)   (d)

FIGURE 7.20
2D compaction is better than two 1D compactions. (a) Input; (b) output of a 2D compactor; (c) (left, bottom); (d) (bottom, left).

A layout is given and its minimum layout is shown in Figure 7.20a,b. The result of x-direction followed by y-direction compaction is shown in Figure 7.20c. Also, the result of y-direction followed by x-direction compaction is shown in Figure 7.20d.

A 2D compaction algorithm was presented in [387]. The coordinates of the southwest corner of each block B_i are represented by (a_i, b_i), where a_i is its x-coordinate and b_i is its y-coordinate. Similarly, (c_i, d_i) denote x- and y-coordinates of the northwest corner of B_i. There are two kinds of constraints: base constraints, which are concerned with size of the elements and integrity of connection; and distance constraints, which are design rules and user-specified constraints.

- **Base Constraints.**

  Base constraints that force the size of elements. Consider a block B_i of height h_i and width w_i. There are the following constraints:

  $$a_i + w_i \leq c_i$$

  $$c_i \leq a_i + w_i$$

  $$b_i + h_i \leq d_i$$

  $$d_i \leq b_i + h_i.$$

  If B_i is a vertical wire, the last constraint should be deleted because it can be stretched in the y-direction. Similarly, if B_i is a horizontal wire, the second constraint should be deleted because it can be stretched in the x-direction.

  Base constraints that preserve the integrity of connections. If a wire B_j is connected to the right of a block B_i, the following inequalities must be satisfied:

  $$a_j \leq c_i$$

  $$c_i \leq c_j$$

  $$a_i \leq a_j$$

  $$b_i + \alpha_{ij} \leq b_j$$

  $$d_j + \beta_{ij} \leq d_i,$$

  where α_{ij} and β_{ij} dictate how far the wire can slide. See Figure 7.21.
- **Distance Constraints.** These are the constraints examined in 1D compaction. They are either spacing constraints or user-specified constraints.

A layout is legal if all base and distance constraints are satisfied. The technique proposed in [312] is a branch-and-bound method. The initial solution is, in general, an illegal one satisfying only the base constraints. Then distance constraints are added one by one, and the solution is changed to satisfy them. Consequently, the area is increased.

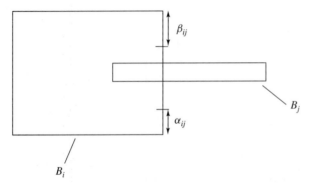

FIGURE 7.21
Integrity-preserving constraints.

The general procedure is demonstrated in the flowchart given in Figure 7.22. Consider a set of constraints C including the base constraint B. First test to see if there are any conflicting constraints. If there are (yes), the procedure backtracks. Otherwise, a minimum-area layout based on C is obtained. Specifically, two constraint graphs corresponding to x-constraints and y-constraints are stored. Then x-direction 1D compaction is performed, followed by y-direction 1D compaction. This dictates the position of all elements satisfying C. If its area is smaller than the previously obtained minimum area, then check to see if there has been any violation of constraints. Otherwise, do not continue this solution (bound the solution space).

7.3 DISCUSSION

A 1D compaction algorithm, minimizing the total wire length, was proposed in [129]. Their algorithm generates a minimum-width layout among all layouts with minimum total length. The algorithm was generalized to minimize $w + \lambda l$, where w is the layout width, l is the total wire length, and λ is a constant specified by the user.

A 1D compaction algorithm considering forbidden regions was proposed in [128]. The modules are allowed to move left and right but cannot be placed (partly or entirely) in a set of prespecified forbidden regions. Forbidden regions represent, for example, modules whose positions are fixed.

In addition to traditional spacing rules, conditional rules were considered in [49, 201]. These rules are caused by three major factors: step coverage, photolithographic resolution tolerance, and defect considerations. See [49] for details.

Two-dimensional compaction, if solved optimally, produces minimum-area layouts. However, it is very time-consuming to solve the problem optimally. Thus $1\frac{1}{2}$D compaction techniques have been proposed. The aim is to perform x-direction compaction moves while making small moves in the y-direction.

The first class of algorithms (e.g., see [330, 386]) performs 1D compaction in a preferred direction. The main goal is to minimize the width in a preferred dimension and, on the way, permit movement in the other direction.

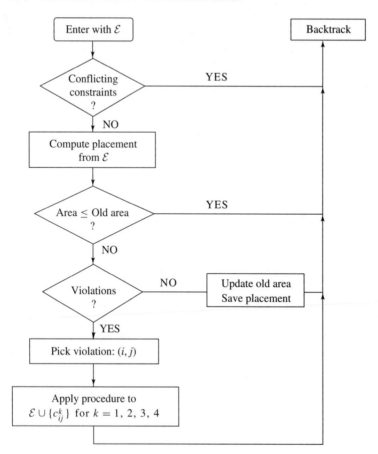

FIGURE 7.22
Flowchart of the branch-and-bound algorithm.

Another technique was proposed in [332]. At each stage, the elements are partitioned into two classes, called floor and ceiling, separated by a region called the zone. An element is moved from the ceiling, across the zone, to the floor.

With a one-dimensional compactor, the layout needs to be compacted alternately in the x- and y-directions. The final layout may depend on which direction the compaction is done first. During the one-dimensional compaction, objects move in one direction at a time. This restriction prevents objects from moving in the situation of x-y interlocks [385]. In this case, one can obtain a better compaction if a simultaneous movement perpendicular to the direction of compaction is allowed. [385] proposes using critical paths in the compaction direction to find the interlocks and to improve the layout area by shearing the interlocks. This iterative improvement method is refined by [395], in which both x- and y-constraint graphs are considered when choosing the shearing candidates. However, these iterative methods to remove interlocks may get stuck in a local optimum solution.

To overcome this problem, a truly two-dimensional compactor is needed. So far, there are two different approaches, branch-and-bound and simulated annealing.

[180] proposes the use of decision variables to determine whether a dual constraint between two interlocked objects can be realized as an x-constraint or a y-constraint. Then the problem is cast in the form of a mixed integer problem, for which a branch-and-bound algorithm is employed for solution. The main problem here is the costly run time. On the other hand, [253] applies simulated annealing techniques to the two-dimensional compaction problem. The wires are regarded as rubber bands with objects moving at the ends. [370] simplifies the annealing moves by removing all the wires and replacing them with the routability constraints described in [237]. Then a Monte-Carlo algorithm is applied, followed by a post-routing phase in which the missing wires are added back. Both compactors bring out a truly compacted layout with spaghettilike wiring. Such layouts are hard to manufacture in today's IC technology. Also, the run time is prohibitive. [80] modifies the Monte-Carlo approach so that the layout can be compacted into an octagon geometry. Such layouts can be manufactured more easily. However, the run time is still very long.

The successful application of flat compactors (a flat compactor considers the entire layout with all details at once) is generally limited to smaller layout problems, such as leaf cells and wiring channels. For large macros such as bit-stacks, PLAs, and ROMs, a hierarchical approach is necessary since it can manage the growth of run time and maintain the layout hierarchy more efficiently. In the bottom-up hierarchical method, compaction starts with the lowest level cells. Then patch-up routings [331] are performed to reconnect these cells before the compaction proceeds to the next level of hierarchy. This method usually does not lead to a globally optimized layout. Also, the patch-up routing is a very time-consuming process. [185] presents a pitch-matching compactor, which consists of two phases. The bottom-up phase starts with the construction of constraint graphs for the lowest level cells. Then port abstraction graphs are extracted from these cells and passed to the next level cells. The top-down phase starts with solving the constraint graph at the topmost level, and the port positions are passed to its children cells. Because of the cell-to-cell spacing problem, a half-maximum-design-rule width of empty space is reserved around the cell boundary to form a protective frame. [289] finds the following remedy to this cell spacing problem. First, a doughnut-shaped region around each cell's boundary is cut. Then a cell abstraction graph for objects inside the doughnut is generated. For both methods, that in [289] and that in [185], there is no guarantee that one will find a solution that keeps the cell repetitions identical. [249], [92], and [7] consider the cell repetition problem by studying the special case of an array composed of single cells, as in a systolic array or a memory array. However, these methods cannot be applied to general macro layouts. [244] formulates the general hierarchical compaction problem as an LP problem. The constraints derived from cell repetitions, mirroring, and rotations are represented by linear equations. But all the positional variables must take integer values. [249] then uses a simple rounding of real-valued solutions to get an approximate solution. An exact solution to the hierarchical compaction problem will require solving an integer linear programming (ILP) problem. However, both LP and ILP are computationally expensive. Therefore, it is essential to reduce the problem size of hierarchical compaction. [15] presents a method to reduce the LP size of the hierarchical compaction by using a minimum design, but it makes the assumption that each cell is enclosed by a protective frame that contains a space of one-half of the maximum design rule on each side. In the more general case without protective frames, [204] presents the dependency tree and a bipartite graph algorithm to reduce the hierarchical constraints to a minimum set. Then the Simplex method coupled with a branch-and-bound search is used to solve the associated compaction and wire-length minimization problems. The hierarchical compaction problem of a large layout is an active research area. There are still many open issues in this area, for example, how to efficiently handle the feed-through wires (i.e., wires at a top level cell overlaying some of the descendent cells) and cell rotations of 90 or 270, which couple the x- and y-coordinates and make a two-dimensional hierarchical compaction necessary.

EXERCISES

7.1. Discuss the advantages and disadvantages of using an abstract representation in the layout process.

7.2. Show an example where an x-direction 1D compaction followed by one in the y-direction does not produce a minimum-area layout.

7.3. Show an example where the compression-ridge method does not produce a minimum-width layout.

7.4. Apply the compression-ridge algorithm to the layout shown in Figure E7.4.

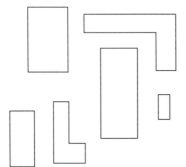

FIGURE E7.4
A layout to be compacted.

7.5. Design an algorithm (different from the algorithms described in the text) for 1D compaction. Apply your algorithm to the layout shown in Figure E7.4.

7.6. Apply a 2D compactor to the layout shown in Figure E7.4.

7.7. Consider the channel shown in Figure E7.7. Shift the two rows (shores) first to minimize the width of the channel and then to minimize the number of jogs (bends). Now shift the rows to minimize the number of bends. Design a general algorithm for this problem: in a channel involving only two-terminal nets, a net may have both terminals on the same row.

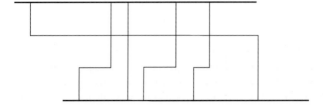

FIGURE E7.7
A channel to be compacted.

7.8. Construct a layout that results in a positive cycle without using user constraints (hint: use rigid elements in the layout). Give the constraint graph of the layout. Suggest alternatives to fix the layout.

7.9. Extend the graph construction in Section 7.1.2 so that objects in the same net can move past each other. Are additional algorithmic changes necessary? Explain.

7.10. Give an example demonstrating that the longest-path algorithm does not work when the graph has cycles. Does the algorithm work when there are special types of cycles? Explain.

7.11. Design an algorithm for compacting a set of three-dimensional cubes.

COMPUTER EXERCISES

7.1. Consider a set of nonoverlapping rectangles in the plane. Implement the compression-ridge method to obtain a minimum-width layout.
Input format.

SWx SWy NEx NEy   (* southwest and northeast coordinate
of each rectangle *)
(* no commas between the entries of the same or different rectangles *)

Output format. The output consists of a drawing of each application of the method (a sequence of figures like the ones shown in Figures 6.3a,b).

7.2. Consider a directed acyclic graph. Implement an algorithm for finding the longest path in the graph.
Input format.

v1 v2 v3 v4 v5   (* set of vertices *)
v1 v2, v3 v4, v2 v5   (* set of directed edges *)

Output format. The output is a drawing of the graph. A longest path is to be highlighted.

7.3. Implement a longest-path algorithm on a cyclic directed graph. Input and output formats are the same as CE 7.2.

7.4. Consider an x-monotone layout as shown in Figure CE7.4. Design an algorithm that compacts the layout. The order of horizontal segments should be preserved; if a segment is originally above another segment, it should remain that way.
Input format. The input is given by the starting point of each net on the x-axis, followed by the number of units of upward, downward, and right moves (there

is no left move since the layout is x-monotone). With reference to Figure CE7.4, the input is represented as

$$1\ 5\ 3\ -2\ 5\ -2\ 4\ -1\ ,\ \ldots$$

This means that a net (its name is not specified in the input) starts at x-coordinate 1, moves 5 units upward and then 3 units to the right, followed by 2 units downward, and so on. The specifications of two nets are separated by a comma.

Output format. The output is as shown in Figure CE7.4. Show the total length before and after the compaction. Note that the horizontal length does not change in the compaction process.

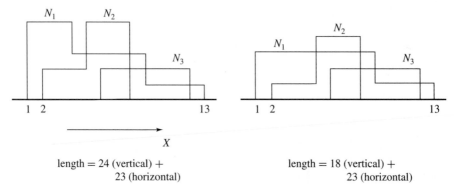

length = 24 (vertical) +
       23 (horizontal)

length = 18 (vertical) +
       23 (horizontal)

FIGURE CE7.4
Input and output formats, an x-monotone layout.

7.5. Given a set of standard cells and a net list, implement an efficient algorithm for placing the cells, routing the nets, and compacting the final result. The number of rows is given, and the width of each row is a given number. The goal is to route all nets, minimizing the height of the chip. First find a loose placement and routing, and then compact the result. Note that in each cell the same terminal appears on both the top side and the bottom side of the cell, and there are no terminals on the left or the right sides of the cell. You are not allowed to use the area over the cells.
   Input format.

   3 7 200
   1 3 0 4, ...,
   0 2 3 1 0 0

   The above example contains a number of cells to be placed in three rows, with each cell row having height 7 and the total number of columns being 200.
   Two of the modules are shown, the first and the last one. The first module occupies four columns: net 1 has a terminal at the first column, net 3 has a terminal at the second column, the third column is empty, and net 4 has a terminal at the fourth column. Note that modules should be placed at least one unit apart.
   Output format. Show a placement (containing the name of the modules) and a complete routing of the nets. Indicate the height of your routing (including the height of the cells and the total number of tracks used for routing) before and

after compaction. For example, your output can be similar to the one shown in Figure CE7.5.

7.6. Implement the previous algorithm. This time after compaction in the y-direction, perform a compaction in the x-direction. How much improvement is obtained? Is this what you expected?

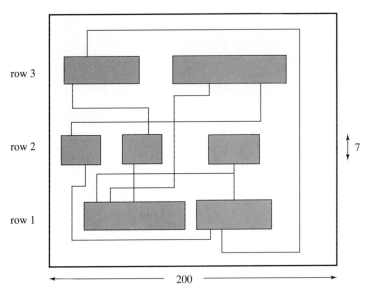

Number of tracks = 21 (cells) + 9 (routing)

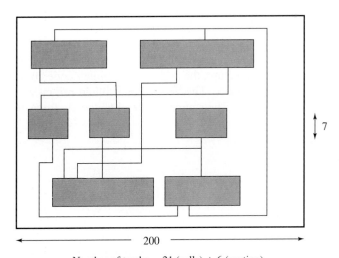

Number of tracks = 21 (cells) + 6 (routing)

FIGURE CE7.5
Output for placement, routing, and compaction of standard cells.

APPENDIX

SOFTWARE AVAILABLE FOR COMPUTER EXERCISES

There is software available for displaying under X windows. If you have access to a Sunworkstation with sun4 architecture, the code will be binary code compatible. If not, you are to recompile the source. The binary code should run on X11R4 or X11R5, with any window manager. The source can be obtained via ftp as follows:

```
$ ftp cad2.eecs.nwu.edu
user: vlsi
passwd: xxxx
ftp> binary
ftp> get display
ftp> quit
```

DISPLAY is a program that reads input files containing graphic commands and draws them on the computer screen. The current version runs under the X window system. The program shields you from the complicated process of developing graphics in X window system. All graphic commands are described in

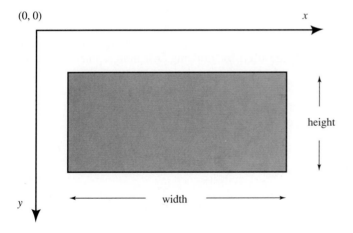

FIGURE A.1
DISPLAY coordinate system.

ASCII files. For example, a command line SL 10 20 30 40 will draw a line
from (10, 20) to (30, 40). To run DISPLAY, type

```
display input_file
```

where input_file is a text file containing graphic commands. If the filename
is omitted, the default filename display.dat is assumed.

A.1 USING DISPLAY

DISPLAY assumes a rectangular graphic display area. Following the convention
defined by the X window system, the origin of the coordinate system is on the
top left corner. The term "width" is associated with the x coordinate and the term
"height" is associated with the y coordinate, as shown below.

The coordinate values specified by the command lines are called *world coordinates*; it is the coordinate system you are working with. The rectangular area
for displaying graphics is a 2D array of pixels. A coordinate (x, y) uniquely specifies a pixel location. Each pixel is typically $\frac{1}{20}$ inch and is capable of displaying
a colored or black/white dot. This coordinate system for pixel location is called
the *device coordinate*. The device coordinate is limited by the size of screen (the
hardware). A full screen is approximately 1150 pixels wide and 880 pixels high.

The world coordinate (specified in your graphic command file) is multiplied
by a magnification factor to yield device coordinate. The magnification factor is
set to 1.0 by default; thus the world coordinate and device coordinate are identical.
The user is allowed to set this factor while displaying. It has a range of [0.1 to
10.0] with 0.1 step increments.

Each command line is terminated by a newline character (\n in C language). If the first character of a line is %, the line is ignored. All coordinate values must be non-negative. There are two types of commands, graphic commands and special commands. Graphic commands are used for drawing. All nongraphic commands are special commands.

A graphic command consists of three alpha characters, and a space, followed by some integers to specify coordinates. The third alpha character may be omitted.

The first character specifies line type:

S Solid lines.

D Dashed lines.

The second character specifies the object to be drawn:

L Straight line. The command must be followed by four integers x_1, y_1, x_2, y_2 specifying the coordinates of the end points of the line.

R Rectangle. The command must be followed by four integers x_1, y_1, x_2, y_2 specifying the coordinates of the four sides of the rectangle.

C Circle. The command must be followed by three integers x, y, r specifying the coordinates (x, y) of the center of the circle and the radius r.

E Ellipse. The command must be followed by four integers x, y, r_x, r_y specifying the coordinates of the center of the ellipse, x-radius and y-radius.

T Text. The command must be followed by two integers x, y specifying the coordinates of the reference point and a text string. The string starts at the first nonspace character after the integers and is terminated by a newline character. The text string can be left-, center- or right-justified. The default is center-justified.

The third character is optional. It specifies some characteristics associated with the drawn object:

F Fill the area inscribed by the object.

L Left-justify a text string.

R Right-justify a text string.

C Center-justify a text string.

If the character does not apply to the object, it is ignored. For example, it is meaningless to fill a line or a text string.

The special commands are as follows:

MAGNIFICATION f This command, if it appears, must be on the beginning of the input file. It sets the magnification factor for mapping world coordinate to device coordinate. The value of f must be [0.1, 10.0]. The default value is 1.0.

WORLD *width height* This command, if it appears, must be on the beginning of the input file. It sets the range of your world coordinate to be displayed. The default values are *width = 800, height = 600*. To limit memory usage, the maximum world coordinate is currently limited to 5000×5000. The actual memory usage is directly proportional to *world_width* $\times$ *world_height* $\times$ *magnification_factor*.

BACKGROUND *colorname* This command, if it appears, must be in the beginning of the input file. It sets the background color of the rectangular drawing area. The default background color is white.

#IN *filename* This command allows you to include a file into current file for processing. When using this command, make sure that there is no loop in file inclusion, such as f_A includes f_B and f_B includes f_A.

COLOR *colorname* This command changes the color of drawing pencil. See below.

END This command terminates a command file prematurely. All text after this command is ignored.

DISPLAY automatically determines the capability of your screen. If it is a monochrome screen, all color commands will be ignored. While drawing objects, DISPLAY acts like drawing on a piece of paper using colored pencils. The initial color of the paper is specified by BACKGROUND *colorname* command. The default background color is white and, the first "pencil" selected by DISPLAY is black.

The command COLOR *colorname* is used to replace the pencil. All subsequent graphic commands after a COLOR command will be drawn in the specified color until the next COLOR command is encountered. *colorname* is taken from a list of standard text string defined by X window environment. The list consists of some 700 strings. All commonly used color names like "red", "yellow", "blue" are in the list. Some less common names such as "ivory", "gold", "khaki" are also found. The complete list of names is in the file /usr/lib/X11/rgb.txt. An easier way to choose a color is to use the xcolors command on a color monitor to look at the colors and their names. If the string is an invalid name, a warning message is generated and the COLOR command is ignored. At any time, one and only one color pencil is active. As mention above, the first pencil is black.

A.2 EXAMPLES

- SL 10 20 30 40 draws a solid line from (10, 20) to (30, 40).
- DL 10 20 30 40 draws a dashed line from (10, 20) to (30, 40).
- SR 10 20 30 40 draws a rectangle with solid line as boundary. The diagonal of the rectangle is from (10, 20) to (30, 40).
- SRF 10 20 30 40 fills the region of a rectangle. The diagonal of the rectangle is from (10, 20) to (30, 40).

- `SC 10 20 30` draws a solid, hollow circle with (10, 20) as center and radius 30 units.
- `SEF 10 20 30 40` draws an ellipse with (10, 20) as the center, an *x*-radius of 30 units, and a *y*-radius of 40 units. The ellipse will be filled with current color.
- `ST 10 20 abcd efgh` puts string "abcd efgh" with the center of the string at (10, 20).
- `STL 10 20 abcd efgh` puts string "abcd efgh" left justified, that is, with the right most end of the string at (10, 20).
- `#IN foo.dat` includes the file `foo.dat` in current line.

A.3   RUNNING DISPLAY

To run DISPLAY, type:

```
display input_file
```

If there are errors in the input file, error messages will appear in a popup. When the window frame appears, click the left mouse button to realize the window. All operations use the left mouse button.

The command buttons are self-explanatory. The `redraw` button redraws the objects.

The `file-new` button allows you to discard the current file and draw another input file.

The `file-include` command allows you to include another file at the end of current file.

The `view` popup allows you to change the magnification factor.

The `home` button of the `view` popup sets the magnification to 1.0. Scrollbars automatically appears when the drawing area is smaller than window display area.

Click on `quit` to exit DISPLAY.

A.4   BUGS

- DISPLAY does not detect the situation where file A includes file B and file B includes file A. However, it will detect the case where file A includes file A.
- Please report bugs, comments and suggestions to `majid@eecs.nwu.edu`.

BIBLIOGRAPHY

[1] Afghahi, M. and C. Svensson: "Performance of Synchronous and Asynchronous Schemes for VLSI Systems," *IEEE Transactions on Computers* 41(7):858–872, 1992.

[2] Agarwal, P. K. and M. T. Shing: "Algorithm for Special Cases of Rectilinear Steiner Trees: I. Points on the Boundary of a Rectilinear Rectangle," *Networks* 20(4):453–485, 1990.

[3] Aho, A. V., M. R. Garey, and F. K. Hwang: "Rectilinear Steiner Trees: Efficient Special-Case Algorithm," *Networks* 7:35–58, 1977.

[4] Aho, A. V., J. Hopcroft, and J. Ullman: *The Design and Analysis of Computer Algorithms*. Addison-Wesley, Reading, MA, 1974.

[5] Akama, T., H. Suzuki, and T. Nishizeki: "Finding Steiner Forests in Planar Graphs," in *The First Annual ACM-SIAM Symposium on Discrete Algorithms*, pp. 444–453, ACM, 1990.

[6] Akers, S. B., M. E. Geyer, and D. L. Roberts: "IC Mask Layout with a Single Conductor Layer," in *Design Automation Conference*, pp. 7–16, IEEE/ACM, 1970.

[7] Anderson, R., S. Kahan, and M. Schlag: "An O(n log n) Algorithm for 1-D Tile Compaction," in *International Conference on Computer-Aided Design*, pp. 144–147, IEEE, 1989.

[8] Antreich, K. J., F. M. Johannes, and F. H. Kirsch: "A New Approach for Solving the Placement Problem Using Force Models," in *International Symposium on Circuits and Systems*, pp. 481–486, IEEE, 1982.

[9] Anway, H., G. Farnham, and R. Reid: "Plint Layout System for VLSI Chip Design," in *Design Automation Conference*, pp. 449–452, IEEE/ACM, 1985.

[10] Aoshima, A., and E. Kuh: "Multi-Channel Optimization in Gate Array LSI Layout," in *International Conference on Computer-Aided Design*, IEEE, 1983.

[11] Asano, T., and H. Imai: "Partioning a Polygon Region into Trapezoids," *Association for Computing Machinery* 33(2):290–312, 1986.

[12] Baker, B. S., S. N. Bhatt, and F. T. Leighton: "An Approximation Algorithm for Manhattan Routing," in *Proc. 15th Annual Symp. Theory of Computing*, pp. 477–486, ACM; 1983.

[13] Bakoglu, H. B.: *Circuits, Interconnections, and Packaging for VLSI*, Addison-Wesley, Reading, MA, 1990.

[14] Bamji, C. S., C. E. Hauck, and J. Allen: "A Design By Example Regular Structure Generator," in *Design Automation Conference*, pp. 16–22, IEEE/ACM, 1985.

307

[15] Bamji, C. S., and R. Vandarajan: "Hierarchical Pitch-matching Compaction Using Minimum Design," in *Design Automation Conference*, pp. 311–317, IEEE/ACM, 1992.

[16] Barnes, E. R.: "An Algorithm for Partitioning the Nodes of a Graph," Technical report, IBM T. J. Watson Research Center, Dept. Comput. Sci., 1981.

[17] Bentley, J. L., and T. Ottmann: "Algorithm for Reporting and Counting Geometric Intersections," *IEEE Transactions on Computers*, C-28:643–647, 1979.

[18] Berger, B., M. L. Brady, D. J. Brown, and T. Leighton: "Nearly Optimal Algorithms and Bounds for Multilayer Channel Routing," unpublished paper, 1986.

[19] Bhasker, J., and S. Sahni: "A Linear Algorithm to Find a Rectangular Dual of a Planar Triangulated Graph," *Algorithmica* 3(2):274–278, 1988.

[20] Bhat, N., and D. Hill: "Routable Technology Mapping for LUT-Based FPGA's," in *International Conference on Computer Design*, pp. 95–98, IEEE, 1992.

[21] Blanks, J. P.: "Near Optimal Placement Using a Quadratic Objective Function," in *Design Automation Conference*, pp. 609–615, IEEE/ACM, 1985.

[22] Blodgett, A. J.: "Microelectronic Packaging," *Scientific American*, (July):86–96, 1983.

[23] Bondy, J. A., and U. S. R. Murty: *Graph Theory with Applications*, American Elsevier, New York, 1976.

[24] Boppana, R. B.: "Eigenvalues and Graph Bisection: An Average-Case Analysis," in *IEEE Symp. on Foundations of Computer Science*, pp. 280–285, IEEE, 1987.

[25] Brady, M. L., and D. J. Brown: "Optimal Multilayer Channel Routing with Overlap," in *Fourth MIT Conference on Advanced Research in VLSI*, pp. 281–296, MIT Press, Cambridge, MA, 1986.

[26] Brayton, R. K., G. D. Hachtel, and L. A. Hemachandra: "A Comparison of Logic Minimization Strategies Using EXPRESSO: An APL Program Package for Partitioned Logic Minimization," in *IEEE Intl. Conf. on Circuits and Computers*, IEEE, 1982.

[27] Brayton, R. K., C. McMullen, G. D. Hachtel, and A. Sangiovanni-Vincentelli: *Logic Minimization Algorithms for VLSI Synthesis*, Kluwer Academic Publishers, Boston, MA, 1984.

[28] Breuer, M. A.: "A Class of Min-cut Placement Algorithms," in *Design Automation Conference*, pp. 284–290, IEEE/ACM, 1977.

[29] Breuer, M. A.: "Min-cut Placement," *J. Design Automation and Fault-Tolerant Computing* 1(4) pp. 343–382, 1977.

[30] Brown, D. J., and R. L. Rivest: "New Lower Bound for Channel Routing," in *CMU Conference on VLSI*, pp. 178–185, CMU, 1981.

[31] Brown, S., J. Rose, and Z. Vranesic: "A Detail Router for Field-Programmable Gate Arrays," in *International Conference on Computer-Aided Design*, pp. 382–385, IEEE/ACM, 1990.

[32] Brown, S., J. Rose, and Z. Vranesic: "A Detailed Router for Field-Programmable Gate Arrays," *IEEE Transactions on Computer Aided Design* 11:620–628, 1992.

[33] Brown, S., J. Rose, and Z. Vranesic: "A Stochastic Model to Predict the Routability of Field-Programmable Gate Arrays," *IEEE Transactions on Computer Aided Design* 12:1827–1838, 1993.

[34] Buckingham, M. A.: "Circle Graphs," PhD thesis, Courant Institute of Mathematical Sciences, Computer Science Department, October 1980. Report No. NSO-21.

[35] Burkard, R. E., and T. Bonniger: "A Heuristic for Quadratic Boolean Programs with Applications to Quadratic Assignment Problems," *European Journal of Operations Research* pp. 377–386, 1983.

[36] Burman, S., C. Kamalanathan, and N. Sherwani: "New Channael Segmentation Model and Associated Routing Algorithm for High Performance FPGAs," in *International Conference on Computer-Aided Design*, pp. 22–25, IEEE, 1992.

[37] Burstein, M., and R. Pelavin: "Hierarchical Channel Router," *Integration: The VLSI Journal*, 1, 1983. (Also pub. in *Proc. 20th Design Automation Conf.*, 1, 1983).

[38] Burstein, M., and R. Pelavin: "Hierarchical Wire Routing," *IEEE Transactions on Computer-Aided Design*, CAD-2(4):223–234, 1983.

[39] Cai, H.: "Multi-Pads, Single Layer Power Net Routing in VLSI Circuits," in *Design Automation Conference*, pp. 183–188, IEEE/ACM, 1988.

[40] Carden, R. C., IV, and C. K. Cheng: "A Global Router Using an Efficient Approximate Multicommodity Multiterminal Flow Algorithm," in *Design Automation Conference*, pp. 316–321, IEEE/ACM, 1991.

[41] Chan, P. K., J. Y. Zien, and M. D. F. Schlag: "On Routability Prediction for Field-Programmable Gate Arrays," in *Design Automation Conference*, pp. 326–330, ACM/IEEE, 1993.

[42] Chao, T. H., Y. C. Hsu, and J. M. Ho: "Zero Skew Clock Net Routing," in *Design Automation Conference*, pp. 518–523, IEEE/ACM, 1992.

[43] Charney, H. R., and D. L. Plato: "Efficient Partitioning of Components," in *Design Automation Workshop*, pp. 16.0–16.21, IEEE, 1968.

[44] Chaudhary, K., and M. Pedram: "A Near Optimal Technology Mapping Minimizing Area Under Delay Constraints," in *Design Automation Conference*, pp. 492–498, ACM/IEEE, 1992.

[45] Chen, C. C., and S. L. Chow: "The Layout Synthesizer: An Automatic Netlist-To-Layout System," in *Design Automation Conference*, pp. 232–238, IEEE/ACM, 1989.

[46] Chen, H. H., and E. Kuh: "A Variable-Width Gridless Channel Router," in *International Conference on Computer-Aided Design*, pp. 304–306, IEEE/ACM, 1985.

[47] Chen, H. H., and C. K. Wong: "Wiring and Crosstalk Avoidance in Multi-Chip Module Design," in *IEEE Custom Integrated Circuits Conference*, IEEE, 1992.

[48] Chen, H. H., and C. K. Wong: "XY and Z Direction Coupled Noise Minimization in Multichip Module Layout Design," in *Proc. ACM/SIGDA Physical Design Workshop, Lake Arrowhead, CA*, pp. 68–79, IEEE, 1993.

[49] Cheng, C. K., X. Deng, Y. Z. Liao, and S. Z. Ya: "Symbolic Layout Compaction under Conditional Design Rules," *IEEE Transactions on Computer Aided Design* 2(4):475–486, 1992.

[50] Cheng, C. K., and E. S. Kuh: "Module Placement Based on Resistive Network Optimization," *IEEE Transactions on Computer Aided Design* 3(3):218–225, 1984.

[51] Cheng, C. K., and Y. C. Wei: "An Improved Two-Way Partitioning Algorithm with Stable Performance," *IEEE Transactions on Computer Aided Design* 10(12):1502–1511, 1991.

[52] Cheng, W. T., J. L. Lewandowski, and E. Wu: "Optimal Diagnostic Methods for Wiring Interconnects," *IEEE Transactions on Computer Aided Design* 11(9):1161–1165, 1992.

[53] Chiang, C., and M. Sarrafzadeh: "On Wiring Overlap Layouts," in *First Great Lakes Symposium on VLSI*, IEEE, 1991.

[54] Chiang, C., M. Sarrafzadeh, and C. K. Wong: "A Weighted-Steiner-Tree-Based Global Router with Simultaneous Length and Density Minimization," *IEEE Trans. on CAD/ICS* 13(12):1461–1469, 1994.

[55] Chiang, C., M. Sarrafzadeh, and C. K. Wong: "An Optimal Algorithm for Constructing a Steiner Tree in a Switchbox (Part 1: Fundamental Theory and Application)," *IEEE Transactions on Circuits and Systems* 39(6):551-563, 1992.

[56] Cho, J. D., S. Raje, M. Sarrafzadeh, M. Sriram, and S. M. Kang: "A Multilayer Assignment Algorithm for Interference Minimization," in *Proc. 4th ACM/SIGDA Physical Design Workshop, Lake Arrowhead, CA*, pp. 63–67, 1993.

[57] Cho, J. D., S. Raje, M. Sarrafzadeh, M. Sriram, and S. M. Kang: "Crosstalk Minimum Layer Assignment," in *Proc. IEEE Custom Integr. Circuits Conf., San Diego, CA*, pp. 29.7.1–29.7.4, IEEE, 1993.

[58] Cho, J. D., and M. Sarrafzadeh: "A Buffer Distribution Algorithm for High-Speed Clock Routing," in *Design Automation Conference*, pp. 537–543, IEEE/ACM, 1993.

[59] Chowdhury, S.: "An Automated Design of Minimum-Area IC Power/Ground Nets," in *Design Automation Conference*, pp. 223–229, IEEE/ACM, 1987.

[60] Chowdhury, S., and M. A. Breuer: "Optimum Design of IC Power/Ground Nets Subject to Reliability Constraint," *IEEE Transactions on Computer Aided Design* 7:787–796, 1988.

[61] Chyan, D., and M. A. Breuer: "A Placement Algorithm for Array Processors," in *Design Automation Conference*, pp. 182–188, IEEE/ACM, 1983.

[62] Cle, R., and A. R. Siegel: "River Routing Every Which Way but Loose," in *Proceedings of the 25th Annual Symposium on Foundations of Focs*, pp. 65–73, IEEE, 1984.

[63] Cohoon, J. P.: "Distributed Genetic Algorithms for the Floorplan Design Problem," *IEEE Transactions on Computer Aided Design* 10(4):483–492, 1991.

[64] Cohoon, J. P., and P. L. Heck: "Beaver: A Computational Geometry Based Tool for Switchbox Routing," *IEEE Transactions on Computer Aided Design* 7(6):684–697, 1988.

[65] Cohoon, J. P., et al: "Floorplan Design Using Distributed Genetic Algorithms," in *International Conference on Computer-Aided Design,* pp. 452–455, IEEE, 1988.

[66] Cohoon, J. P., D. S. Richards, and J. S. Salowe: "An Optimal Steiner Tree Algorithm for a Net Whose Terminals Lie on the Perimeter of a Rectangle," *IEEE Transactions on Computer Aided Design* 9(4):398–407, 1990.

[67] Cong, J., and Y. Ding: "An Optimal Technology Mapping Algorithm for Delay Optimization in Lookup-Table Based FPGA Design," Technical Report CSD-920022, University of California at Los Angeles, May 1992. (Also pub. in *Proceedings of the ICCAD,* 1992.)

[68] Cong, J., L. Hagen, and A. Kahng: "Net Partitions Yield Better Module Partitions," in *Design Automation Conference,* pp. 47–52. IEEE, 1992.

[69] Cong, J., A. Kahng, and G. Robins: "Matching-based Methods for High-Performance Clock Routing," unpublished paper, 1992.

[70] Cong, J., A. Kahng, G. Robins, M. Sarrafzadeh, and C. K. Wong: "Provably Good Performance-Driven Global Routing," *IEEE Transactions on Computer Aided Design* 11(6):739–752, 1992.

[71] Cong, J., K.-S. Leung, and D. Zhou: "Performance-Driven Interconnect Design Based on Distributed RC Delay Model," UCLA Computer Science Department Technical Report CSD-920043, Oct. 1992.

[72] Cong, J. and C. L. Liu: "Over-the-Cell Channel Routing," *IEEE Transactions on Computer Aided Design* 9(4):408–418, 1990.

[73] Cong, J., and C. L. Liu: "On the k-layer Planar Subset Problem and Topological Via Minimization Problems," *IEEE Transactions on Computer Aided Design* 10(8):972–981, 1991.

[74] Cong, J., B. Preas, and C. L. Liu: "General Models and Algorithms for Over-the-Cell Routing in Standard Cell Design," in *Design Automation Conference,* pp. 709–715, IEEE/ACM, 1990.

[75] Cormen, T. H., C. E. Leiserson, and R. L. Rivest: *Introduction to Algorithms,* McGraw-Hill, New York, 1991.

[76] Dai, W. M.: "Performance Driven Layout of Thin-film Substrates for Multichip Modules," in *Proceedings of Multichip Module Workshop,* pp. 114–121, IEEE, 1990.

[77] Dai, W. M.: "Topological Routing in SURF: Generating a Rubber-Band Sketch," in *Proceedings of IEEE Design Automation Conference,* pp. 39–48, IEEE, 1991.

[78] Dai, W. M., and E. S. Kuh: "Global Spacing of Building Block Layout," in *VLSI '87* (C. H. Sequin, ed.), pp. 193–205, Elsevier Science Publishers, Amsterdam, 1987.

[79] Dai, W. M., and E. S. Kuh: "Simultaneous Floor Planning and Global Routing for Hierarchical Building-Block Layout," *IEEE Transactions on Computer Aided Design* 6(5):828–837, 1987.

[80] de Dood, P., J. Wawrzynek, E. Liu, and R. Suaya: "A Two-Dimensional Topological Compactor with Octagonal Geometry," in *Design Automation Conference,* pp. 727–731, IEEE/ACM, 1991.

[81] de Gyvez, J. P., and C. Di: "IC Defect Sensitivity for Footprint-Type Spot Defects," *IEEE Transactions on Computer Aided Design* 11(5):638–658, 1992.

[82] Desoer, C. A., and E. S. Kuh: *Basic Circuit Theory,* McGraw-Hill, New York, 1969.

[83] Deutsch, D. N.: "A Dogleg Channel Router," in *Design Automation Conference,* pp. 425–433, IEEE/ACM, 1976.

[84] Deutsch, D. N.: "Compacted Channel Routing," in *International Conference on Computer-Aided Design,* pp. 223–225, IEEE, 1985.

[85] Dijkstra, E. W.: "A Note on Two Problems in Connection with Graphs," *Numerische Mathematik* 1:269–271, 1959.

[86] Du, D. Z., and F. K. Hwang: "A Proof of Gilbert-Pollak Conjecture on Steiner Ratio," *Algorithmica* 7:121–135, 1992.

[87] Dunlop, A. E., V. D. Agrawal, D. N. Deutsch, M. F. Jukl, P. Kozak, and M. Wiesel: "Chip Layout Optimization Using Critical Path Weighting," in *Design Automation Conference*, pp. 133–136, IEEE/ACM, 1984.

[88] Dunlop, A. E., and B. W. Kernighan: "A Procedure for Placement of Standard Cell VLSI Circuits," *IEEE Transactions on Computer Aided Design* 4(1):92–98, 1985.

[89] Ecker, J. G.: "Geometric Programming: Methods, Computations and Applications," *SIAM Review* 22(3):338–362, 1980.

[90] Edmonds, J.: "Maximum Matching and a Polyhedron with 0,1-vertices," *J. Res. Nat. Bur. Stand.* 69:125, 1965.

[91] Egan, J. R., and C. L. Liu: "Optimal Bipartite Folding of PLA," in *Design Automation Conference*, pp. 141–146, IEEE/ACM, 1982.

[92] Eichenberger, P., and M. Horwitz: "Toroidal Compaction of Symbolic Layouts for Regular Structures," in *International Conference on Computer-Aided Design*, pp. 142–145, IEEE, 1987.

[93] El Gamal, A. et al.: "An Architecture for Electrically Configurable Gate Arrays," *IEEE Journal of Solid-State Circuits* 24(2):394–398, 1989.

[94] El Gamal, A., J. Greene, and V. Roychowdhury: "Segmented Channel Routing is Nearly as Efficient as Channel Routing," *Proc. Advanced Research in VLSI*, pp. 193–211, 1991.

[95] Enbody, R. J., and H. C. Du: "Near Optimal n-layer Channel Routing," in *Design Automation Conference*, pp. 708–714, IEEE/ACM, 1986.

[96] Even, S.: *Graph Algorithms,* Computer Science Press, Potomac, MD, 1979.

[97] Farrahi, A. H., and M. Sarrafzadeh: "Complexity of the Look-up Table Minimization Problem for FPGA Technology Mapping," *IEEE Transactions on Computer Aided Design* 13(11):1319–1332, 1994.

[98] Farrahi, A. H., and M. Sarrafzadeh: "FPGA Technology Mapping for Power Minimization," in *International Workshop on Field-Programmable Logic and Applications*, pp. 66–77, Springer-Verlag, Berlin, 1994.

[99] Feldman, J., I. Wagner, and S. Wimer: "An Efficient Algorithm for Some Multirow Layout Problems," in *International Workshop on Layout Synthesis*, 1992.

[100] Fiduccia, C. M., and R. M. Mattheyses: "A Linear Time Heuristic for Improving Network Partitions," in *Design Automation Conference*, pp. 175–181, IEEE/ACM, 1982.

[101] Ford, L. R., and D. R. Fulkerson: *Flows in Network,* Princeton University Press, Princeton, NJ, 1962.

[102] Francis, R., J. Rose, and K. Chung: "Chortle: A Technology Mapping Program for Lookup Table-based Field Programmable Gate Arrays," in *Design Automation Conference*, pp. 613–619, IEEE/ACM, 1990.

[103] Francis, R., J. Rose, and Z. Vranesic: "Technology Mapping for Lookup Table-based FPGAs for Performance," in *International Conference on Computer-Aided Design*, pp. 568–571, IEEE, 1991.

[104] Frankle, J., and R. M. Karp: "Circuit Placement and Cost Bounds by Eigenvector Decomposition," in *International Conference on Computer-Aided Design*, pp. 414–417, IEEE, 1986.

[105] Franzon, P. D.: Private communication, Electronic MCM Clearing House in North Carolina State University, 1992.

[106] Friedman, E. G.: "Clock Distribution Design in VLSI Circuits—an Overview," in *International Symposium on Circuits and Systems*, pp. 1475–1478, IEEE, 1993.

[107] Funabiki, N., and Y. Takefuji: "A Parallel Algorithm for Channel Routing Problems," *IEEE Transactions on Computer Aided Design* 11(4):464–474, 1992.

[108] Gabbe, J. D., and P. A. Subrahmanyam: "A Note on Clustering Modules for Floorplanning," in *Design Automation Conference* pp. 594–597, IEEE/ACM, 1989.

[109] Gao, S., and S. Hambrush: "Two-Layer Channel Routing with Vertical Unit-Length Overlap," *Algorithmica* 1(2):223–233, 1986.

[110] Gao, S., and M. Kaufmann: "Channel Routing of Multiterminal Nets," in *Proceedings of 28th Annual Symposium on the Foundations of Computer Science*, pp. 316–325, IEEE, 1987.

[111] Gao, T., P. M. Vaidya, and C. L. Liu: "A Performance Driven Macro-Cell Placement Algorithm," in *Design Automation Conference*, pp. 147–152, IEEE/ACM, 1992.

[112] Garey, M. R., and D. S. Johnson: "The Rectilinear Steiner Tree Problem is NP-Complete," *SIAM Journal on Applied Mathematics* 32(4):826–834, 1977.

[113] Garey, M. R., and D. S. Johnson: *Computers and Intractability: A Guide to the Theory of NP-Completeness,* Freeman, New York, 1979.

[114] Garey, M. R., D. S. Johnson, and H. C. So: "An Application of Graph Coloring to Printed Circuit Testing," *IEEE Transactions on Circuits and Systems* CAS-23:591–599, 1976.

[115] Gee, P., S. M. Kang, and I. N. Hajj: "Automatic Synthesis of Metal-Metal Matrix (M^3) Layout," *International Journal of Computer Aided VLSI Design* 2:83–104, 1990.

[116] Ghosh, A., S. Devadas, K. Keutzer, and J. White: "Estimation of Average Switching Activity in Combinational and Sequential Circuits," in *Design Automation Conference*, pp. 253–259, ACM/IEEE, 1992.

[117] Gonzalez, T. F., and S. Kurki-Gowdara: "Minimization of the Number of Layers for Single Row Routing with Fixed Street Capacity," *IEEE Transactions on Computer Aided Design* 7:420–424, 1988.

[118] Greene, J., V. Roychowdhury, S.. Kaptanoglu, and A. El Gamal: "Segmented Channel Routing," in *Design Automation Conference*, pp. 567–572, IEEE/ACM, 1990.

[119] Gudmundsson, G., and S. Ntafos: "Channel Routing with Superterminals," in *Proc. 25th Allerton Conference on Computing, Control and Communication*, pp. 375–376, University of Illinois Press, Champaign-Urbana, IL, 1987.

[120] Hachtel, G. D., A. R. Newton, and A. Sangiovanni-Vincentelli: "Techniques for PLA Folding," in *Design Automation Conference*, pp. 147–155, IEEE/ACM, 1982.

[121] Hadlock, F.: "Finding a Maximum Cut of a Planar Graph in Polynomial Time," *SIAM Journal on Computing* 4(3):221–225, 1975.

[122] Hagen, L., and A. B. Kahng: "Fast Spectral Methods for Ratio Cut Partitioning and Clustering," in *International Conference on Computer-Aided Design*, pp. 10–13, IEEE, 1991.

[123] Hagen, L., and A. B. Kahng: "A New Approach to Effective Circuit Clustering," *IEEE Transactions on Computer Aided Design* 11(9):422–427, 1992.

[124] Hall, K. M.: "An R-Dimensional-Quadratic Placement Algorithm," *Management Science* 17(3):219–229, 1970.

[125] Hamachi, G. T., and J. K. Ousterhout: "A Switchbox Router with Obstacle Avoidance," in *Design Automation Conference*, pp. 173–179, IEEE/ACM, 1984.

[126] Hambrusch, S. E.: "Using Overlap and Minimizing Contact Points in Channel Routing," in *Proc. 21th Annual Allerton Conference on Communication, Control and Computing*, pp. 256–257, University of Illinois Press, Champaign-Urbana, IL, 1983.

[127] Hambrusch, S. E.: "Channel Routing Algorithm for Overlap Models," *IEEE Transactions on Computer Aided Design* CAD-4(1):23–30, 1985.

[128] Hambrusch, S. E., and H. Y. Tu: "A Framework for 1-D Compaction with Forbidden Region Avoidance," *Computational Geometry: Theory and Application* 1:203–226, 1992.

[129] Hambrusch, S. E., and H. Y. Tu: "Minimizing Total Wire Length during 1-Dimensional Compaction," *Integration: The VLSI Journal* 14:113–144, 1992.

[130] Hanan, M.: "On Steiner's Problem with Rectilinear Distance," *SIAM Journal on Applied Mathematics* 14(2):255–265, 1966.

[131] Hanan, M., and J. M. Kurtzbert: "Placement Techniques," in *Design Automation of Digital System* (M. A. Breuer, ed.) vol. 1, pp. 213–282, Prentice Hall, Englewood Cliffs, NJ, 1972.

[132] Harkness, C. L., and D. P. Lopresti: "VLSI Placement Using Uncertain Costs," in *International Conference on Computer-Aided Design*, pp. 340–343, IEEE, 1990.

[133] Haruyama, S., and D. Fussell: "A New Area-Efficient Power Routing Algorithm for VLSI Layout," in *Design Automation Conference*, pp. 38–41, IEEE/ACM, 1987.

[134] Hashimoto, A., and J. Stevens: "Wire Routing by Optimizing Channel Assignment within Large Apertures," in *Proc. 8th Design Automation Workshop, Atlantic City, NJ*, pp. 155–169, ACM, 1971.

[135] Hauge, P. S., R. Nair, and E. J. Yoffa: "Circuit Placement for Predictable Performance," in *International Conference on Computer-Aided Design*, pp. 88–91, IEEE, 1987.

[136] Hightower, D. W.: "A Solution to Line Routing Problems on the Continuous Plane," in *Sixth Design Automation Workshop*, pp. 1–24, IEEE, 1969.

[137] Hill, D. D.: "Sc2; A Hybrid Automatic Layout System," in *International Conference on Computer-Aided Design*, pp. 172–174, IEEE, 1985.

[138] Hill, D. D.: "A CAD System for the Design of Field Programmable Gate Arrays," in *Design Automation Conference*, pp. 187–192, IEEE/ACM, 1991.

[139] Ho, J. M., M. Sarrafzadeh, and A. Suzuki: "An Exact Algorithm for Single-Layer Wire-Length Minimization," in *International Conference on Computer-Aided Design*, pp. 424–427, IEEE, 1990.

[140] Ho, J. M., M. Sarrafzadeh, G. Vijayan, and C. K. Wong: "Layer Assignment for Multi-Chip Modules," *IEEE Transactions on Computer Aided Design* CAD-9(12):1272–1277, 1990.

[141] Ho, J. M., M. Sarrafzadeh, G. Vijayan, and C. K. Wong: "Pad Minimization for Planar Routing of Multiple Power Nets," *IEEE Transactions on Computer Aided Design* CAD-9(4):419–426, 1990.

[142] Ho, J. M., G. Vijayan, and C. K. Wong: "New Algorithm for the Rectilinear Steiner Tree Problem," *IEEE Transactions on Computer Aided Design* 9(2):185–193, 1990.

[143] Ho, K. C., and S. Sastry: "Flexible Transistor Matrix," in *Design Automation Conference* pp. 475–480, IEEE/ACM, 1991.

[144] Holmes, N., N. Sherwani, and M. Sarrafzadeh: "New Algorithm for Over-the-Cell Channel Routing Using Vacant Terminals," in *Design Automation Conference*, pp. 126–131, IEEE/ACM, 1991.

[145] Horowitz, M. A.: "Timing Models for MOS Pass Networks," in *International Symposium on Circuits and Systems*, pp. 202–211, IEEE, 1983.

[146] Hsieh, H., W. Carter, J. Ja, E. Cheung, S. Schreifels, C. Erickson, P. Freidin, L. Tinkey, and R. Kanazawa: "Third Generation Architecture Boosts Speed and Density of Field-Programmable Gate Arrays," in *IEEE Custom Integrated Circuits Conference*, pp. 31.2.1–31.2.7, IEEE, 1990.

[147] Hsieh, Y. C, and C. C. Chang: "A Modifed Detour Router," in *International Conference on Computer-Aided Design*, pp. 301–303, IEEE, 1985.

[148] Hsieh, Y. C., C. Y. Hwang, Y. L. Lin, and Y. C. Hsu: "LiB: A CMOS Cell Compiler," *IEEE Transactions on Computer Aided Design* CAD-10(8):994–1005, 1991.

[149] Hsu, C. P.: "General River Routing Algorithm," in *Design Automation Conference*, pp. 578–583, IEEE/ACM, 1983.

[150] Hsu, C. P.: "Minimum-Via Topological Routing," *IEEE Transactions on Computer Aided Design* CAD-2(2):235–246, 1983.

[151] Hsu, C. P., et al: "APLS2: A Standard Cell Layout System for Double-layer Metal Technology," in *Design Automation Conference*, pp. 443–448, IEEE/ACM, 1985.

[152] Hsueh, M. Y.: "Symbolic Layout and Compaction of Integrated Circuits," Technical Report UCB/ERL-M79/80, University of California, Berkley, December 1979.

[153] Hu, T. C., and E. Kuh: *VLSI Layout: Theory and Design,* IEEE, 1985.

[154] Hu, T. C., and M. T. Shing: "A Decomposition Algorithm For Circuit Routing," *Mathematical Programming Study* 24:87–103, 1985.

[155] Hwang, C. Y., Y. C. Hsieh, Y. L. Lin, and Y. C. Hsu: "A Fast Transistor Chaining Algorithm for CMOS Cell Layout," *IEEE Transactions on Computer Aided Design* CAD-9:781–786, 1990.

[156] Hwang, F. K.: "On Steiner Minimal Trees with Rectilinear Distance," *SIAM Journal on Applied Mathematics* 30(1):104–114, 1976.

[157] Hwang, L.-T., D. Nayak, I. Turlik, and A. Reisman: "Thin-Film Pulse Propagation Analysis Using Frequency Techniques," *IEEE Transactions on CHMT* 14:192–198, 1991.

[158] Hwang, L.-T., and I. Turlik: "Calculation of Voltage Drops in the Vias of a Multichip Package," MCNC Technical Reports, Technical Report Series TR90-41, 1990.

[159] Hwang, L.-T., and I. Turlik: "The Skin Effect in Thin-Film Interconnections for ULSI/VLSI Packages," MCNC Technical Reports, Technical Report Series TR91-13, 1991.

[160] Hwang, L.-T., I. Turlik, and A. Reisman: "A Thermal Module Design for Advanced Packaging," *J. Electron. Mats.* 16:347–355, 1987.

[161] Hwang, S. Y., R. W. Dutton, and T. Blank: "A Best-First Search Algorithm for Optimal PLA Folding," *IEEE Transactions on Computer Aided Design* CAD-5(3):433–442, 1986.

[162] Ihler, E., D. Wagner, and F. Wagner: "Modeling Hypergraphs by Graphs with the Same Mincut Properties," Technical Report B 92-22, Freie Universitat, Berlin, October 1992.

[163] Imai, H., and T. Asano: "Efficient Algorithms for Geometric Graph Search Problems," *SIAM Journal on Computing* 15(2):478–494, 1986.

[164] Ishikawa, M., and T. Yoshimura: "A New Module Generator with Structural Routers and a Graphical Interface," in *International Conference on Computer-Aided Design*, pp. 436–439, IEEE, 1987.

[165] Jackson, M. A. B., and E. S. Kuh: "Performance-Driven Placement of Cell-Based IC's," in *Design Automation Conference*, pp. 370–375, IEEE/ACM, 1989.

[166] Jackson, M. A. B., E. S. Kuh, and M. Marek-Sadowska: "Timing-Driven Routing for Building Block Layout," in *International Symposium on Circuits and Systems*, pp. 518–519, IEEE, 1987.

[167] Jackson, M. A. B., A. Srinivasan, and E. S. Kuh: "Clock Routing for High-Performance ICs," in *Design Automation Conference* pp. 573–579, IEEE/ACM, 1990.

[168] Johnson, D. S., A. Demers, J. D. Ullman, M. R. Garey, and R. L. Graham: "Worst-case Performance Bounds for Simple One-Dimensional Packing Algorithms," *SIAM Journal on Computing* (3):299–325, 1974.

[169] Joobbani, R.: *An Artificial Intelligence Approach to VLSI Routing*, Kluwer Academic Publishers, Boston, MA, 1986.

[170] Joobbani, R.: *Weaver: A Knowledge-Based Routing Expert*, Kluwer Academica Press, Dordrecht, 1986.

[171] Kahng, A. B.: "Fast Hypergraph Partition," in *International Conference on Computer-Aided Design*, pp. 762–766, IEEE, 1989.

[172] Kahng, A. B., and G. Robins: "A New Class of Steiner Tree Heuristics with Good Performance: The Iterated 1-Steiner Approach," in *International Conference on Computer-Aided Design*, pp. 428–431, IEEE, 1990.

[173] Kahng, A. B., and G. Robins: *On Optimal Interconnections for VLSI*, Kluwer Academic Publishers, Boston, MA, 1995.

[174] Kang, S. M.: "Metal-Metal Matrix (M^3) for High-Speed MOS VLSI Layout," *IEEE Transactions on Computer Aided Design* CAD-6(5):886–891, 1987.

[175] Kang, S. M.: "Performance-Driven Layout of CMOS VLSI Circuits," in *International Symposium on Circuits and Systems*, pp. 881–884, IEEE, 1990.

[176] Kang, S. M., and M. Sriram: "Binary Formulations for Placement and Routing Problems," in *Algorithmic Aspects of VLSI Layouts* (M. Sarrafzadeh and D. T. Lee, eds.), pp. 25–68, World Scientific, Singapore, 1993.

[177] Kang, S. M., and W. M. van Cleemput: "Automatic PLA Synthesis from a DDL-P Description," in *Design Automation Conference*, pp. 391–397, 1981.

[178] Kaplan, D.: "Routing with a Scanning Window," in *Design Automation Conference*, pp. 629–632, IEEE/ACM, 1987.

[179] Karp, R. M., F. T. Leighton, R. L. Rivest, C. D. Thompson, U. V. Vazirani, and V. V. Vazirani: "Global Wire Routing in Two-Dimensional Arrays," *Algorithmica* 2(1):113–129, 1987.

[180] Kedem, G., and H. Watanabe: "Graph-Optimization Techniques for IC Layouts and Compaction," *IEEE Transactions on Computer Aided Design* (3):12–19, 1983.

[181] Kernighan, B. W., and S. Lin: "An Efficient Heuristic Procedure for Partitioning Graphs," *Bell System Technical Journal* 49:291–307, 1970.

[182] Khoo, K. Y., and J. Cong: "A Fast Multilayer General Area Router for MCM and Dense PCB Designs," Technical Report CSD-920011, University of California at Los Angeles, March 1992.

[183] Khoo, K. Y., and J. Cong: "Four Vias Are Sufficient in Mutilayer MCM and Dense PCB Routing," unpublished paper, 1992.

[184] Kim, S., R. M. Owens, and M. J. Irwin: "Experiments with a Performance Driven Module Generator," in *Design Automation Conference*, pp. 687–690, IEEE/ACM, 1992.

[185] Kingsley, C.: "A Hierachical Error-Tolerant Compactor," in *Design Automation Conference*, pp. 126–132, IEEE/ACM, 1984.

[186] Kirkpatrick, S., C. D. Gelatt, Jr., and M. P. Vecchi: "Optimization by Simulated Annealing," *Science* 220:671–680, 1983.

[187] Kleinhans, J. M., G. Sigl, F. M. Johannes, and K. J. Antreich: "GORDIAN: VLSI Placement by Quadratic Programming and Slicing Optimization," *IEEE Transactions on Computer Aided Design* 10(3):365–365, 1991.

[188] Kollaritsch, P. W., and N. H. E. Weste: "TOPOLOGIZER: An Expert System Transistor of Transistor Connectivity to Symbolic Cell Layout," *IEEE Journal of Solid-State Circuits* 20:799–804, 1985.

[189] Kozminski, K., and E. Kinnen: "Rectangular Dual of Planar Graphs," *Networks* 15:145–157, 1985.

[190] Krishnamurthy, B.: "An Improved Min-Cut Algorithm for Partitioning VLSI Networks," *IEEE Transactions on Computers* C-33:438–446, 1984.

[191] Krohn, H. E.: "An Over-the-Cell Gate Array Channel Router," in *Design Automation Conference,* pp. 665–670, IEEE/ACM, 1983.

[192] Kruskal, J. B., Jr.: "On the Shortest Spanning Subtree of a Graph and the Traveling Salesman Problem," *American Mathematical Society* (November):48–50, 1956.

[193] Kuhn, H. W., and A. W. Tucker: "Nonlinear Programming," in *Proceedings of the Second Berkley Symposium on Mathematical Statistics and Probability*, pp. 481–492, University of California Press, Berkeley, 1951.

[194] Kuo, Y. S., T. C. Chern, and W. K. Shih: "Fast Algorithm for Optimal Layer Assignment," in *Design Automation Conference*, pp. 554–559, IEEE/ACM, 1988.

[195] Kurtzberg, J. M.: "Algorithms for Backplane Formation," in *Microelectronics in Large Systems*.

[196] Lai, Y. T., and S. M. Leinwand: "Algorithms for Floor-plan Design via Rectangular Dualization," *IEEE Transactions on Computer Aided Design* 7(12):1278–1289, 1988.

[197] LaPaugh, A. S., F. L. Heng, W. W. Lin and R. Y. Pinter: "Decreasing Channel Width Lower Bounds by Channel Lengthening," Technical Report CS-TR-218-89, Princeton University, 1989.

[198] Lathi, B. P.: *Linear Systems and Signals,* Berkeley-Cambridge Press, Carmichael, CA, 1993.

[199] Lawler, E. L.: *Combinatorial Optimization: Networks and Matroids,* Holt, Rinehart and Winston, New York, 1976.

[200] Lee, C. Y.: "An Algorithm for Path Connection and Its Application," *IRE Transaction on Electronic Computer* EC-10:346–365, 1961.

[201] Lee, J. F.: "A New Framework of Design Rules for Compaction of VLSI Layouts," *IEEE Transactions on Computer Aided Design* 7(11):1195–1204, 1988.

[202] Lee, J. F.: "A Layout Compaction Algorithm with Multiple Grid Constraints," in *International Conference on Computer Design*, pp. 30–33, IEEE, 1991.

[203] Lee, J. F., and D. T. Tang: "VLSI Layout Compactor with Grid and Mixed Constraints," *IEEE Transactions on Computer Aided Design* CAD-6(5):903–910, 1987.

[204] Lee, J. F., and D. T. Tang: "HIMALAYAS—A Hierarchical Compactor System with a Minimized Constraint Set," in *International Conference on Computer-Aided Design*, pp. 150–157, 1992.

[205] Lee, J. F., and C. K. Wong: "A Performance-Aimed Cell Compactor with Automatic Jogs," *IEEE Transactions on Computer Aided Design* CAD-11(12):1495–1507, 1992.

[206] Lee, K.-W., and C. Sechen: "A New Global Router for Row-Based Layout," in *International Conference on Computer-Aided Design*, pp. 180–183, IEEE, 1988.

[207] Leighton, F. T., T. N. Bui, S. Chauduri, and M. Siper: "Graph Bisection Algorithm with Good Average Case Behavior," *Combinatorica* 7:171–191, 1987.

[208] Leighton, T., and S. Rao: "An Approximate Max-Flow Min-Cut Theorem for Uniform Multi-Commodity Flow Problems with Applications to Approximation Algorithms," in *Annual Symposium on Foundations of Computer Science*, pp. 422–431, IEEE, 1988.

[209] Leiserson, C. E., and F. M. Maley: "Algorithms for Routing and Testing Routability of Planar VLSI Layouts," in *Symposium on the Theory of Computation*, pp. 69–78, ACM, 1985.

[210] Leiserson, C. E., and R. Y. Pinter: "Optimal Placement for River Routing," *SIAM Journal on Computing* 12(3):447–462, 1983.

[211] Lek, N., R. W. Thaik, and S.-M. Kang: "A New Global Router Using Zero-One Integer Linear Programming Techniques for Sea-of-Gates and Custom Logic Arrays," *IEEE Transactions on Computer Aided Design* CAD-11(12), 1992.

[212] Lemieux, G., and S. Brown: "A Detailed Routing Algorithm for Allocating Wire Segments in FPGAs," in *Proc. 4th ACM/SIGDA Physical Design Workshop*, ACM, 1993.

[213] Lengauer, T.: *Combinatorial Algorithms for Integrated Circuit Layout*. John Wiley & Sons, New York, 1990.

[214] Lewandowski, J. L., and C. L. Liu: "A Branch and Bound Algorithm for Optimal PLA Folding," in *Design Automation Conference*, pp. 426–433, IEEE/ACM, 1982.

[215] Liao, K. F., D. T. Lee, and M. Sarrafzadeh: "Planar Subset of Multi-terminal Nets," *Integration: The VLSI Journal* 10(1):19–37, 1990.

[216] Liao, K. F., and M. Sarrafzadeh: "Boundary Single-Layer Routing with Movable Terminals," *IEEE Transactions on Computer Aided Design* CAD-10(11):1382–1391, 1991.

[217] Liao, K. F., M. Sarrafzadeh, and C. K. Wong: "Single-Layer Global Routing," *IEEE Trans. on CAD/ICS* 13(1):38–47, 1994.

[218] Liao, Y. Z., and C. K. Wong: "An Algorithm to Compact a VLSI Symbolic Layout with Mixed Constraints," *IEEE Transactions on Computer Aided Design* 2(2):62–69, 1983.

[219] Lie, M., and C. S. Horng: "A Bus Router for IC Layout," in *Design Automation Conference*, pp. 129–132, IEEE/ACM, 1982.

[220] Lin, I., and D. H. C. Du: "Performance-Driven Constructive Placement," in *Design Automation Conference*, pp. 103–106, IEEE/ACM, 1990.

[221] Lin, M. S., H. W. Perng, C. Y. Hwang, and Y. L. Lin: "Channel Density Reduction by Routing Over the Cells," in *Design Automation Conference*, pp. 120–125, IEEE/ACM, 1991.

[222] Lin, P. F., and K. Nakajima: "A Linear Time Algorithm for Optimal CMOS Functional Cell Layouts," in *International Conference on Computer-Aided Design*, pp. 449–453, IEEE, 1990.

[223] Lin, R.-B., and E. Shragowitz: "Fuzzy Logic Approach to Placement Problem," in *Design Automation Conference*, pp. 153–158, IEEE/ACM, 1992.

[224] Lin, S. L., and J. Allen: "Minplex—A Compactor that Minimizes the Bounding Rectangle and Individual Rectangles in a Layout," in *Design Automation Conference*, pp. 123–130, IEEE/ACM, 1986.

[225] Lin, S. P., M. Marek-Sadowska, and E. S. Kuh: "Delay and Area Optimization in Standard Cell Design," in *Design Automation Conference*, pp. 349–352, IEEE/ACM, 1990.

[226] Lin, T. M., and C. A. Mead: "Signal Delay in General RC Networks," *IEEE Transactions on Computer Aided Design* 3(4):331–349, 1984.

[227] Longway, C., and R. Siferd: "A Doughnut Layout Style for Improved Switching Speed with CMOS VLSI Gates," *IEEE Journal of Solid-State Circuits* 24(1):194–198, 1989.

[228] Lopez, A. D., and H. S. Law: "A Dense Gate Matrix Layout Method for MOS VLSI," *IEEE Trans. on Electronic Devices* ED-27(8):1671–1675, 1980.

[229] Lou, R. D., K. F. Liao, and M. Sarrafzadeh: "Planar Routing Around a Rectangle," *Journal of Circuits, Systems and Computers,* 2(1):27–38, 1992.

[230] Lou, R. D., M. Sarrafzadeh, and D. T. Lee: "An Optimal Algorithm for the Maximum Two-Chain Problem," *SIAM Journal on Discrete Mathematics* 5(2):284–304, 1992.

[231] Luby, M., V. Vazirani, U. Vazirani, and A. L. Sangiovanni-Vincentelli: "Some Theoretical Results on PLA Folding," in *IEEE International Conference on Circuits and Computers*, pp. 165–170, 1982.

[232] Luenberger, D. G.: *Linear and Nonlinear Programming*, Addison-Wesley, Reading, MA, 1984.

[233] Luhukay, J., and W. J. Kubitz: "A Layout Synthesis System for NMOS Gate-Cells," in *Design Automation Conference*, pp. 307–314, IEEE/ACM, 1982.

[234] Luk, W. K.: "A Greedy Switchbox Router," Technical Report CMU-CS-84-148, Carnegie-Mellon University, 1984.

[235] Luk, W. K., P. Sipala, M. Tamminen, D. Tang, L. S. Woo, and C. K. Wong: "A Hierarchical Global Wiring Algorithm for Custom Chip Design," *IEEE Transactions on Computer Aided Design* CAD-6(4):518–533, 1987.

[236] Makedon, F., and S. Tragoudas: "Approximate Solutions for Graph and Hypergraph Partitioning," in *Algorithmic Aspects of VLSI Layouts* (M. Sarrafzadeh and D. T. Lee, eds.), pp. 133–166, World Scientific, Singapore, 1993.

[237] Maley, F. M.: "Single Layer Routing," PhD thesis, MIT/LCS/TR-403, 1987.

[238] Manber, U.: *Introduction to Algorithms—A Creative Approach*, Addison-Wesley, Reading, MA, 1989.

[239] Marek-Sadowska, M.: "An Unconstraint Topological Via Minimization," *IEEE Transactions on Computer Aided Design* CAD-3(3):184–190, 1984.

[240] Marek-Sadowska, M.: "Two-Dimensional Router for Double-Layer Layout," in *Design Automation Conference*, pp. 117–123, IEEE/ACM, 1985.

[241] Marek-Sadowska, M.: "Issues in Timing Driven Layout," in *Algorithmic Aspects of VLSI Layouts* (M. Sarrafzadeh and D. T. Lee, eds.), pp. 1–24, World Scientific, Singapore, 1993.

[242] Marek-Sadowska, M., and E. S. Kuh: "A New Approach to Channel Routing," in *International Symposium on Circuits and Systems*, pp. 764–767, IEEE, 1982.

[243] Marek-Sadowska, M., and T. T.-K. Tarng: "Single-Layer Routing for VLSI : Analysis and Algorithms," *IEEE Transactions on Computer Aided Design* CAD-2(4):246–259, 1983.

[244] Marple, D.: "A Hierarchy Preserving Hierarchical Compactor," in *Design Automation Conference*, pp. 375–381, IEEE/ACM, 1990.

[245] McCoy, B. A., K. D. Boese, A. B. Kahng, and G. Robins: "Fidelity and Near-Optimality of Elmore-Based Routing Constructions," Technical Report TR-CS-93-14, University of Virginia, 1993.

[246] McGeoch, L. A., D. S. Johnson, C. R. Aragon and C. Schevon: *Optimization by Simulated Annealing: An Experimental Evaluation (Part 1)*, AT&T Bell Lab., Murray Hill, NJ, 1985.

[247] Mead, C., and L. Conway: *Introduction to VLSI System*, Addison-Wesley, Reading, MA, 1980.

[248] Mehlhorn, K., F. P. Preparata, and M. Sarrafzadeh: "Channel Routing in Knock-Knee Mode: Simplified Algorithms and Proofs," *Algorithmica* 1(2):213–221, 1986.

[249] Mehlhorn, K., and W. Ruling: "Compaction on the Torus," *IEEE Transactions on Computer Aided Design* (4):389–397, 1990.

[250] Mirzaian, A.: "River Routing in VLSI," *Journal of Computer and System Sciences* 31(1):43–54, 1987.

[251] Moore, E. F.: "Shortest Path through a Maze," in *Annals of Computation Laboratory*, pp. 285–292, Harvard University Press, Cambridge, MA, 1959.

[252] Moreno, H.: "OSS Physical Design CAD Tools Specifications," MCC Technical Report P/I-250-90, 1990.

[253] Mosteller, R. C.: "Monte Carlo Methods for 2D Compaction," Technical Report 5230, California Institute of Technology, 1986.

[254] Murgai, R., Y. Nishizaki, N. Shenoy, R. K. Brayton, and A. Sangiovanni-Vincentelli: "Logic Synthesis Algorithms for Programmable Gate Arrays," in *Design Automation Conference*, pp. 620–625, IEEE/ACM, 1990.

[255] Murgai, R., N. Shenoy, R. K. Brayton, and A. Sangiovanni-Vincentelli: "Performance Directed Synthesis for Table Look Up Programmable Gate Arrays," in *International Conference on Computer-Aided Design*, pp. 572–575, IEEE, 1991.

[256] Muroga, S.: *VLSI System Design*. John Wiley & Sons, New York, 1982.

[257] Nagel, W.: "SPICE2, A computer program to simulate semiconductor circuits," Technical Report UCB/ERL-M520, Electronics Research Lab, University of California, Berkeley, May 1975.

[258] Nair, R.: "A Simple Yet Effective Technique for Global Wiring," *IEEE Transactions on Computer Aided Design* CAD-6(2):165–172, 1987.

[259] Nair, R., C. L. Berman, P. S. Hauge, and E. J. Yoffa: "Generation of Performance Constraints for Layout," *IEEE Transactions on Computer Aided Design* CAD-8(8):860–874, 1989.

[260] Najm, F.: "Transition Density: A New Measure of Activity in Digital Circuits," *IEEE Transactions on Computer Aided Design* 12(2):310–323, 1992.

[261] Nakatani, K., T. Fujii, T. Kikuno, and N. Yoshita: "A Heuristic Algorithm for Gate Matrix Layout," in *International Conference on Computer-Aided Design*, pp. 324–327, IEEE, 1986.

[262] Natarajan, S., N. Holmes, N. A. Sherwani, and M. Sarrafzadeh: "Over-the-Cell Channel Routing for High Performance Circuits," in *Design Automation Conference* pp. 600–603, IEEE/ACM, 1992.

[263] Obermeier, F. W., and R. H. Katz, "An Electrical Optimizer That Considers Physical Layout," in *Design Automation Conference*, pp. 453–459, IEEE/ACM, 1988.

[264] Ohtsuki, T.: *Layout Design and Verification*, North-Holland, Amsterdam, 1986.

[265] Ohtsuki, T., H. Mori, E. S. Kuh, T. Kashiwabara, and T. Fujisawa: "One-Dimensional Logic Gate Assignment and Interval Graphs," *IEEE Transactions on Circuits and Systems* 26: 675–684, 1979.

[266] Ohtsuki, T., T. Sudo, and S. Goto: "CAD Systems for VLSI in Japan," *Information and Control*, Vol. 59, 1983.

[267] Ong, C. L., J. T. Li, and C. Y. Lo: "GENAC: An Automatic Cell Synthesis Tool," in *Design Automation Conference*, pp. 239–234, IEEE/ACM, 1989.

[268] Osterhout, J. K.: "A Switch-Level Timing Verifier for Digital MOS VLSI," *IEEE Transactions on Computer Aided Design*, 4:336–349, 1985.

[269] Otten, R. H. J. M.: "Automatic Floorplan Design," in *Design Automation Conference*, pp. 261–267, IEEE/ACM, 1982.

[270] Otten, R. H. J. M.: "Efficient Floorplan Optimization," in *International Conference on Computer Design*, pp. 499–503, IEEE/ACM, 1983.

[271] Palczewski, M.: "Plane Parallel A* Maze Router and Its Application," in *Design Automation Conference*, pp. 691–697, IEEE/ACM, 1992.

[272] Papadimitriou, C. H., and K. Steiglitz: *Combinatorial Optimization—Algorithms and Complexity*. Prentice-Hall, Englewood Cliffs, NJ, 1982.

[273] Pedram, M., B. Nobandegani, and B. Preas: "Design and Analysis of Segmented Routing Channels for Row-Based FPGA's," *IEEE Transactions on Computer Aided Design* 13:1470–1479, 1994.

[274] Pinter, R. Y.: "River Routing: Methodology and Analysis," in *The Third Caltech Conference on VLSI*, MIT Press, Cambridge, MA, 1983.

[275] Pinter, R. Y., R. Bar-Yehuda, J. A. Feldman, and S. Wimer: "Depth-First-Search and Dynamic Programming Algorithm for Efficient CMOS Cell Generation," *IEEE Transactions on Computer Aided Design* 8(7):737–743, 1989.

[276] Prasitjutrakul, S., and W. J. Kubitz: "Path Delay Constrained Floorplanning: A Mathematical Programming Approach for Initial Placement," in *Design Automation Conference*, pp. 364–367, IEEE/ACM, 1989.

[277] Preas, B., and M. Lorenzetti: *Physical Design Automation of VLSI Systems,* Benjamin/Cummings, Menlo Park, CA, 1988.

[278] Preas, B., M. Pedram, and D. Curry: "Automatic Layout of Silicon-on-Silicon Hybrid Packages," in *Design Automation Conference*, pp. 394–399, IEEE/ACM, 1989.

[279] Preparata, F. P., and W. Lipski, Jr.: "Optimal Three-Layer Channel Routing," *IEEE Transactions on Computer Aided Design* C-33(5):427–437, 1984.

[280] Preparata, F. P., and M. I. Shamos: *Computational Geometry: An Introduction,* Springer-Verlag, Berlin, 1985.

[281] Prim, R. C.: "Shortest Connection Networks and Some Generalizations," *Bell System Technical Journal* (36):1389–1401, 1957.

[282] Pullela, S., N. Menezes, J. Omar, and L. T. Pillage: "Skew and Delay Optimization for Reliable Buffered Clock Trees," in *International Conference on Computer-Aided Design,* pp. 556–562, IEEE, 1993.

[283] Raghavan, P.: "Probabilistic Construction of Deterministic Algorithms: Approximating Packing Integer Programs," *Journal of Computer and System Sciences* 37(2):130–143, 1988.

[284] Raghavan, R., J. Cohoon, and S. Sahni: "Single Bend Wiring," *Journal of Algorithms* 7(2):232–257, 1986.

[285] Ramanathan, P., and K. G. Shin: "A Clock Distribution Scheme for Non-Symmetric VLSI Circuits," in *International Conference on Computer-Aided Design,* pp. 398–401, IEEE/ACM, 1989.

[286] Rao, S.: "Finding Near Optimal Separators in Planar Graphs," in *Annual Symposium on Foundations of Computer Science,* pp. 225–237, IEEE, 1987.

[287] Ratzlaff, C. L., N. Gopal, and L. T. Pillage: "'RICE: Rapid Interconnect Circuit Evaluator," in *Design Automation Conference,* pp. 555–560, IEEE/ACM, 1991.

[288] Ravi, S. S., and E. L. Lloyd: "The Complexity of Near-Optimal PLA Folding," *J. Comput.* 17(4):698–710, 1988.

[289] Reichelt, M., and W. Wolf: "An Improved Cell Model for Hierarchical Constraint-Graph Compaction," in *International Conference on Computer-Aided Design,* pp. 482–485, IEEE, 1986.

[290] Richards, D.: "Complexity of Single-Layer Routing," *IEEE Transactions on Computer Aided Design* C-33(3):286–288, 1984.

[291] Richards, D.: "Fast Heuristic Algorithms for Rectilinear Steiner Trees," *Algorithmica* 4:191–207, 1989.

[292] Rivest, R. L.: "The PI (Placement and Interconnect) System," in *Design Automation Conference,* pp. 475–481, IEEE/ACM, 1982.

[293] Rivest, R. L., A. E. Baratz, and G. Miller: "Provably Good Channel Routing Algorithms," in *VLSI Systems and Computations* (H. T. Kung, R. Sproull, and G. Steel, eds.), pp. 178–185, Computer Science Press, Rockville, MD, 1981.

[294] Rivest, R. L., and C. M. Fiduccia: "A Greedy Channel Router," in *Design Automation Conference,* pp. 418–424, IEEE/ACM, 1982.

[295] Rose, J., and S. Brown: "The Effect of Switch Box Flexibility on Routability of Field-Programmable Gate Arrays," in *IEEE Custom Integrated Circuits Conference,* pp. 27.5.1–27.5.4, IEEE, 1990.

[296] Rose, J., and S. Brown: "Flexibility of Interconnection Structures in Field-Programmable Gate Arrays," *IEEE Journal of Solid-State Circuits* 26:277–282, 1991.

[297] Roy, K., and C. Sechen: "A Timing Driven N-Way Chip Partitioner," in *International Workshop on Layout Synthesis* pp. 189–190, MCNC, 1992.

[298] Roychowdhury, V., J. Greene, and A. El Gamal: "Segmented Channel Routing," *IEEE Transactions on Computer Aided Design* 12:79–95, 1993.

[299] Rubinstein, J., P. Penfield, and M. A. Horowitz: "Signal Delay in RC Tree Networks," *IEEE Transactions on Computer Aided Design* CAD-2(3):202–211, 1983.

[300] Sakurai, T.: "Approximation of Wiring Delay in MOSFET LSI," *IEEE Journal of Solid-State Circuits* 18(4):418–426, 1983.

[301] Sanchis, L. A.: "Multi-Way Network Partitioning," *IEEE Transactions on Computers* 38(1):62–81, 1989.

[302] Sangiovanni-Vincentelli, A., and M. Santomauro: "YACR: Yet Another Channel Router," in *Proc. Custom Integr. Circuits Conf., Rochester, NY,* pp. 460–466, IEEE, 1982.

[303] Sangiovanni-Vincentelli, A., M. Santomauro, and J. Reed: "A New Gridless Channel Router: Yet Another Channel Router the Second (YACR-II)," in *International Conference on Computer-Aided Design*, pp. 72–75, IEEE, 1984.

[304] Sarrafzadeh, M.: "Channel-Routing Problem in the Knock-Knee Mode Is NP-Complete," *IEEE Transactions on Computer Aided Design* 6(4):503–506, 1987.

[305] Sarrafzadeh, M., and D. T. Lee: "A New Approach to Topological Via Minimization," *IEEE Transactions on Computer Aided Design* CAD-8(8):890–900, 1989.

[306] Sarrafzadeh, M. and D. T. Lee, (eds.): *Algorithmic Aspects of VLSI Layouts*, World Scientific, Singapore, 1993.

[307] Sarrafzadeh, M., and R. D. Lou: "Maximum k-Covering of Weighted Transitive Graphs with Applications," *Algorithmica* 9(1):84–100, 1993.

[308] Sarrafzadeh, M., and C. K. Wong: "Hierarchical Steiner Tree Construction in Uniform Orientations," *IEEE Transactions on Computer Aided Design* 11(8):1095–1103, 1992.

[309] Sawkar, P., and D. Thomas: "Performance Directed Technology Mapping for Look-Up Table Based FPGAs," in *Design Automation Conference*, pp. 208–212, IEEE/ACM, 1993.

[310] Schiele, W. L.: "Improved Compaction by Minimized Length of Wires," in *Design Automation Conference*, pp. 121–127, IEEE/ACM, 1983.

[311] Schlag, M., J. Kong, and P. K. Chan: "Routability-Driven Technology Mapping for Lookup Table-Based FPGA's," *IEEE Transactions on Computer Aided Design* 13:13–26, 1994.

[312] Schlag, M., Y. Z. Liao, and C. K. Wong: "An Algorithm for Optimal Two Dimensional Compaction of VLSI Layouts," *Integration: The VLSI Journal*, 1983.

[313] Schlag, M., F. Luccio, P. Maestrini, D. T. Lee, and C. K. Wong: "A Visibility Problem in VLSI Layout Compactor," in *VLSI Theory*, pp. 259–282, JAI Press Inc., Greenwich, CT, 1984.

[314] Schrage, L.: *Linear, Integer, and Quadratic Programming with LINDO*, Scientific Press, Palo Alto, CA, 1986.

[315] Schuler, D. M., and E. G. Ulrich: "Clustering and Linear Placement," in *Proc. 9th Design Automation Workshop*, pp. 50–56, ACM, 1972.

[316] Schweikert, D. G.: "A Two-Dimensional Placement Algorighm for the Layout of Electrical Circuits," in *Proceedings of IEEE Design Automation Conference*, pp. 408–416, IEEE/ACM, 1976.

[317] Schweikert, D. G., and B. W. Kernighan: "A Proper Model for the Partitioning of Electrical Circuits," in *Design Automation Conference*, pp. 57–62, IEEE/ACM, 1972.

[318] Sechen, C.: "The TimberWolf3.2 Standard Cell Placement and Global Routing Program: User's Guide for Version 3.2, Release 2," unpublished paper, 1986.

[319] Sechen, C.: "Chip-Planning, Placement, and Global Routing of Macro/Custom Cell Integrated Circuits Using Simulated Annealing," in *Design Automation Conference*, pp. 73–80, IEEE/ACM, 1988.

[320] Sechen, C.: *VLSI Placement and Global Routing Using Simulated Annealing*, Kluwer, Deventer, The Netherlands, 1988.

[321] Sechen, C., and A. Sangiovanni-Vincentelli: "TimberWolf3.2: A New Standard Cell Placement and Global Routing Package," in *Design Automation Conference*, pp. 432–439, IEEE/ACM, 1986.

[322] Shahookar, K., and P. Mazumder: "A Genetic Approach to Standard Cell Placement Using Meta-Genetic Parameter Optimization," *IEEE Transactions on Computer Aided Design* 9(5):500–511, 1990.

[323] Shahookar, K., and P. Mazumder: "VLSI Cell Placement Techniques," *ACM Computing Surveys* 23(2):143–220, 1991.

[324] Shapiro, J. F.: *Mathematical Programming: Structures and Algorithms,* John Wiley & Sons, New York, 1979.

[325] Shargowitz, E., and J. Keel: "A Global Router Based on Multicommodity Flow Model," *Integration: The VLSI Journal* 5:3–16, 1987.

[326] Sherwani, N. A.: *Algorithms For VLSI Physical Design Automation,* Kluwer Academic Publishers, Boston, MA, 1993.

[327] Sherwani, N. A., and B. Wu: "Effective Buffer Insertion of Clock Tree for High Speed VLSI Circuits," *Microelectronics Journal* 23:291–300, 1992.

[328] Shih, M., and E. S. Kuh: "Circuit Partitioning under Capacity and I/O Constraints," in *IEEE Custom Integrated Circuits Conference*, IEEE, 1994.

[329] Shih, M., E. S. Kuh, and R. S. Tsay: "Performance Driven System Partitioning on Multi-Chip Modules," in *Design Automation Conference* pp. 53–56, IEEE/ACM, 1992.

[330] Shin, H., and L. Chi-Yuan: "An Efficient Two-Dimensional Layout Compaction Algorithm," in *Design Automation Conference*, pp. 290–295, IEEE/ACM, 1989.

[331] Shin, H., A. Sangiovanni-Vincentelli, and C. Sequin: "Two-Dimensional Module Compactor Based on Zone-Refining," in *International Conference on Computer Design*, pp. 201–203, IEEE, 1987.

[332] Shin, H., A. Sangiovanni-Vincentelli, and C. Sequin: "Two-Dimensional Compaction by 'Zone Refining' " in *Design Automation Conference*, pp. 115–122, IEEE/ACM, 1989.

[333] Shiraishi, Y., and Y. Sakemi: "A Permeation Router," *IEEE Transactions on Computer Aided Design* CAD-6(3):462–471, 1987.

[334] Sigl, G., K. Doll, and F. M. Johannes: "Analytical Placement: A Linear or a Quadratic Objective Function," in *Design Automation Conference*, pp. 427–431, IEEE/ACM, 1991.

[335] Singh, S., J. Rose, D. Lewis, K. Chung, and P. Chow: "Optimization of Field-Programmable Gate Array Logic Block Architecture for Speed," in *IEEE Custom Integrated Circuits Conference*, pp. 6.1.1–6.1.6, 1991.

[336] Singh, U., and C. Y. Roger: "From Logic to Symbolic Layout for Gate Matrix," *IEEE Transactions on Computer Aided Design* CAD-11(2):216–227, 1992.

[337] Smith, J. M., D. T. Lee, and J. S. Liebman: "An $O(n \log n)$ Heuristic Algorithm for the Rectilinear Steiner Minimal Tree Problem," *Engineering Optimization* 4(4):179–192, 1980.

[338] Soukup, J.: "Circuit Layout," in *Proc. of the IEEE*, pp. 1281–1305, IEEE, 1981.

[339] Srinivasan, A.: "An Algorithm for Performance-Driven Initial Placement of Small-Cell ICs," in *Design Automation Conference*, pp. 636–639, IEEE/ACM, 1991.

[340] Srinivasan, A., K. Chaudhary, and E. S. Kuh: "RITUAL: An Algorithm for Performance Driven Placement if Cell-Based IC's," in *Third Physical Design Workshop*, 1991.

[341] Sriram, M., and S. M. Kang: "A Modified Hopfield Network for Two-Dimensional Module Placement," in *International Symposium on Circuits and Systems*, pp. 1664–1667, IEEE, 1990.

[342] Sriram, M., and S. M. Kang: "Detailed Layer Assignment for MCM Routing," in *International Conference on Computer-Aided Design*, pp. 386–389, IEEE, 1992.

[343] Stallmann, M., T. Hughes, and W. Liu: "Unconstrainted Via Minimization for Topological Multilayer Routing," *IEEE Transactions on Computer Aided Design* CAD-9(9):970–980, 1990.

[344] Stockmeyer, L.: "Optimal Orientation of Cells in Slicing Floorplan Designs," *Information and Control* 57(2):91–101, 1983.

[345] Su, C. C., and M. Sarrafzadeh: "Optimal Gate-Matrix Layout of CMOS Functional Cells," *Integration: The VLSI Journal* 10:3–23, 1990.

[346] Sutanthavibul, S., and E. Shragowitz: "An Adaptive Timing-Driven Layout for High Speed VLSI," in *Design Automation Conference*, pp. 90–95, IEEE/ACM, 1990.

[347] Sutanthavibul, S., E. Shargowitz, and J. B. Rosen: "An Analytical Approach to Floorplan Design an Optimization," in *Design Automation Conference*, pp. 187–192, IEEE, 1990.

[348] Sutanthavibul, S., E. Shragowitz, and J. B. Rosen: "An Analytical Approach to Floorplan Design and Optimization," *IEEE Transactions on Computer Aided Design* 10(6):761–769, 1991.

[349] Syed, Z. A., and A. El Gamal: "Single Layer Routing of Power and Ground Networks in ICs," *Journal of Digital Systems* VI(1):53–63, 1982.

[350] Szepieniec, A. A.: "Integrated Placement/Routing in Sliced Layouts," in *Design Automation Conference*, pp. 300–307, IEEE/ACM, 1986.

[351] Szymanski, T. G.: "Dogleg Channel Routing is NP-Complete," *IEEE Trans. on CAD* 4(1):31–41, 1985.

[352] Tang, H., and W. K. Chen: "Generation of Rectangular Duals of a Planar Triangulated Graph by Elementary Transformations," in *International Symposium on Circuits and Systems*, pp. 2857–2860, IEEE, 1989.

[353] Téllez, G. E., and M. Sarrafzadeh: "Clock Period Constrained Buffer Insertion in Clock Trees," in *International Conference on Computer-Aided Design*, IEEE/ACM, pp. 219–223, 1994.

[354] The, K. S., D. F. Wong, and J. Cong: "A Layout Modification Approach to Via Minimization," *IEEE Transactions on Computer Aided Design* 10(4):536–541, 1991.

[355] Ting, B. S., and B. N. Tien: "Routing Techniques for Gate Arrays," *IEEE Transactions on Computer Aided Design* CAD-2(4):301–312, 1983.

[356] Tragoudas, S.: "VLSI Partitioning Approximation Algorithms Using Multicommodity Flow and Other Techniques," PhD thesis, University of Texas at Dallas, 1991.

[357] Tsay, R. S.: "Exact Zero Skew," in *International Conference on Computer-Aided Design*, pp. 336–339, IEEE/ACM, 1991.

[358] Tsay, R. S., and J. Koehl: "An Analytic Net Weighting Approach for Performance Optimization in Circuit Placement," in *Design Automation Conference*, pp. 620–625, IEEE/ACM, 1991.

[359] Tsay, R. S., E. S. Kuh, and C. P. Hsu: "PROUD: A Sea-of-Gates Placement Algorithm," *IEEE Design and Test of Computers*, :44–56, 1988.

[360] Tseng, B., J. Rose, and S. Brown: "Improving FPGA Routing Architecture Using Architecture and CAD Interactions," in *International Conference on Computer Design*, pp. 99–104, IEEE, 1992.

[361] Tsui, C., M. Pedram, and A. Despain: "Efficient Estimation of Dynamic Power Dissipation under a Real Delay Model," in *International Conference on Computer-Aided Design*, pp. 224–228, IEEE, 1993.

[362] Tsukiyama, S., K. Koike, and I. Shirakawa: "An Algorithm to Eliminate All Complex Triangles in a Maximal Planar Graph for Use in VLSI Floor-plan," in *International Symposium on Circuits and Systems*, pp. 321–324, IEEE, 1986.

[363] Tsukiyama, S., and E. S. Kuh: "On the Layering Problem of Multilayer PWB Wiring," in *Design Automation Conference*, pp. 738–745, 1981.

[364] Tsukiyama, S., K. Tani, and T. Maruyama: "A Condition for a Maximal Planar Graph to Have a Unique Rectangular Dual and Its Application to VLSI Floor-plan," in *International Symposium on Circuits and Systems*, pp. 931–934, IEEE, 1989.

[365] Uehara, T., and W. M. van Cleemput: "Optimal Layout of CMOS Functional Arrays," *IEEE Transactions on Computers* C-30(5):305–312, 1981.

[366] Ullman, J. D.: *Computational Aspects of VLSI*, Computer Science Press, Rockville, MD, 1984.

[367] Upton, M., K. Samii, and S. Sugiyama: "Integrated Placement for Mixed Macro Cell and Standard Cell Designs," in *Design Automation Conference*, pp. 32–35, IEEE/ACM, 1990.

[368] Vaidya, P. M., S. S. Sapatnekar, V. B. Rao and S.-M. Kang: "An Exact Solution to the Transistor Sizing Problem for CMOS Circuits using Convex Optimization," *IEEE Transactions on Computer Aided Design* 12:1621–1634, 1993.

[369] Vaishnav, H., and M. Pedram: "A Performance Driven Placement Algorithm for Low Power Designs," in *European Design Automation Conference*, 1993.

[370] Valainis, J., S. Kaptanoglu, E. Liu, and R. Suaya: "Two-Dimensional IC Layout Compaction Based on Topological Design Rule Checking," *IEEE Transactions on Computer Aided Design* CAD-9(3):260–275, 1990.

[371] van Ginneken, L. P. P. P.: "Buffer Placement in Distributed RC-tree Networks for Minimal Elmore Delay," in *International Symposium on Circuits and Systems*, pp. 865–868, IEEE, 1990.

[372] Varga, R. S.: *Matrix Iterative Analysis*, Prentice-Hall, Englewood Cliffs, NJ, 1962.

[373] Vecchi, M. P., and S. Kirkpatrick: "Global Wiring by Simulated Annealing," *IEEE Transactions on Computer Aided Design* CAD-2(4):215–222, 1983.

[374] Vijayan, G.: "Generalization of Min-Cut Partitioning to Tree Structure and Its Applications," *IEEE Transactions on Computer Aided Design* 10(3):307–314, 1991.

[375] Vijayan, G., N. Hasan, and C. K. Wong: "A Neighborhood Improvement Algorithm for Rectilinear Steiner Trees," in *International Symposium on Circuits and Systems*, pp. 2869–2872, IEEE, 1990.

[376] Wang, T. C., and D. F. Wong: "An Optimal Algorithm for Floorplan Area Optimization," in *Design Automation Conference*, pp. 180–186, IEEE/ACM, 1990.

[377] Wang, T. C., and D. F. Wong: "A Note on the Complexity of Stockmeyer's Floorplan Optimization Technique," unpublished paper, 1991.

[378] Wei, Y. C., and C. K. Cheng: "Ratio-Cut Partitioning for Hierachical Designs," *IEEE Transactions on Computer Aided Design*, 40(7):911–921, 1991.

[379] Weinberger, A.: "Large Scale Integration of MOS Complex Logic: A Layout Method," *IEEE Journal of Solid-State Circuits* Sc-2(4):182–190, 1967.

[380] Weis, B. X., and D. A. Mlynski: "A Graph-Theoretic Approach to the Relative Placement Problem," *IEEE Transactions on Circuits and Systems* 35(3):286–293, 1988.

[381] Weste, N., and K. Eshraghian: *Principles of CMOS VLSI Design—A Systems Perspective*, Addison-Wesley, Reading, MA, 1985.

[382] Wimer, S., I. Koren, and I. Cederbaum: "Optimal Aspect Ratios of Building Blocks in VLSI," *IEEE Transactions on Computer Aided Design* 8(2):139–145, 1989.

[383] Wimer, S., R. Y. Pinter, and J. Feldman: "Optimal Chaining of CMOS Transistors in a Functional Cell," in *International Conference on Computer-Aided Design*, pp. 66–69, IEEE, 1986.

[384] Wing, O., S. Huang, and R. Wang: "Gate Matrix Layout," *IEEE Transactions on Computer Aided Design* CAD-4(3):220–231, 1985.

[385] Wolf, W. H., R. G. Mathews, J. Newkirk, and R. Dutton: "Two-Dimensional Compaction Strategies," in *International Conference on Computer-Aided Design*, pp. 90–91, IEEE, 1983.

[386] Wolf, W. H., R. G. Mathews, J. A. Newkirk, and R. W. Dutton: "Algorithms for Optimizing, Two-Dimensional Symbolic Layout Compaction," *IEEE Transactions on Computer Aided Design* 7(4):451–466, 1988.

[387] Wong, C. K.: "An Optimal Two-Dimensional Compaction Scheme," in *VLSI: Algorithms and Architectures* (P. Bertolazzi and F. Luccio, eds.), pp. 205–220, North-Holland, Amsterdam, 1984.

[388] Wong, D. F., H. W. Leong, and C. L. Liu: "Multiple PLA Folding by the Method of Simulated Annealing," in *Custom Integrated Circuits Conf.*, pp. 351–355, 1986.

[389] Wong, D. F., H. W. Leong, and C. L. Liu: *Simulated Annealing for VLSI Design*, Kluwer Academic, Boston, MA, 1988.

[390] Wong, D. F., and C. L. Liu: "Floorplan Design of VLSI Circuits," *Algorithmica* 4:263–291, 1989.

[391] Wong, D. F., and P. S. Sakhamuri: "Efficient Floorplan Area Optimization," in *Design Automation Conference*, pp. 586–589, IEEE/ACM, 1989.

[392] Wu, Y., and M. Marek-Sadowska: "An Efficient Router for 2-D Field-Programmable Gate Arrays," in *European Design Automation Conference*, 1994.

[393] Wu, Y., S. Tsukiyama, and M. Marek-Sadowska: "Graph-Based Analysis of 2-D FPGA Routing," 1994.

[394] Xilinx Inc.: *The Programmable Gate Array Data Book*, 1986.

[395] Xiong, X. M.: "Two-Dimensional Compaction for Placement Refinements," in *International Conference on Computer-Aided Design*, pp. 136–139, IEEE, 1989.

[396] Xiong, X. M., and E. S. Kuh: "The Scan Line Approach to Power and Ground Routing," in *International Conference on Computer-Aided Design*, pp. 6–9, IEEE, 1986.

[397] Xiong, X. M., and E. S. Kuh: "Nutcracker: An Efficient and Intelligent Channel Spacer," in *Design Automation Conference*, pp. 298–304, IEEE/ACM, 1987.

[398] Yang, C. D., D. T. Lee, and C. K. Wong: "On Bends and Lengths of Rectilinear Paths: A Graph-Theoretic Approach," in *Proceedings of Workshop on Algorithms and Data Structure*,

Canada, 1991, Lecture Notes in Computer Science, vol. 519, pp. 320–330, Springer-Verlag, Berlin, 1991.

[399] Yeap, G., M. Sarrafzadeh, and C. K. Wong: "A Global Routing Algorithm for Array Architecture FPGA Environment," unpublished paper, 1993.

[400] Yeap, K. H., and M. Sarrafzadeh: "Floorplanning by Graph Dualization: 2-Concave Rectilinear Modules," 1993.

[401] Yeap, K. H., and M. Sarrafzadeh: "Net-Regular Placement for High Performance Circuits," *Journal of High Speed Electronics,* to appear.

[402] Yu, Q., and O. Wing: "Interval-Graph-Based PLA Folding," in *International Symposium on Circuits and Systems*, pp. 1463–1466, IEEE, 1985.

[403] Zhang, C., and D. A. Mlynski: "VlSI-Placement with a Neural Network Model," in *International Symposium on Circuits and Systems*, pp. 475–478, IEEE, 1990.

[404] Zhu, K., and D. F. Wong: "On Channel Segmentation Design for Row-Based FPGA's," in *International Conference on Computer-Aided Design*, pp. 26–29, IEEE, 1992.

INDEX

Acyclic graph, directed (DAG), 163
Algorithm, *see also* algorithms by name
   k-segment routing, 149
   assignment, 81–82
   BEAVER, 135–136
   benchmark, 126, 145
   bottom_up_routing, 114
   constructive force-directed, 71–72
   efficiency of, 20, 126, 145
   exhaustive search, 97
   Fiduccia-Mattheyses, 37
   finite time, 19
   GPSP-ALG, 210–211
   greedy channel router, 128
   greedy push, 224
   greedy switchbox, 137
   greedy-routing, 217, 218
   greedy-slide, 229
   GREEDY2 push, 227
   hypergraph-transform, 34
   integer linear programming, 117–118
   iterative force-directed, 72
   iterative-improvement-for-wire-length, 283
   Kernighan-Lin, 35
   layout-BRST, 177–178
   layout-WRST, 104
   linear placement, 82
   longest-path, 278
   max-cut, 192
   MNP-ALG, 209–210
   multicommodity flow, 115–117
   NODE_OP, 211
   NP-complete, 20
   one-step approach, 118–119
   order notation, 19
   performance-driven placement, 174
   PLA-redundancy-removal, 251
   placement, 71
   polynomial time, 19
   ratio-cut, 40–42
   resistive network, 78–80
   routing, 111
   Simplex, 285
   simulated annealing, 58–59
   top_down_routing, 113
   vertical node sizing, 63
   weight-based algorithm, 172
   wiring, 2
   zero-slack, 165–169
Algorithmic paradigms, 21
   in bin packing, 22–24
   branch and bound, 22, 24
   dynamic programming, 21, 23
   exhaustive search, 21, 23
   genetic algorithm, 22, 24
   greedy approach, 21, 23
   hierarchical approach, 21, 23
   mathematical programming, 22–24
   neural-network alogrithm, 22
   simulated annealing, 22, 24
Analog simulation, 156
Aspect ratio, 65
Assignment algorithm, 80–82
   cost function, 80, 84
   relaxed phase, 80
   removing overlaps, 81
   running time, 84
Automated generation, 199

Backtrace, 239
Basic cell, 7–8
BEAVER algorithm, 133–136
   corner router, 133–134
   line sweep router, 135
   thread router, 135
Benchmark, 126, 145, 199
Bend minimization, 201, 227–230
   greedy slide algorithm, 227–229
Bin packing, 22–24, 143–145
Block Gauss-Seidel optimization, 84
Boolean function, 248
   prime implicant, 249
   tautology, 250
Boolean programming, quadratic, 45
Bottom-up matching algorithm, 181–182
Bottom_up_routing algorithm, 114
Branch, 127
Branch-and-bound approach, 22, 24, 150, 252, 263, 294, 297, 298
   compaction algorithm, 296
BRMRT, *see* Routing tree, bounded radius minimum
BRST, *see* Routing tree, bounded radius spanning
BSLR, *see* Single-layer routing, boundary
Buffer delay model, 186
Bus routing, 241

Capacitance
interconnect, 159
output loading, 157
parasitic, 159
wire loading, 157
Capacity constraint, 43–44
Cell, 60 *see also* Basic cell;
Regular cell
fixed, 60
variable, 60
Cell generation
layout, *see* Layout style
random, 8, 247
regular, 8, 247
Cell layout generation, *see*
Layout style
Channel, 3, 98, 121
Channel routing, 123–133,
138–139
compaction algorithms, 279
directional multilayer model,
139
divide-and-conquer approach,
130–133
free-hand model, 138
greedy algorithm, 128–129
greedy channel router, 126
gridless routing, 138
hierarchical approach, 126,
129
L/k-layer model, 139
left-edge algorithm, 126–127
multilayer model, 138–139
neural net algorithm, 139
segmented, 138
shift-right-1, 124–125
single-layer routing, 233
unit-vertical overlap model,
139
YACR-II, 126
yet another channel router,
127–128
Channel routing algorithm, 287
Channel width, 123
Channel-based thin-film wiring,
235
methodology, 200
Circle graph, 193
Circuit graph, 36
Circuit partitioning, 31
Clock
distribution network, 184–185
latency, 184

phase delay, 184
rise time, 184
skew, 180, 184
skew rate, 184
Clock power-up tree, *see* Clock
tree, buffered
Clock skew
Elmore delay model, 182–184
zero clock skew algorithm,
182–184
Clock tree, 184
buffered, 185
wiring-oriented, 185
zero skew, 186
Clock tree optimization, 200
Cluster graph, 189
CMOS functional arrays, *see*
Transistor chaining
Compaction
one and one-half dimensional,
295
one-dimensional, 272–293,
295–296
two-dimensional, 272,
293–295, 297
Complex triangle, 50
Compression ridge, 274
Compression-ridge algorithms,
273–276
flow-based, 274–275
maximum-flow algorithm,
275
Conductance, shunt, 159
Conducting layer, 7
Confronting graph, 144
Confronting nets, 144
Congestion map (CM), 213
Connection graph, 260
Constrained layer assignment
problem, 234
Constraint
base, 294
capacity, 43–44
distance, 294
grid, 289–293
I/O, 43–44
two-dimensional compaction,
294
Constructive placement, 69
Constructive placement
algorithm
adaptive control approach,
174

window, 174
Contour compaction algorithm,
287, 288
Convex programming, 174
Corner router, 133–134
Cost function, 70–71
CPSP, *see* Planar subset
problem, circle
Critical area, 201
Critical nets, 163
Crossing graph, 188
Crosstalk, 139
CVM, *see* Via minimization,
constrained

Decomposition tree (DT), 255
binary (BDT), 262
Delaunay triangulation,
constrained, 220
Delaunay triangulation-based
algorithm, 100
Delay model, 155–163
analog simulation, 156
fall time, 156
gate, 156–158
interconnect, 159–161
RC network, 161–163, 176
rise time, 156
timing simulation, 156
Delay time, 156
Density
global, 124
local, 124
Design for manufacturability,
269
Design for testability, 269
Detailed routing, 17, 91
area routing, 121
bounded region, 215–219
channel, 121
channel routing, 122,
123–133, 138–139
channel width minimization,
123
delay minimization, 175
fixed terminal, 121
floating terminal, 121
FPGA, *see* FPGA routing
free-hand layer model, 123
general region, 219–221
greedy-routing algorithm,
217–219
gridless routing, 138

Capacitance (*continued*)
  hierarchical approach, 133
  Lee-Moore algorithm, 121
  Manhattan model, 122
  multilayer model, 138–139
  overlap model, 138
  placement, 220
  planar routing, 138
  reserved layer model,
    122–123
  segmented channel routing,
    138
  shift-right-1, 124–125
  single-layer, 207, 215–221,
    see also Single-layer
    routing
  STACK algorithm, 216–219
  Steiner tree, 123, 131–133
  straight cut, 219
  switchbox, 121
  switchbox routing, 133–138
Diffusion, 3
Dijkstra's algorithm, 95, 116,
    215, *see also* SLGRP
    algorithm
Divide-and-conquer algorithm,
    130–133
  running time, 132–133
Drain, 3
Dynamic programming, 21, 23,
    98–99, 208, 233, 265
  running time, 98

Edge-cut model, 36
Elmore delay
  interconnect design, 161
  RC network, 161
  real delays, 161
  timing-driven critical sink
    routing, 161
  zero-skew clock tree routing,
    161
Elmore delay formula, 162
Elmore delay model, 175,
    182–184
Eulerian path, 254, 255
Exhaustive search, 21, 23, 264
Exhaustive search algorithm,
    97–98
  running time, 97–98

Fall time, 156
Fiduccia-Mattheyses algorithm,
    37–39, 43, 67, 75

  balanced partition, 37
  cut-cost, 37
  data structure, 37
  extensions of, 39
  running time, 37, 38–39
Field-programmable gate arrays,
    see FPGA
Finite time, 19
Fixed terminal, 121
Flexible transistor matrix, 267
Floating terminal, 121
Floorplan
  dual graph, 50
  extended dual graph, 51
  hierarchical, 68
  hierarchically defined, 48
    order 5, 64, 68
  nonsliceable, 48, 68, 122
  in Polish expression, 59
  rectangular, 49
  sliceable, 48, 68, 122
  topology, 67
  tree, 49, 59
Floorplanning, 16–17, 47
  chip area, 47
  cost function, 57
  dual decomposition, 51–53
  dual graph, 50
  dualization algorithm, 54
  Fiduccia-Mattheyses
    algorithm, 67
  floorplan merging, 51–53
  floorplan-sizing, 49–50,
    60–67
  genetic algorithm, 69
  global routing, 68, 107–108
  graph matching, 54
  greedy algorithm, 55
  hierarchical approach, 54–57
  hierarchical floorplan sizing,
    62–64
  hierarchically defined, 50
  Kernighan-Lin algorithm, 67
  linear time algorithm, 54
  LP-based approach, 65–67,
    69
  nonhierarchical floorplan
    sizing, 65–67
  partitioning, 56
  performance, 47
  rectangular dissection, 47
  routing, 47
  routing channel ordering, 50

  simulated annealing, 57–61,
    68–69
  splitting path, 52
  vertical node sizing
    algorithm, 63–64
  weighted complex triangle
    elimination, 54
Flux, 125
Force-directed algorithms
  constructive, 71–72
  iterative, 72
FPGA, 12, 14, 15, 122, 140, 149
  array-based, 140
  diagonal S block, 143
  flexibility, 141
  long wire segment, 141
  one-segment routing, 147
  row-based, 140
FPGA routing, 141–151
  array-based, 141–145, 150
  branch-and-bound algorithm,
    150
  detailed routing, 150
  global routing, 150
  graph-coloring problem, *see*
    Graph, chromatic
    number
  greedy bin-packing algorithm,
    143–145
  issues in, 150–151
  k-segment routing algorithm,
    147–149
  left-edge algorithm, 146–147
  one-segment routing, 147,
    148
  problem, 142
  row-based, 145–149, 150
  segmented channel routing,
    146–147
Full custom design, 12
Fuzzy logic algorithm, 175

Gate, 3
Gate array, 12–13, 15, 122
Gate delay analysis, 158
Gate delay model, 156–158
  capacitance, 157
  on-resistance model, 157
  waveform, 157
Gate matrix layout, 10–11,
    257–263
  advantages, 258
  area minimization, 257,
    258–259

Gate matrix layout (*continued*)
  bounding algorithm, 261
  connection graph, 260
  disadvantages, 258
  gate assignment function, 259
  gate line, 259
  greedy algorithm, 261
  interval graph, 259
  interval graph-based
      algorithm, 269
  net assignment function, 259
  structure, 257–258
  vertical diffusion run, 261
GENAC, 263
Genetic algorithm, 22, 24, 69, 83
Geometric design rules, *see*
      Layout rules
Global density, 124
Global routing, 17, 68, 91
  bottom-up matching
      algorithm, 181–182
  bottom-up routing, 114
  clock skew problem, 180–184
  clock tree, buffered, 184–187
  delay minimization, 175–180
  to detailed routing, 121, 219
  Dijkstra's algorithm, 116
  after floorplanning, 107–108
  FPGA, *see* FPGA routing
  GPSP, 210
  hierarchical approach,
      112–114, 120
  ILP-based algorithm, 114
  integer linear programming,
      119–120
  integer programming, 112
  integer programming
      algorithm, 113, 121
  JSK algorithm, 180–181
  k-layer planar subset, 195
  k-layer unconstrained via
      minimization, 195
  layout environment, 108
  layout-BRST algorithm,
      176–180
  Lee-Moore algorithm, 110,
      121
  linear delay model, 182
  maze running, 110, 180
  multicommodity flow
      algorithm, 114–117,
      120

  one-step approach, 118–119,
      121
  output, 108–110
  after placement, 108, 109
  Prim-based algorithm, 110
  randomized routing, 117–118
  routing algorithm, 111–112
  sequential approach,
      110–112, 120
  simulated annealing, 121
  single-layer, 207, 212–215,
      see also Single-layer
      routing
  Steiner tree, 100, 110–111
  tapping points, 182
  timing-driven, 175–187
  top-down-routing algorithm,
      113
  unconstrained via
      minimization, 193–194
Global-stack, 281
GPSP, *see* Planar subset
      problem, general
GPSP-ALG algorithm, 210–212
Graph
  chromatic number, 28, 143,
      201
  circle, 193
  circuit, 7, 36
  cluster, 189
  conducting layer, 7
  configuration, 57
  confronting, 144
  connection, 260
  constraint, 276, 277
  crossing, 188
  directed acyclic (DAG), 61,
      163, 276
  dual, 50
  extended dual, 51
  floorplan, 50
  horizontal constraints, 124
  horizontal dependency, 61
  and hypergraph, 32, 33
  independent set, 29
  intersection, 124, 252
  interval, 259
  maximum independent set
      (MIS), 29
  minimum vertex cover
      (MVC), 30
  net interference (NIG), 201

  planar triangulated (PTG), 50
  routing, 112
  search, 101
  series-parallel, 62, 255
  track, 95–96
  vertex cover, 30
  vertical constraints, 124
  vertical dependency, 62
  wiring, 7
Graph-based algorithm, 255,
      276–279
  constraint graph, 276, 277
  directed acyclic graph, 276
  longest-path algorithm,
      276–279
Graph-coloring problem, *see*
      Graph, chromatic
      number
Greedy algorithm, 21, 23,
      55–56, 261
  channel routing, 128–129
Greedy bin-packing algorithm,
      143–145
  track domain, 143
Greedy clustering, *see*
      Floorplanning, greedy
      algorithm
Greedy embedding algorithm,
      100
Greedy push algorithm, 222–227
  running time, 224
Greedy switchbox algorithm,
      136–138
Greedy-routing algorithm,
      217–219
  boundary-dense solution, 218
  running time, 218
Greedy-slide algorithm, 227–229
  full-slide, 228, 229
  one-unit shift, 228, 229
  partial-slide, 228, 229
  stopped-point, 228
Grid, 7
Grid node, 291
Grid searching algorithm, 94
  Lee-Moore, 110
Grid-graph, 6, 7
Gridless routing, 138
GRP, *see* Global routing,
      single-layer

H-flipping, 180
H-tree, 180

Hanan grid, 97, 106
Heuristic, *see* Algorithm
Hierarchical approach, 21, 23
  bottom-up, 21, 55–56, 68,
    129
  to channel routing, 129
  floorplanning, 54–57
  global routing, 112–114, 120
  to layout, 31
  to switchbox routing, 133
  top-down, 21, 31, 56–57, 68,
    129
Hierarchical clustering-based
    algorithm, 198
Hierarchical construction
    algorithm, 100
Hierarchical placement
    algorithm, 169
Homotropic transformation, 222
Homotropy, 213
Horizontal constraints graph,
    124
Hypergraph
  bipartitioning, 33
  and graph, 32, 33
  multiway partitioning, 33
Hypergraph-transform algorithm,
    33–34

ILP-based algorithm, *see*
    Integer linear
    programminng
Inductance, interconnect, 159
Input-output constraint, 43–44
Integer linear programming, 114,
    117, 119–120, 298,
    see also Randomized
    routing
  constrained, 120
  unconstrained, 120
Integer programming algorithm,
    112, 113, 117, 118, 121
Interconnect delay model,
    159–161
  capacitance, 159
  conductance, 159
  inductance, 159
  RC network, 160–161
  resistance, 159
Interconnection delay, 155
Intersection graph, 124, 252
Interval graph, 259
Inverter, 4

Iterated one-Steiner algorithm,
    99
Iterative improvement, 69
Iterative-improvement-for-wire-
    length algorithm,
    283
  cost function, 283

Job scheduling problem, 126
Jog, automatic, 286–289
  contour compaction
    algorithm, 287–288
  force-directed algorithm, 287
  noncritical paths, 288–289
JSK algorithm, 180–181

k-layer planar subset problem
    (k-PSP), 195
k-layer topological via
    minimization problem
    (k-UVM), 195
k-segmented routing, *see*
    Segmented channel
    routing
Kernighan-Lin algorithm, 34–37,
    40, 46, 67, 75, 198
  bipartitioning, 35
  edge-cut model, 36
  running time, 35
Knock-knee model, 138
Kruskal-based algorithm, 100
Kuhn-Tucker conditions, 78–79,
    175
  gradient, 171
  Hessian matrix, 171
  Lagrange multiplier, 171

L_1 metric, 96
L-shaped optimal embedding
    algorithm, 99
Lagrange multiplier, 171
Lagrangian optimization
    algorithm, 200
λ-geometry, 104–106
Laplace transform, 161
Layer assignment algorithm,
    200–201
Layout
  abstract, 6, 271
  automated, *see* Algorithm
  compacted, 271
  constraints, *see* Layout rules
  environment, 108

grid, 8
gridless, 6
legal, 6
mask, 6, 264
measures of quality, 1–2
optimization, 17
physical, 271, 272
symbolic, 264, 271, 272
verification, 17
Layout architecture
  FPGA, 14, 15, 122
  gate array, 12–13, 15, 122
  module, 11
  PLA, 15
  sea-of-gates, 13–14, 122
  standard cell, 12, 13, 15, 122
Layout methodology
  detailed routing, 17
  floorplanning, 16–17
  global routing, 17
  hierarchical approach, 2
  layout optimization, 17
  layout verification, 17
  partitioning, 14, 16
Layout optimization, 17
Layout rules
  assumptions, 6
  contact, 6
  wire separation, 5
  wire width, 5
Layout style
  design for manufacturability,
    269
  design for testability, 269
  free-hand layer model, 123
  gate matrix, 10–11, 257–263
  PLA, 8, 248–253
  reserved layer model,
    122–123
  transistor chaining, 9, 10,
    253–256
  Weinberger array, 9–10,
    256–257
Layout synthesis, 199, 263
Layout-BRST algorithm,
    176–180
  routing tree, 179
Layout-WRST algorithm,
    100–104, 111
  in λ-geometry, 107
  running time, 104
  search graph, 101

Lee's algorithm, *see* Lee-Moore algorithm
Lee-Moore algorithm, 92, 94, 110, 121
Left-edge algorithm (LEA), 126–127, 146–147, 259
Level gain, 47
LiB
  chain formation, 263–264
  chain placement, 264
  clustering, 263
  compaction, 264
  detailed routing, 264
  geometry transfer, 264
  net assignment, 264
  pairing, 263
  routing, 264
  transistor folding, 264
Line probing algorithm, *see* Line searching algorithm
Line searching algorithm, 94–95, 110
  line probes, 94
  sink probes, 94
  source probes, 94
  track graph, 95–96
Line sweep connection
  dogleg, 134
  horseshoe, 134
  single bend, 134
  stairstep, 134
  straight-line wire, 134
Line sweep router, 134–135
Linear delay model, 175
Linear placement algorithm, 82–83
  cost function, 82
  running time, 82
Linear programming, 22, 80, 281, 297
  cost function, 67, 173
  module size, 67
  timing-driven placement, 172–174
Linear programming constraints
  cost, 67
  module size, 66–67
  nonoverlap, 65–66
Linear time algorithm
  horizontal contact offsetting, 265
  symbolic layout optimization, 265

transistor orientation, 265
vertical contact relocation, 265
Load matching, 186
Local density, 124
Local-slack, 281
Logic minimization, 9, 248–251
  personality, 250
Longest-path algorithm, 276–279, 291–293
  running time, 278

Manhattan model, 122
Matching, 81
Mathematical programming, 23–24, 44, 198, *see also* Linear programming
Max-cut (L) algorithm, 190–192
  running time, 190
Max-cut problem, 190
Maximum-flow algorithm, 275
Maximum-weighted pairwise nonoverlapping polygon (MNP), 208
Maximum-weighted planar subset, *see* Planar subset problem
Maze running, 91–94, 110, 239, *see also* Lee-Moore algorithm
  bidirectional search, 93
  cost, 93–94
  memory optimization, 93
  and multilayer routing, 94
  and multiterminal routing, 94
  sink terminal, 92
  source terminal, 91
  thread router, 135
MCM, 19, 234
  layers of, 19
  packaging, 19
MCM routing, 233–240
  channel-based thin-film wiring, 235
  concurrent maze router, 239
  constrained layer assignment problem, 234
  cost function, 240
  cross-talk, 235–236
  maze running, 239
  model, 237–238

pin redistribution problem, 238–239
propagation delay, 235
routing requirements, 236
rubber-band sketches, 234
signal distribution problem, 239–240
skin effect, 236
topological multilayer assignment, 235
MCTP, *see* Partitioning, min-cost tree
Merging, 211
Metal-metal matrix (M^3), 266
Min-cut algorithm, 74–75, 174, 264
  block-oriented, 75
Minimum weight matching algorithm
  running time, 189
MNP, *see* Maximum-weighted pairwise nonoverlapping polygon
MNP-ALG algorithm, 208–210
  running time, 209, 210
Module placement algorithm, *see* Resistive network algorithm
Monotone section, 227, 228
Monte-Carlo algorithm, 297
MST, *see* Steiner tree, minimum spanning; Routing tree, minimum spanning
Multichip module, *see* MCM
Multicommodity flow, 115
Multicommodity flow algorithm, 114–117, 120
Multilayer model, 138–139
  L/k-layer, 139
  directional, 139
  unit-vertical overlap, 139
  unrestricted overlap, 139
Multilayer routing, 94
Multiplicity, 110
Multiterminal routing, 94

NAND gate, 4
Neighboring Steinerization algorithm, 100
Net interference graph, 201

Net-cut model, 37
Neural networks, 22, 83–84, 139
Node-edge incidence matrix, 115
NOR gate, 4
Normalized, 60
NP-complete (NP-hard), 20

On-resistance model
  analysis, 158
  capacitance, 157
  delay equation, 158
  gate delay, 157
  gate delay analysis, 158
One-dimensional compaction,
    272–293, 295–296
  automatic jog, 286–289
  branch-and-bound approach,
    296
  channel routing, 279, 280
  compression-ridge algorithms,
    273–276
  contour compaction
    algorithm, 287–288
  effective cycle length, 292
  force-directed algorithm, 287
  graph-based algorithms,
    276–279
  grid constraints, 289–293
  maximum-flow algorithm,
    275
  noncritical-path jog, 288–289
  $1\frac{1}{2}D$, 295
  standard cell layout, 289
  wire-length minimization,
    279–286
One-segment routing, *see*
    Segmented channel
    routing
One-step approach, 118–119,
    121
Optimal linear arrangement, 82
Order notation, 19
Over-the-cell routing (OTC)
  channel routing, 233
  channel segment assignment,
    233
  CPSP algorithm, 233
  dynamic programming, 233
  model, 230–231
  net classification, 232
  net decomposition, 231–232
  single-layer, 230–233

  vacant terminal/abutment
    assignment, 233
Overlap model, 138

Packaging, 18
  DIP, 18
  MCM, 19, 233, 234
  PGA, 18
Partitioning, 14, 16, 84
  aggregation algorithm, 45
  balance factor, 33
  balanced, 37
  capacity constraint, 43–44
  circuit, 31
  clustering, 45
  constructive methods, 45
  eigenvector decomposition,
    45, 46–47
  Fiduccia-Mattheyses
    algorithm, 37–39
  floorplanning, 56
  I/O constraint, 43–44
  Kernighan-Lin algorithm,
    34–37
  mathematical programming,
    44
  maximum-flow minimum-cut
    algorithm, 45
  min-cost tree (MCTP)
    running time, 46
  multiway
    graph, 47
    network, 47
  quadratic Boolean
    programming, 45
  and quadratic optimization
    techniques, 197–198
  ratio-cut algorithm, 39–43, 47
  simulated annealing, 45, 47
Partitioning placement, *see*
    Placement, min-cut
    algorithm
Partitioning resistive network, 84
Pattern router, 127
PCUB algorithm, 196–197
Performance-driven layout,
    265–269
  area minimization, 200
  automated generation, 199
  bend minimization, 201
  channel-based thin-film
    wiring, 200
  clock tree optimization, 200

  delay model, 155–163
  doughnut layout style, 268
  electrical performance, 200
  flexible transistor matrix
    (FTM), 267
  Lagrangian optimization
    algorithm, 200
  layer assignment algorithm,
    200–201
  layout synthesis, 199
  metal-metal matrix (M^3), 266
  module generation, 268
  multilayer area determination,
    201
  over-the-cell routing, 199
  placement, 199
  power minimization, 196–198
  power pad minimization, 200
  regularity, 199
  timing-driven floorplanning,
    198–199
  timing-driven partitioning,
    198
  timing-driven placement,
    163–175
  timing-driven routing,
    175–187
  via minimization, 187–196
Performance-driven routing,
    MCM, 235
Photolithography, 3
Pin redistribution problem
    (PRP), 238–239
PLA, 8, 12, 15, 248–253
  folding, 248–249, 251–253
  interval graph-based
    algorithm, 269
  logic minimization, 248–251
  personality, 250, 251
  redundancy-removal
    algorithm, 251
  tautology test algorithm,
    250–251
PLA column-folding
  breakpoint, 252
  intersection graph, 252
  ordered folding pair, 252
  topological sort, 253
PLA folding, 9, 248–249,
    251–253
  best-first search algorithm,
    252
  branch-and-bound search, 252

PLA folding (*continued*)
  column-folding, 252
  graph-based algorithm, 252
  optimal bipartite folding, 252
  PLA decomposition
    technique, 252
  row-folding, 252
  simulated annealing, 252
PLA-redundancy-removal
    algorithm, 251
Placement, 17, *see also*
    Partitioning;
    Floorplanning
  assignment algorithm, 80–82,
    84
  block Gauss-Seidel optimiza-
    tion, 84
  cell-based, 197
  cost function, 70–71, 84
  delay optimization, 84
  Fiduccia-Mattheyses
    algorithm, 75
  force-directed algorithm,
    71–72, 84
  fuzzy logic algorithm, 175
  genetic algorithm, 83
  and global routing, 108, 109
  Kernighan-Lin algorithm, 75
  Kuhn-Tucker conditions,
    78–79, 175
  linear placement algorithm,
    82–83
  linear programming, 172–174
  min-cut algorithm, 74–75
  module placement algorithm,
    see Resistive network
    algorithm
  neural networks, 83–84
  partitioning, 84, 108
  PCUB algorithm, 196–197
  resistive network algorithm,
    75, 78–80, 84
  simulated annealing, 22, 24,
    46, 47, 73–74, 83
  terminal propagation
    technique, 75, 76
  timing constraint, 84
  timing-driven, 163–175
  weight-based, 169–172
Placement algorithm, 69–71
  cost function, 70–71
Placement topology
  nonslicing, 122

  slicing, 122
Planar
  pattern, 213
  solution, 213
  tile, 213
Planar routing, 138, 200,
    241–242
Planar subset problem (PSP),
    207–212
  circle (CPSP), 208, 233
  general (GPSP), 210–212
  general region, 208–210
  running time, 211–212
Polish expression, 59
Polynomial time, 19
Porosity, 290
Power minimization, 196–198
  objective function, 197
  PCUB algorithm, 196–197
  quadratic optimization
    techniques, 197
Power pad minimization, 200
Power/ground net sizing, 241
Prim-based algorithm, 100, 103,
    110
  running time, 103
Procedure, *see* Algorithm
Programmable logic array, *see*
    PLA
PTG, *see* Graph, planar
    triangulated

Randomized routing, 117–118
Ratio gain, 41
Ratio-cut algorithm, 39–43
  group swapping, 41–42
  initialization, 41
  iterative shifting, 41
  running time, 42–43
  spectral method, 43
RC, *see* Capacitance; Resistance
RC network, 160–163
  asymptotic waveform
    evaluation (AWE), 161
  Elmore delay, 161
RC tree, 162
RC tree delay model, 175
Rectangular dual graph, 50
Rectangular floorplan, 49
Regular cell, 8, 247
Regularity, 199
Relaxed placement, 80
Removing overlap, 80
Residual net, 193

Resistance
  interconnect, 159
  on-resistance, 157
  parasitic, 159
Resistive network algorithm, 75,
    78–80, 84
  n-terminal, 77
  cost function, 75–76
  power dissipation, 77
  running time, 75
  scaling, 79
Right-shifting operation, 41
Rise time, 156
River routing, 241
River-routing problem, 215
Routing, *see also* Detailed
    routing; FPGA routing;
    Global routing
  Dijkstra's algorithm, 95
  Lee-Moore algorithim, 92
  line searching algorithm,
    94–95
  maze running, 91–94
  optimal-bend, 227
  Steiner tree, 96–107
Routing algorithm, 111–112
Routing graph, 112
Routing tree, *see also* Steiner
    tree
  bounded radius minimum
    (BRMRT), 176
  bounded radius spanning
    (BRST), 176
  clock, 184
  minimum spanning (MST),
    176, 177
  RC, 162
  shortest path (SPT), 176
  wire, 213
RST (rectilinear Steiner tree),
    see Steiner tree
Rubber-band sketch, 220

Sc2, 263
Sea-of-gates layout, 13–14, 122,
    290
Search graph, 101
Segmented channel routing, 138
  k-segment routing algorithm,
    147–149
    assignment tree, 148
    frontier, 148
    running time, 148

Segmented channel routing
(*continued*)
left-edge algorithm, 146–147
one-segment routing, 147,
148
Semicustom design, 12
Semiperimeter method, 70
Sequential approach, 110–112,
120
Lee-Moore algorithm, 110
rip up and reroute, 112
Series-parallel multigraph, 255
Series-parallel tree, 261
Signal distribution problem
(SDP), 239–240
Simplex algorithm, 284–286,
298
cost function, 284–285
tree transformation, 284
Simulated annealing, 22, 24,
45–47, 68–69, 73–74,
83, 121, 252, 297
cost function, 73–74
in floorplanning, 57–61
TimberWolf, 73
Single-layer routing
bend minimization, 227–230
boundary (BSLR), 215
detailed, 216
topological, 216
bounded region, 215–219
bus, 241
degenerate cut, 220
detailed, 215–221
fixed module, 210–212
general region, 208–210,
219–221
global, 212–215
graph algorithm, 242
greedy push algorithm,
222–227
greedy slide algorithm,
227–229
greedy-routing algorithm,
217–219
MCM, 222, 233–240
nondegenerate cut, 220
optimal-bend, 227
over-the-cell, *see*
Over-the-cell routing
planar, 241–242
planar subset problem,
207–212

power/ground net sizing, 241
river routing, 241
river-routing problem, 215
rubber-band sketch, 220
single-row, 241
STACK algorithm, 216–219
straight cut, 219
wire-length minimization
problem (SLWP),
222–227
Single-row routing, 241
SLGRP, *see* Global routing,
single-layer
SLGRP algorithm, 213–215
cost function, 215
directional cost, 214
Slicing, 122
SLWP, *see* Single-layer
routing, wire-length
minimization problem
SMT, *see* Steiner tree, minimal
Solid net, 194
Source, 3
Spot defect, 201
SPT, *see* Routing tree, shortest
path
STACK algorithm, 216–219
Standard cell layout, 12, 13,
122, 289
Steiner approximation, 198
Steiner point, 97
Steiner tree
channel, 98
Delaunay triangulation-based
algorithm, 100
detailed routing, 123,
131–133
dynamic programming, 98–99
exhaustive search algorithm,
97–98
in global routing, 100,
110–111
greedy embedding algorithm,
100
Hanan grid, 97
hierarchical construction
algorithm, 100
iterated one-Steiner
algorithm, 99
k-comb, 151
Kruskal-based algorithm, 100
L-shaped optimal embedding
algorithm, 99

in λ-geometry, 96, 104–107
layout-WRST algorithm,
100–104, 111
maze running, 110
minimal (SMT), 96
minimum length, 96, 97–100
boundary, 98, 99
corner, 98, 99
cross, 98, 99
earthworms, 98, 99
interior, 98, 99
minimum spanning tree
(MST), 97
minimum weight, 100
MST-based, 179
neighboring Steinerization,
100
Prim-based algorithm, 100,
103
rectilinear (RST), 179
switchbox, 98
weighted, 96, 100–104
Z-shaped optimal embedding
algorithm, 99
Switchbox, 98, 121
boundary segment, 98
interior segment, 98
Switchbox routing, 133–138
BBLSR, 136
BEAVER algorithm, 133–136
DETOUR, 136
greedy switchbox algorithm,
136–138
hierarchical approach, 133
MDR, 136
SWR, 136
WEAVER, 133, 136

Target cell, 80
Tautology test algorithm,
250–251
Temperature, 73
Tesselation, 6
Thread router, 135
Three-layer over-the-cell routing
algorithm, 199
Tile, 6
Timing critical path, 168
Timing simulation, 156
Timing-driven floorplanning,
198–199
Timing-driven layout, 265–269

Timing-driven partitioning
  hierarchical clustering-based
    algorithm, 198
  Kernighan-Lin algorithm, 198
Timing-driven placement
  constructive approach, 174
  convex programming, 174
  critical nets, 163
  fuzzy logic algorithm, 175
  hierarchical placement
    algorithms, 169
  Kuhn-Tucker conditions, 171
  Lagrange multiplier, 171
  linear programming, 172–174
  min-cut algorithm, 174
  weight-based algorithm,
    169–172
  zero-slack algorithm,
    163–168
Timing-driven routing
  clock skew, 180–184
  clock tree, buffered, 184–187
  delay minimization, 175–180
  linear delay model, 175
Top-down-routing algorithm,
    113
  integer programming, 113
  sequencing of nets, 113
TOPOLOGIZER, 263
Track, 143
Track graph, 95
Transistor chaining, 9, 10,
    253–256
  branch-and-bound approach,
    263
  doughnut layout style, 269
  dynamic programming, 265
  Eulerian path, 254, 255
  exhaustive search, 264
  flexible transistor matrix
    (FTM), 269
  GENAC, 263
  graph-based algorithm, 255
  Layout synthesizer, 263
  LiB, 263
  linear time algorithm, 265
  logic diagram, 254
  metal-metal matrix (M^3), 269
  min-cut algorithm, 264
  one-dimensional, 265

Sc2, 263
TOPOLOGIZER, 263
Trunk, 127
Two-dimensional compaction,
    293–295, 297–298
  base constraints, 294
  branch-and-bound approach,
    294, 297
  distance constraints, 294
  flat, 297
  hierarchical approach,
    297–298
  integer linear programming,
    298
  linear programming, 297
  Monte-Carlo algorithm, 297
  simulated annealing, 297

U, 222
  box, 225
  fixed, 223
  nes, 225
Universal design
  FPGA, 12
  PLA, 12
UVM, *see* Via minimization,
    unconstrained

**Vertical constraints graph
  (VG), 124**
Vertical node sizing algorithm,
    63–64
  running time, 63–64
Via, 240
Via minimization
  constrained (CVM), 187–192
  max-cut (L) algorithm,
    190–192
  max-cut problem, 190
  minimum weight matching
    algorithm, 189
  other issues, 195–196
  unconstrained (UVM),
    193–195
  weighted constrained, 202
VLSI technology
  CMOS, 2, 4–5
  NAND gate, 4
  NMOS, 2–5

NOR gate, 4
PMOS, 2, 4

Wave propagation, 239
Waveform, 157
WCVM, *see* Via minimization,
    weighted-constrained
Weight-based placement
    algorithm, 169–172
Weinberger array, 9–10,
    256–257
Wire tree, 213
Wire-length minimization,
    279–286
  branch-and-bound approach,
    298
  cost function, 280–281
  global optimization, 284
  global-slack, 281
  greedy push algorithm,
    222–227
  iterative-improvement-for-
    wire-length algorithm,
    283
  linear programming, 281
  local-stack, 281
  longest-path algorithm, 279
  MCM, 222
  Simplex algorithm, 284–286,
    298
  single-layer routing, 222–227
Wiring, 7
WRST, *see* Steiner
    tree, weighted;
    Layout-WRST
    algorithm

x-y monotone path, 103

Yet another channel router
  branch, 127
  trunk, 127
  YACR, 127
  YACR-II, 127–128

**Z-shaped optimal embedding
  algorithm, 99**
Zero clock skew algorithm, 182,
    184–185
Zero-skew clock tree, 186
Zero-slack algorithm, 163–168